Not by Genes Alone

Not by Genes Alone

How Culture Transformed Human Evolution

Peter J. Richerson
and Robert Boyd

The University of Chicago Press
Chicago and London

Peter J. Richerson is professor of environmental science and policy at the University of California–Davis. The author of scores of articles, he has edited *Human by Nature* (with Peter Weingart, Sandra Mitchell, and Sabine Maasen). His previous book, *Culture and the Evolutionary Process,* was written with Robert Boyd and published by the University of Chicago Press. A volume of their important papers, *The Origin, an Evolution of Cultures,* is forthcoming from Oxford University Press.

Robert Boyd is professor in the Department of Anthropology at the University of California–Los Angeles. In addition to numerous articles and the books written or edited with Professor Richerson, he is the author, with J. B. Silk, of *How Humans Evolved.* He is coeditor of two books, *Foundations of Human Sociality* (with J. Henrich, S. Bowles, C. Camerer, E. Fehr, and H. Gintis) and *Moral Sentiments and Material Interests* (with Gintis, Bowles, Camerer, and Fehr) from MIT Press.

The University of Chicago Press, Chicago 60637
The University of Chicago Press, Ltd., London
© 2005 by The University of Chicago
All rights reserved. Published 2005
Printed in the United States of America

14 13 12 11 10 09 08 07 06 05   2 3 4 5
ISBN: 0-226-71284-2 (cloth)

Library of Congress Cataloging-in-Publication Data
Richerson, Peter J.
  Not by genes alone : how culture transformed human evolution / Peter J. Richerson and Robert Boyd.
      p.   cm.
  Includes bibliographical references and index.
  ISBN 0-226-71284-2 (hardcover : alk. paper)
  1. Social evolution.   2. Culture—Origin.   3. Human evolution.   4. Human behavior.   5. Sociobiology.   I. Boyd, Robert, Ph.D.   II. Title.
  GN360.R5 2005
  306—dc22

                                                    2004006601

♾ The paper used in this publication meets the minimum requirements of the American National Standard for Information Sciences—Permanence of Paper for Printed Library Materials, ANSI Z39.48-1992.

Contents

Acknowledgments

The intellectual odyssey that led to this book began with some after-dinner conversations in the early 1970s. Over the years we have accumulated debts to many people, beginning with our families. Their indulgence of our preoccupation with matters of cultural evolution we much appreciate. The strong support of Don Campbell, Gerry Edelman, Ralph Burhoe, and Harvey Wheeler mattered a lot in the early days. Others who have provided substantial aid, comfort, and intellectual stimulation over the years include Robert Aunger, Howard Bloom, Chris Boehm, Sam Bowles, L. L. Cavalli-Sforza, Tom Dietz, Marc Feldman, Russ Genet, John Gillespie, Herb Gintis, Tatsuya Kameda, Hillard Kaplan, Kevin Laland, John Odling Smee, Alan Rogers, Eric Smith, Michael Turelli, Polly Wiessner, David Sloan Wilson, and Bill Wimsatt.

Our graduate students are and have been a source of much inspiration and feedback on the ideas in this book. Thanks to Alpina Begossi, Mika Cohen, Ed Edsten, Charles Efferson, Francisco Gil-White, Joe Henrich, Jen Mayer, Richard McElreath, Brian Paciotti, Karthik Panchathan, Lore Ruttan, Joseph Soltis, Bryan Vila, and Tim Waring.

PJR owes a large debt to his departmental colleagues, past and present, who have steadfastly supported his rather unusual research program. Thanks especially to chairs Francisco Ayala, Charles Goldman, Paul Sabatier, and Alan Hastings for extra help here and there. Other colleagues at

Davis that have provided magnificent intellectual stimulation and other forms of support include Billy Baum, Bob Bettinger, Monique Borgerhoff Mulder, Larry Cohen, Bill Davis, Jim Griesemer, Sarah Hrdy, Bob and Mary Jackman, Mark Lubell, Richard McElreath, Jim McEvoy, and Aram Yengoyan. I am grateful to Lesley Newson and the School of Psychology at Exeter University for hosting me during the final preparation of the manuscript.

RB thanks his colleagues in the anthropology departments at Emory and UCLA for much intellectual stimulation, and for being broad-minded about what constitutes anthropology. Bradd Shore, Peter Brown, and Bruce Knauft provided generous help and advice to an anthropological novice when he was starting out. Many thanks are also due to colleagues at UCLA, especially Clark Barrett, Nick Blurton-Jones, Dan Fessler, Alan Fiske, Allen Johnson, Nancy Levine, Joan Silk, and Tom Weisner, and in the Center for Behavior, Evolution and Culture, including Martie Haselton, Jack Hirshleifer, Susanne Lohman, Neil Malamuth, Derek Penn, and John Schumann. A joint progam with UCSB has afforded stimulating interactions with Leda Cosmides, John Tooby, Donald Symons, and their students. I am also grateful to the many students in Anthropology 186P and Anthropology 120G who had to put up with the less polished and poorly copyedited versions of this manuscript that they were assigned to read. Moreover, I benefited from the time and intellectual stimulation of a year at the Wissenschaftskolleg zu Berlin; many thanks to the kolleg staff and the fellows, especially John Breuilly, Martin Daly, Örjan Ekeberg, Alex Kacelnik, John McNamara, and Margo Wilson. I have also benefited from many interactions with my colleagues in the MacArthur Preferences Network, including Sam Bowles, Colin Camerer, Catherine Eckel, Ernst Fehr, Herb Gintis, David Laibson, and Paul Romer. I worked on early drafts of this book at Baboon Camp, Maun, Botswana, and am thankful to Dorothy Cheney and Robert Seyfarth for their many courtesies there. Finally, of course, there are Joan, Sam, and Ruby, whose love and support is the foundation of everything.

Many people read all or part of this book and helped us improve the various drafts. Sandy Hazel, Christie Henry, Joan Silk, and Eric Alden Smith provided particularly thorough editorial comments. Other readers included Sam Bowles, Peter Corning, Richard McElreath, Russ Genet, Peter Godfrey-Smith, Ed Hagen, Kim Hill, Robert Hinde, Daniel Fessler, Gary Marcus, Lesley Newson, John Odling Smee, Luke Rendell, Kim Sterelny, Hal Whitehead, and several anonymous reviewers. A UCD Animal Behavior 270 class, co-taught with Monique Borgerhoff Mulder, Tim Caro, and John Eadie, read and critiqued a complete draft of the book.

We began this book during The Biological Foundations of Culture, a project directed by Peter Weingart at the Center for Interdisciplinary Research at the University of Bielefeld. Participants provided us with much useful feedback, particularly Monique Borgerhoff Mulder, Leda Cosmides, Bill Durham, Bernd Giesen, Peter Hejl, Sabine Maasen, Alexandra Maryanski, Sandra Mitchell, Wulf Schievenhovel, Ullica Segerstråle, Peter Meyer, Nancy Thornhill, John Tooby, Jonathan Turner and Peter Weingart.

Christie Henry and her team at the University of Chicago Press provided all the editorial support authors could hope for, and under trying circumstances too.

Chapter One

Culture Is Essential

The American South has long been more violent than the North. Colorful descriptions of duels, feuds, bushwhackings, and lynchings feature prominently in visitors' accounts, newspaper articles, and autobiographies from the eighteenth century onward. Statistics bear out these impressions. For example, over the period 1865–1915, the homicide rate in the South was ten times the current rate for the whole United States, and twice the rate in our most violent cities. Modern homicide statistics tell the same story.

In their book, *Culture of Honor,* psychologists Richard Nisbett and Dov Cohen argue that the South is more violent than the North because southern people have culturally acquired beliefs about personal honor that are different from their northern counterparts.[1] Southerners, they argue, believe more strongly than Northerners that a person's reputation is important and worth defending even at great cost. As a consequence, arguments and confrontations that lead to harsh words or minor scuffles in Amherst or Ann Arbor often escalate to lethal violence in Asheville or Austin.

What else could explain these differences? Some feature of the southern environment, such as its greater warmth, could explain why Southerners are more violent. Such hypotheses are plausible, and Nisbett and Cohen are at pains to test them. Northerners and Southerners might differ genetically, but this hypothesis is not very plausible. The settlers of the North and

South came mostly from the British Isles and adjacent areas of northwestern Europe.[2] Human populations are quite well mixed on this scale.

Nisbett and Cohen support their hypothesis with an impressive range of evidence. Let's start with statistical patterns of violence. In the rural and small-town South, murder rates are elevated for arguments among friends and acquaintances, but not for killings committed in the course of other felonies. In other words, in the South men are more likely than Northerners to kill an acquaintance when an argument breaks out in a bar, but they are no more likely to kill the guy behind the counter when they knock off a liquor store. Thus, Southerners seem to be more violent than other Americans only in situations that involve personal honor. Competing hypotheses don't do so well: neither white per-capita income nor hot climate nor history of slavery explain this variation in homicide.

Differences in what people say about violence also support the "culture of honor" hypothesis. For example, Nisbett and Cohen asked people to read vignettes in which a man's honor was challenged—sometimes trivially (for example, by insults to his wife), and in other cases seriously (for example, by stealing his wife). Southern respondents were more likely than Northerners to say that violent responses were justified in all cases, and that one would "not be much of a man" unless he responded violently to insults. In the case of more serious affronts, southern respondents were almost twice as likely to say that shooting the perpetrator was justified.

Interestingly, this difference in behavior is not just talk; it can also be observed under the controlled conditions of the psychology laboratory. Working at the University of Michigan, Nisbett and Cohen recruited participants from northern and southern backgrounds, ostensibly to participate in an experiment on perception. As part of the procedure, an experimenter's confederate bumped some participants and muttered "Asshole!" at them. This insult had very different effects on southern and northern participants, as revealed by the next part of the experiment. Sometime after being bumped, participants encountered another confederate walking toward them down the middle of a narrow hall, setting up a little game of chicken. This confederate, a six-foot, three-inch, 250-pound linebacker on the UM football squad, was much bigger and stronger than any participant, and had been instructed to keep walking until either the participant stepped aside and let him pass or a collision was immanent. Northerners stepped aside when the confederate was six feet away, whether or not they had been insulted. Southerners who had not been insulted stepped aside when they were nine feet away from the confederate, while previously insulted Southerners continued walking until they were just three feet away. Polite, but prepared to be violent, uninsulted Southerners take more care, presumably because they attribute a sense of honor to the football player and are careful not to test it. When their own honor is challenged, however, they are willing to challenge someone at considerable risk to their own safety. These behavioral differences have physiological correlates. In a similar confederate-insulter experiment, Nisbett and Cohen measured levels of two hormones, cortisol and testosterone, in participants before and after they had been insulted. Physiologists know that cortisol levels increase in response to stress, and testosterone levels rise in preparation for violence. Insulted Southerners showed much bigger jumps in cortisol and testosterone than insulted Northerners.

Nisbett and Cohen argue that the difference in beliefs between northern and southern people can be understood in terms of their cultural and economic histories. Scots-Irish livestock herders were the main settlers of the South, while English, German, and Dutch peasant farmers populated the North. States historically have had considerable difficulty imposing the rule of law in the sparsely settled regions where herding is the dominant occupation, and livestock are easy to steal. Hence in herding societies a culture of honor often arises out of necessity as men seek to cultivate reputations for willingly resorting to violence as a deterrent to theft and other predatory behavior. Of course, bad men may also subscribe to the same code, the better to intimidate their victims. As this arms race escalates, arguments over trivial acts can rapidly get out of hand if a man thinks his honor is at stake. This account is supported by the fact that Southern white homicide rates are unusually high in poor regions with low population density and a historically weak presence of state institutions, not in the richer, more densely settled, historically slave-plantation districts. In such an environment the Scots-Irish honor system remained adaptive until recent times.

This fascinating study illustrates the two main points we want to make in this book.

Culture is crucial for understanding human behavior. People acquire beliefs and values from the people around them, and you can't explain human behavior without taking this reality into account. Murder is more common in the South than in the North. If Nisbett and Cohen are right, this difference can't be explained in terms of contemporary economics, climate, or any other external factor. Their explanation is that people in the South have acquired a complex set of beliefs and attitudes about personal honor that make them more polite, but also more quick to take offense than people in the North. This complex persists because the beliefs of one generation are learned by the next. This is not an isolated example. We will present several other similar well-studied examples demonstrating that culture plays an important role in human behavior. These are only the tip of the iceberg—a complete scholarly rehearsal of the evidence would try the patience of all but the most dedicated reader. Culturally acquired ideas are crucially important for explaining a wide range of human behavior—opinions, beliefs, and attitudes, habits of thought, language, artistic styles, tools and technology, and social rules and political institutions.

Culture is part of biology. An insult that has trivial effects in a Northerner sets off a cascade of physiological changes in a southern male that prepare him to harm the insulter and cope with the likelihood that the insulter is prepared to retaliate violently. This example is merely one strand in a skein of connections that enmesh culturally acquired information in other aspects of human biology. Much evidence suggests that we have an evolved psychology that shapes what we learn and how we think, and that this in turn influences the kind of beliefs and attitudes that spread and persist. Theories that ignore these connections cannot adequately account for much of human behavior. At the same time, culture and cultural change cannot be understood solely in terms of innate psychology. Culture affects the success and survival of individuals and groups; as a result, some cultural variants spread and others diminish, leading to evolutionary processes that are every bit as real and important as those that shape genetic variation. These culturally evolved environments then affect which genes are favored by natural selection. Over the evolutionary long haul, culture has shaped our innate psychology as much as the other way around.

Few who have thought much about the problem would dispute either of these claims *in principle*. Beliefs and practices that we learn from one another are clearly important, and like all human behavior, culture must in some way be rooted in human biology. However, *in practice* most social scientists ignore at least one of them. Some scholars, including most economists, many psychologists, and many social scientists influenced by evolutionary biology, place little emphasis on culture as a cause of human behavior. Others, especially anthropologists, sociologists, and historians, stress the importance of culture and institutions in shaping human affairs, but usually fail to consider their connection to biology. The success of all these disciplines suggests that many questions can be answered by ignoring culture or its connection to biology. However, the most fundamental questions of how humans came to be the kind of animal we are can *only* be answered by a theory in which culture has its proper role *and* in which it is intimately intertwined with other aspects of human biology. In this book we outline such a theory.

Culture can't be understood without population thinking

Eminent biologist Ernst Mayr has argued that "population thinking" was Charles Darwin's key contribution to biology.[3] Before Darwin, people thought of species as essential, unchanging types, like geometric figures and chemical elements. Darwin saw that species were populations of organisms that carried a variable pool of inherited information through time. To explain the properties of a species, biologists had to understand how the day-to-day events in the lives of individuals shape this pool of information, causing some variant members of the species to persist and spread, and others to diminish. Darwin famously argued that when individuals carrying some variants were more likely to survive or have more offspring, these would spread through a process of natural selection. Less famously, he also thought that beneficial behaviors and morphologies acquired during an individual's lifetime were transmitted to the offspring, and that this process, which he called the "inherited effects of use and disuse," also shaped which variants were present. We now know that the latter process is unimportant in organic evolution, and that many processes Darwin never dreamed of are important in molding populations, including mutation, segregation, recombination, genetic drift, gene conversion, and meiotic drive. Nonetheless, modern biology is fundamentally Darwinian, because its explanations of evolution are rooted in population thinking; and if through some miracle of cloning Darwin were to be resurrected from his grave in Westminster Abbey, we think that he would be quite happy with the state of the science he launched.

Population thinking is the core of the theory of culture we defend in this book. First of all, let's be clear about what we mean by *culture*:

> *Culture is information capable of affecting individuals' behavior that they acquire from other members of their species through teaching, imitation, and other forms of social transmission.*[4]

By *information* we mean any kind of mental state, conscious or not, that is acquired or modified by social learning and affects behavior. We will use everyday words like *idea, knowledge, belief, value, skill,* and *attitude* to describe this information, but we do not mean that such socially acquired information is always consciously available, or that it necessarily corresponds to folk-psychological categories. Our definition is rooted in the conviction that most cultural variation is caused by information stored in human brains—information that got into those brains by learning from others.[5]

People in culturally distinct groups behave differently, mostly because they have acquired different skills, beliefs, and values, and these differences persist because the people of one generation acquire their beliefs and attitudes from those around them. Hence Southerners are more likely to kill than Northerners because they hold different attitudes about personal honor. The same is true of many other aspects of culture. Different populations exhibit persistent variation in language, social customs, moral systems, practical skills and devices, and art. These and all the other dimensions of culture exist because people possess different socially acquired skills, beliefs, or values.

Population thinking is the key to building a causal account of cultural evolution. We are largely what our genes and our culture make us. In the same way that evolutionary theory explains why some genes persist and spread, a sensible theory of cultural evolution will have to explain why some beliefs and attitudes spread and persist while others disappear. The processes that cause such cultural change arise in the everyday lives of individuals as people acquire and use cultural information. Some moral values are more appealing and thus more likely to spread from one individual to another. These will tend to persist, while less attractive alternatives tend to disappear. Some skills are easy to learn accurately, while others are more difficult and are likely to be altered as we learn them. Some beliefs make people more likely to be imitated, because the people who hold those beliefs are more likely to survive or more likely to achieve social prominence. Such beliefs will tend to spread, while beliefs that lead to early death or social stigma will disappear. In the short run, a population-level theory of culture has to explain the net effect of such processes on the distribution of beliefs and values in a population during the previous generation. Over the longer run, the theory explains how these processes, repeated generation after generation, account for observed patterns of cultural variation. The heart of this book is an account of how the population-level consequences of imitation and teaching work.

Taking a population approach does not imply that cultural evolution is closely analogous to genetic evolution. For example, population thinking that does not require cultural information takes the form of *memes*, discrete, faithfully replicating, genelike bits of information. A range of models are consistent with the facts of cultural variation as they are presently understood, including models in which cultural information is not discrete and is never replicated. The same goes for the processes that give rise to cultural change. Natural selection−like processes are sometimes important, but processes that have no analog in genetic evolution also play important roles. Culture is interesting and important because its evolutionary behavior is distinctly different from that of genes. For example, we will argue that the human cultural system arose as an adaptation, because it can evolve fancy adaptations to changing environments rather more swiftly than is possible by genes alone. Culture would never have evolved unless it could do things that genes can't!

Population thinking makes it easy to link cultural and genetic evolution

Many social scientists have treated culture as a "superorganic" phenomenon. As one of the founders of modern anthropology, A. L. Kroeber, put it,

> [P]articular manifestations of culture find their primary significance in other cultural manifestations, and can be most fully understood in terms of these manifestations; whereas they cannot be specifically explained from the generic organic endowment of the human personality, even though cultural phenomena must always conform to the frame of this endowment.[6]

Social scientists in Kroeber's tradition have long dismissed the need to incorporate biology in any serious way into their study of human behavior. Humans cannot fly by flapping their arms or breathe underwater, but outside of such obvious constraints, biology has little to do with culture. On this view, biology is important, of course, because we need bodies and brains to have culture. But biology just furnishes the blank slate on which culture and personal experience write.[7]

Superorganicism is wrong because it ignores the rich interconnections between culture and other aspects of our behavior and anatomy. Culture is as much a part of human biology as walking upright. Culture causes people to do many weird and wonderful things. Nonetheless, the equipment in human brains, the hormone-producing glands, and the nature of our bodies play a fundamental role in how we learn and why we prefer some ideas to others. Culture is taught by motivated human teachers, acquired by motivated learners, and stored and manipulated in human brains. Culture is an evolving product of populations of human brains, brains that have been shaped by natural selection to learn and manage culture. Culture-making brains are the product of more than two million years of more or less gradual increases in brain size and cultural complexity. During this period, culture must have increased the reproductive success of our ancestors; otherwise, the features of our brain that make culture possible would not have evolved.[8] The operational products of this evolution are innate predispositions and organic constraints that influence the ideas that we find attractive, the skills that we can learn, the emotions that we can experience, and the very way we see the world. To take an exceedingly simple example, why are the doorways of houses in many cultures usually a little above head high? Because the human skull, for obvious adaptive reasons, is rather well endowed with pain sensors. Those who emphasize the role that organic evolution plays in explaining human behavior are surely correct to emphasize that a plethora of such innate adaptations strongly affect how culture evolves, although we still know little about the details. Why did Southerners need a *culture* of honor? Perhaps because on average, human males are *neither* innately sufficiently sensitive to insults *nor* sufficiently ready to respond violently to them in an environment where self-help violence is the chief means of protecting one's livelihood.

Thinking about culture as something that is acquired, stored, and transmitted by a population of *individuals* enables us to explore interactions between culture and other aspects of human biology. Individual psychologies determine which ideas are likely to be easy to learn and remember and which kinds of people are likely to be imitated. Of course, individuals do not behave in isolation. Individual psychologies may interact in interesting and complex ways, and we have to be careful to make sure that such structure finds its way into our theories. Individuals are also the main locus of genetic variation within the human species; to a first approximation, selection has acted over time to increase the fitness of individuals. A population-based theory of cultural change tells us how the details of individual psychology affect what kinds of skills, beliefs, and values that individuals acquire. In concept, modeling the evolution of the innate psychological machinery that gives rise to social learning is easy—you just allow individual psychology to be genetically variable. Individuals with different psychologies will acquire different beliefs and values that will lead to different fitness outcomes. Of course, many complications can arise, so making such theory can be very hard work indeed. This is, however, straightforward scientific labor—when you use population thinking to conceptualize culture, intriguing questions appear where paradoxes and confusion once reigned.

Culture changes the nature of human evolution in fundamental ways

Although we do not doubt that culture is deeply intertwined with other aspects of human biology, we also believe that the evolution of culture has led to fundamental changes in the way that our species responds to natural selection. Over the last forty years or so, behavioral evolutionists have developed a rich theory predicting how natural selection will shape social behavior under various conditions. This theory explains a great deal about different aspects of behavior—mating and parenting, signaling, and cooperation—and has been fairly successful in explaining the differences between species throughout the animal kingdom. In the 1970s a group of scientists, then called human sociobiologists, created an intense controversy by applying the same body of theory to humans.[9] Two contemporary research traditions have grown out of this work: human behavioral ecology and evolutionary psychology. Human behavioral ecologists typically use evolutionary theory to understand contemporary human behavior. Evolutionary psychologists use it to generate hypotheses about the evolved structure of human psychology. While both traditions have been quite successful, their application of evolutionary theory to humans is still the cause of much debate.[10]

Some of the opposition to evolutionary approaches to human behavior comes from thinking about these issues in terms of nature versus nurture. Biology is about nature; culture is about nurture. Some things, like whether you have sickle-cell anemia, are determined by genes—nature. Other things, like whether you speak English or Chinese, are determined by the environment—nurture. Evolutionary biology, many opponents of evolutionary explanations believe, can explain genetically determined behaviors, but not behaviors that are learned or are the result of contact with the environment. Since most human behavior is learned, they conclude evolutionary theory has little to contribute toward shaping or understanding it.

Although this way of thinking is common, it is deeply mistaken. To ask whether behavior is determined by genes or environment does not make sense. *Every bit* of the behavior (or physiology or morphology, for that matter) of every single organism living on the face of the earth results from the interaction of genetic information stored in the developing organism and the properties of its environment. To think of genes like blueprints that specify the adult properties of the organisms—one gene says you are tall, the other short—is wrong. A much better analogy is that genes are like a recipe, but one in which the ingredients, cooking temperature, and so on are set by the environment. Different traits *do* vary in how sensitive they are to environmental differences. Some traits aren't much affected by the normal range of environments—humans develop five fingers on each hand in almost all environments[11]—while others are highly sensitive—genetically similar people may end up with very different body sizes depending on nutrition and health during their childhood. Asking whether observed *differences* are due to genetic differences, differences in the environment, or some combination of these factors is sensible. However, the answer you get will tell you nothing about whether the traits in question are adaptations shaped by natural selection.

The reason is that natural selection shapes the way that developmental processes respond to environmental variation. Environment plays only a *proximate* role.[12] Differences in the environment may cause genetically identical individuals to behave differently, and in this sense environmental differences are immediate causes of behavior. However, if we want to know why the organism develops one way in one environment and a different way in a different environment, we have to find out how natural selection has shaped the developmental process of the organism so that it responds to the environment as it does. Or, as biologists put it, the *ultimate* determinant of behavior is natural selection on genes. Learning and other developmental processes that cause individuals to respond differently to different environments implement structures built into the genes.[13] In the natural world, proximate causes are typically physiological. Birds migrate toward the equator when days shorten because their brain converts changes in day length to hormonal signals that activate migratory behavior. Ultimate causes are evolutionary. Migration is an evolved strategy to exploit the favorable season at higher latitude while passing the harsh winter in less demanding habitats. Selection has shaped the reaction of the brain to day length and all the downstream physiological and behavioral machinery in order to motivate geese to fly from the Yukon River delta to central California before Arctic winter weather arrives.

While evolutionary social scientists reject the naïve idea that genes and environment can be independent causes, many accept that culture can be lumped with other environmental influences. They think that the psychological mechanisms that govern the acquisition of culture are just another form of behavioral plasticity whose structure can be understood in terms of natural selection acting on genes.[14] As a result, many in the evolutionary social science community rejected the idea that culture makes any *fundamental* difference in the way that evolutionary thinking should be applied to humans. Because the psychological machinery that molds human culture was shaped by natural selection, so, at least in ancestral environments, the machinery *must have* led to fitness-enhancing behavior. If it goes wrong in modern environments, culture is not the culprit but the fact that our evolved, formerly adaptive psychology "misfires" these days. While the sort of adaptationist thinking inherent in this approach has many famous critics, we are not among their number.[15]

Instead, our concern is that lumping culture with other environmental influences leads people to ignore the novel evolutionary processes that are created by culture. Selection shapes individual learning mechanisms so that interaction with the environment produces adaptive behavior. For example, many plants contain toxic substances. Selection makes these chemicals taste bitter to herbivores so that they learn not to consume the toxic plant species. Culture adds something quite new and different to this scenario. Like other animals, humans normally use bitter taste as a signal that a plant is inedible. However, some bitter plant compounds (like salicylic acid in willow bark) have medicinal value, so we also learn from others that we can override the aversive bitter taste of certain plants when we have the need to cure an ailment. The genes making the plant taste bitter don't change at all, but the behavior of a whole population can change anyway as the belief in the bitter plant's medicinal value spreads. We take our medicine in spite of its bitter taste, not because our sensory physiology has evolved to make it less bitter, but because the idea that it has therapeutic value has spread through the population. In the distant past, some inquisitive and observant healer discovered the curative properties of a bitter plant. Then a number of processes that we describe in this book might cause this belief to increase in frequency, despite its horrible taste. You can't understand this process by asking how individuals interact with their environment. Instead, you have to understand how a population of individuals interact with their environments *and* each other over time.

Thus, culture is neither nature nor nurture, but some of both. It combines inheritance and learning in a way that cannot be parsed into genes or environment.[16] This fact has two important consequences for human evolution, consequences to which we now turn.

Culture is a necessary part of the design problem for human psychology

One of the key steps in an adaptationist analysis of human behavior is to decide on the design problem that natural selection had to solve. Most students of human evolution begin by asking, how should evolution have shaped the psychology of a group-living, foraging hominid? From there, they ask how the evolved psychology will shape human culture. The implicit evolutionary scenario seems to be that Pleistocene hominids were just extra-smart chimpanzees, clever social animals in which learning from each other played a negligible role until the evolution of our brain was complete, at which point the souped-up chimpanzee was able to take up culture. *First* we got human nature by genetic evolution; *then* culture arose as an evolutionary byproduct.

This way of thinking neglects the inevitable feedback between the nature of human psychology and the kind of social information that this psychology should be designed to process. For us to take bitter medicine, our psychology must have evolved both to learn from others and to let this culturally acquired information override aversive stimuli. Culture is adaptive because the behavior of other individuals is a rich source of information about which behaviors are adaptive and which are not. We all know that plagiarism is often easier than the hard work of writing something by ourselves; imitating the behavior of others can be adaptive for the same reason. The trick is that once culture becomes important, the nature of the behavior that is available to imitate is itself strongly affected by the psychology that shapes how we learn from others. To take an extreme example, if everyone relied completely on imitation, behavior would become decoupled from the environment. With any environmental change, imitation would no longer be adaptive. To understand the evolution of the psychology that underlies culture, we must take this population-level feedback into account. We want to know how evolving psychology shapes the ideas and behaviors that can be acquired from others, and how natural selection shapes how we think and learn in an environment featuring direct information from personal experience *and* the potential to use the behavior of others at a lower cost but perhaps greater risk of error.

This kind of reasoning leads to conclusions quite different from other evolutionary theories of human behavior. Under the right conditions, selection can favor a psychology that causes most people most of the time to adopt behaviors "just" because the people around them are using those behaviors. The last 800,000 years or so have seen especially large, rapid fluctuations in world climate; the world average temperature sometimes changed more than 10 degrees Celsius in a century, leading to massive shifts in ecosystem structure.[17] A group of hominids living in a habitat something like contemporary Madrid could find themselves in a habitat like Scandinavia one hundred years later. You might think that such rapid and extreme environmental changes would put a premium on individual learning over imitation. Odd as it may seem, in many kinds of variable environments, the best strategy is to rely mostly on imitation, not your own individual learning. Some individuals may discover ways to cope with the new situation, and if the not-so-smart and not-so-lucky can imitate them, then the lucky or clever of the next generation can add other tricks. In this way the ability to imitate can generate the cumulative cultural evolution of new adaptations at blinding speed compared with organic evolution. A population of purely individual learners would be stuck with what little they can learn by themselves; they can't bootstrap a whole new adaptation based on cumulatively improving cultural traditions. This design for human behavior depends on people adopting beliefs and technologies *largely* because other people in their group share those beliefs or use these technologies. When lots of imitation is mixed with a little bit of individual learning, *populations* can adapt in ways that outreach the abilities of any individual genius.

Thinking about the population properties of culture helps us understand the psychology of social learning. For example, we will see that selection can favor a psychology that causes people to conform to the majority behavior even though this mechanism sometimes prevents populations from adapting to a change in the environment. Evolution also favors a psychology that makes people more prone to imitate prestigious individuals and individuals who are like themselves even though this habit can easily result in maladaptive fads. These psychological mechanisms in turn give rise to important patterns of behavior, like the symbolic marking of social groups that would not evolve unless their culture had certain population-level consequences.

Culture is an ultimate cause of human behavior

If the only processes shaping culture arose from our innate evolved psychology, then culture would be a strictly proximate cause of human behavior. Understanding how natural selection gave rise to our psychology would be more complicated than for other forms of behavioral plasticity, but in the end we could, at least in principle, reduce human culture to the actions of evolution by natural selection to increase genetic fitness.[18]

However, not all of the processes shaping culture *do* arise from our innate psychology—culture itself is subject to natural selection. Much as a child resembles her parents, people resemble those from whom they have acquired ideas, values, and skills. Culturally acquired ideas, values, and skills affect what happens to people during their lives—whether they are successful, how many children they have, and how long they live. These events in turn affect whether their behavior will be culturally transmitted to the next generation. If successful people are more likely to be imitated, then those traits that lead to becoming successful will be favored. Even more obviously, if living people are more likely to be imitated than the dead, then ideas, values, and skills that promote survival will tend to spread. Consequently, a culture of honor arises, at least in part, because in lawless societies, men who are not aggressive in protecting their herds and their families tend to fall victim to tough, ruthless predators. If these advantages to a culture of honor have disappeared in the modern South, the higher death rate of those who cling to the custom will eventually extinguish it.

Such selective processes can often favor quite different behaviors from those favored by selection on genes. For example, beliefs and values that lead to prestige and economic success in modern societies may also reduce fertility. Such beliefs spread because the prestigious are more likely to be imitated, even though this lowers genetic fitness. Opening our minds to ideas in the environment allows rapid adaptation, but it also leads to the evolution of pathological cultural maladaptations. Our psychology has a delicately balanced set of mechanisms designed to exclude harmful ideas in the environment yet not attack the beneficial ones.

Natural selection acting on culture is an ultimate cause of human behavior, just like natural selection acting on genes. Consider an example we will return to repeatedly. Much cultural variation exists at the group level. Different human groups have different norms and values, and the cultural transmission of these traits can cause such differences to persist for long periods of time. Now, the norms and values that predominate in a group plausibly affect the probability that the group is successful, whether it survives, and whether it expands. For the purposes of illustration, suppose that groups having norms that promote group solidarity are more likely to survive than groups lacking this sentiment. This creates a selective process that leads to the spread of solidarity. Of course, this process may be opposed by an evolved innate psychology that biases what we learn from others, making us more prone to imitate and invent selfish or nepotistic beliefs rather than ones favoring group solidarity, like patriotism. The long-run evolutionary outcome would then depend on the balance of the processes favoring and disfavoring patriotism. Again for the sake of illustration, let us suppose that net effect of these opposing processes causes patriotic beliefs to predominate. In this case, the population behaves patriotically *because* such behavior promotes group survival, in exactly the same way that the sickle-cell gene is common in malarial areas *because* it promotes individual survival. Human culture participates in ultimate causation.

Cultural scientists, we believe, should not fear a reunion with biology. Culture is a brawny phenomenon and is in no real danger of being "reduced" to genes. Of course genetic elements of our evolved psychology shape culture—how could it be otherwise? But at the same time, natural selection acting on cultural variation shaped the environments in which our psychology evolved (and is evolving). The coevolutionary dynamic makes genes as susceptible to cultural influence as vice versa. We will argue that the phenomenon of group selection on cultural variation described above could have produced institutions encouraging more cooperation with distantly related people than would be favored by our original evolved psychology. These cooperators would have discriminated against individuals who carried genes that made them too belligerent to conform to the new cooperative norms. Then the cultural rules could expand cooperation a bit further, generating selection for still more-docile genes. Eventually, innate elements of human social psychology became tolerably well adapted to promote living in tribes, not just families.

Culture makes us odd

Thinking about cultural evolution at the population level leads to a picture of a powerful adaptive system that is necessarily accompanied by some exotic side effects. Some of our evolutionist friends take a dim view of this notion, seeing it as giving aid and comfort to those who would deny the relevance of evolution to human affairs. We prefer to think that population-based theories of cultural evolution strengthen the Darwinian's grasp on the human species by providing a picture of the engine that powered the furious pace of human evolution over the last few hundred thousand years. Our ape cousins still live in the same tropical forests in the same small social groups, and eat the same fruits, nuts, and bits of meat as our common ancestors did. By the late Pleistocene (say, 20,000 years ago), human foragers already occupied a much wider geographical and ecological range than any other species, using a remarkable range of subsistence systems and social arrangements. Over the last ten millennia we have exploded to become the earth's dominant organism by dint of deploying ever more-sophisticated technology and ever more-sophisticated social systems. The human species is a spectacular evolutionary anomaly, so we ought to expect that the evolutionary system behind it is pretty anomalous as well. Our quest is for the evolutionary motors that drove our divergence from our ancestors, and we believe that the best place to hunt is among the anomalies of cultural evolution. This does not mean that gene-based evolutionary reasoning is worthless. To the contrary, human sociobiologists and their successors have explained a lot about human behavior even though most work ignores the novelties introduced by cultural adaptation. However, there is still much to explain, and we think that the population properties of culture are an essential ingredient of a satisfactory theory of human behavior.

The path not taken

In the preface to the second edition of the *Descent of Man* in 1874, Darwin noted that he

> [took the] opportunity of remarking that my critics frequently assume that I attribute all changes of corporeal structure and mental power exclusively to the natural selection of such variations as are often called spontaneous; whereas, even in the first edition of the *Origin of Species* I distinctly stated that great weight must be attributed to the inherited effects of use and disuse, with respect both to the body and mind.[19]

From the biologists' point of view, Darwin's belief in the inheritance of acquired variation was his greatest error. Darwin thought "inherited habits," by which he meant something very close to human culture, were important in a wide variety of species. In a sense he was correct—simple forms of social learning are widespread in the animal kingdom.[20] However, Darwin imagined that even honeybees had humanlike imitative capacities, whereas the best modern evidence, as we shall see, suggests that all other animals, including our closest ape relatives, have rudimentary capacities for culture compared with ourselves.

Darwin's intuitions about "inherited habits" no doubt came from his observation that humans had such things, combined with his desire to minimize the gap between humans and other animals. He is sometimes said to have biologized human culture, but he is more accurately accused of culturizing biology.[21] Darwin had a sophisticated, if erroneous, picture of the distribution of the inherited effects of use and disuse across traits. He thought that behavior was more susceptible to the inheritance of acquired variation and that anatomy was much more conservative in this regard, so he could account for the fact that human behavior was much more variable from place to place than were human bodies. As "On the Races of Man," chapter 7 of the *Descent,* shows, Darwin was not seduced into thinking that the huge behavioral differences he and other pioneering anthropologists observed among humans could be accounted for by differences in conservative—we would say today genetic—characters. Rather, he attributed them to the more labile characters that we would today label cultural.

We thus have an interesting historical paradox: Darwin's theory was a better starting point for humans than any other species, and required a major pruning to adjust to the rise of genetics. Nevertheless, the *Descent* had no lasting influence on the social sciences that emerged at the turn of the twentieth century.[22] Darwin was pigeonholed as a biologist, and sociology, economics, and history all eventually wrote biology out of their disciplines. Anthropology relegated his theory to a subdiscipline, biological anthropology, behind the superorganic firewall. Since the midtwentieth century, many social scientists have treated Darwinian initiatives as politically tainted threats. If anything, the gulf between the social and natural sciences continues to widen as some anthropologists, sociologists, and historians adopt methods and philosophical commitments that seem to natural scientists to abandon the basic norms of science entirely.

In this book, we follow Darwin's path not taken. Beginning with psychologist Donald T. Campbell's work in the 1960s, we, and a few compatriots,[23] have sought to give cultural evolution its due weight *without* divorcing culture from biology. We hope to convince you that this approach to cultural evolution delivers new and powerful tools to dissect some of the enduring problems of the human sciences: How do genes and culture interact to influence our behavior? Why are humans so extraordinarily successful a species? How do individual processes and the institutional structures and functions of groups articulate? What are the sources of cultural diversity? Why, despite our success as a species, do our actions often seem mildly (or sometimes wildly) dysfunctional? Why does our behavior sometimes lead to colossal catastrophes? Why are we sometimes downright heroic in our concern for others' welfare while in other circumstances indifferent, callous, exploitative, or vicious? As far as we can see, the benefits of such a theory are large compared with the cost of abandoning certain cherished commitments to disciplines, methods, and hypotheses that it casts into doubt. We hope that by the time you finish this book you will agree.

Chapter Two

Culture Exists

Anthropologists, sociologists, and historians express disbelief when told that serious students of human behavior find culture peripheral to their analyses of human behavior. Nevertheless, the truth is that culture plays little role in disciplines like economics and psychology. Scholars working in such traditions usually don't deny that culture is real and important, but maintain that worrying about how it works or why it exists is just not part of their job description.[1] But we suspect that for some in these disciplines, benign neglect is accompanied by a largely unarticulated prejudice against cultural explanations. Confronted with differences in marriage systems, inheritance rules, or economic organization, many scholars prefer economic or ecological explanations, no matter how far-fetched, over those that invoke cultural history.

This view is common (though far from universal) among our colleagues in evolutionary social science. From the beginning, many such scholars have been blunt in their rejection of the idea that culture has any important role in human affairs. As one of the founders of sociobiology, Richard Alexander, puts it, "Cultural novelties do not replicate or spread themselves, even indirectly. They are replicated as a consequence of the behavior of vehicles of gene replication."[2] In the same vein, psychologist David Buss remarks, "'Culture' is not an autonomous causal process in competition with 'biology' for explanatory power."[3] Or, even more directly, anthropologist Laura Betzig says in reaction to claims for the importance of culture: "I, personally, find culture unnecessary."[4]

The main purpose of this chapter is to convince the skeptics that culture *is* necessary, and to show that variation in human behavior cannot be understood without accounting for beliefs, values, and other socially acquired determinants of behavior. Those who would deny a role for culture place the entire burden of explaining human diversity on some mix of genetic and environmental variation—but neither genetic nor environmental differences can bear the explanatory weight this approach places on them. The evidence accords better with the traditional views of cultural anthropologists and kindred thinkers in other disciplines: heritable cultural differences are crucial for understanding human behavior.

Cultural differences account for much human variation

The diversity of the human species is striking especially when you think about peoples in other parts of the world. Consider, for example, the Copper Eskimo and the Trobriand Islanders. In the winter, the Copper Eskimo lived in snow houses built on the frozen sea. They obtained food by spearing seals at breathing holes in the ice, sometimes waiting motionlessly for hours in the bitterly cold darkness. In the summer, they lived in skin houses and hunted from cunningly constructed sealskin kayaks. They dwelled in groups of families linked together by a web of reciprocity without chiefs or councils. On the Trobriand Islands, many families shared a large wooden house. They subsisted on yams and taro grown in gardens that had been cleared and cultivated by hours of backbreaking labor in the humid tropical sun. They were ruled by a hereditary aristocracy with an elaborate system of rights and privileges based on membership in large matrilineally organized clans. Now add to the list nomadic pastoralists living in the starkness of central Arabia, the rice farmers of Java with their intricately nuanced social life, and the teeming economic and ethnic complexity of Los Angeles, and you will be convinced of the magnitude of human variation.

Three things could act as *proximate* causes of this variation. First, people may vary because they inherited different genes from their parents. Second, genetically similar individuals may differ because they have lived in different environments.[5] Finally, people may differ because they have acquired different beliefs, values, and skills through teaching and observational learning. Because the three sources of variation interact so richly in determining our behavior, people sometimes lose track of the important differences.[6] Consider the causes of variation in body weight, a character of concern to many of us. Clearly, environment can have a powerful effect on body weight. Central Europeans were undoubtedly leaner on average in 1918 and 1945 than they are today. Culture can powerfully affect body weight through work habits, ideas about appropriate diet, recreational preferences, innovations in the restaurant industry, and ideas about what constitutes physical beauty. In one West African culture, young girls are secluded for months and force-fed large meals several times a day for the express purpose of making them become extremely fat. In the United States, young girls (among others) avoid desserts and do aerobics to achieve a very different culturally transmitted ideal. At the same time, cheap, calorie-dense foods are heavily promoted by a highly competitive fast food industry. Caught between the gym and the supersized extra value meal, variation in the weight of Americans is enormous. Recent research has also shown that some genetic constitutions are predisposed to be heavier than others even with similar diets.

The "common garden experiment"

So, which is more important in determining people's behavior: genes, environment, or culture? You can calibrate your own position on this question by considering the following thought experiment. Choose two groups of people who live in different environments and behave differently—say, Eskimos and Trobriand Islanders. Next, suppose a population of Eskimos moves to an empty island in Melanesia and a population from the Trobriands moves to the high Arctic. Then, allow enough time for the *individuals* in each group to learn as much as they can about how to best behave in their new environment. Now here's the test: Do you think that the political system, religious practice, or kinship system of the Trobriands living in the Arctic will resemble their Eskimo neighbors more than their Trobriand ancestors? If so, then you are one of those who minimize the importance of culture. Or, will the political system, religious practice, or kinship system of the Trobriands living in the Arctic resemble that of their Trobriand ancestors more closely than their neighbors, the Eskimo? If that is your position, you think that the natural environment was not the source of the original variation in these characters—there must be something else that is transmitted through time. It could be culture, but it also could be genes or a self-replicating social environment.

Much better than a thought experiment would be a real experiment. While such an experiment would be unethical and impractical, its essential elements have been played out in various ways as people with different cultural histories have come to live in the same environment, and as culturally similar people have been challenged by divergent environmental changes. We submit that the following examples provide as strong evidence that some transmitted factor—culture, genes, or transmitted environment—plays an important role in shaping human societies. Then we will present evidence that neither genes nor transmissible environment is likely to be sufficient to explain the variation between human societies, leaving culture as the most likely suspect.

Illinois farmers from different immigrant backgrounds behave differently

The Midwest region of the United States was settled in the nineteenth century by immigrants from many different parts of Europe who brought with them the language, values, and customs of their native lands. Today, most overt traces of ethnic origin are gone—you cannot guess people's origin from their language or dress. But their farming practices are still substantially different. Rural sociologist Sonya Salamon and her colleagues have studied the effect of ethnic background on midwestern farmers, and found that people from different ethnic backgrounds have quite dissimilar beliefs about farming and family, and make very different decisions about farm management even though they have similar farms on nearly identical soils only a few miles apart.

One of Salamon's studies focused on two farming communities in southern Illinois, Freiburg (a pseudonym), inhabited by the descendants of German-Catholic immigrants who arrived during the 1840s, and Libertyville (also a pseudonym), settled in 1870 by people from other parts of the United States, mainly Kentucky, Ohio, and Indiana. These two communities are only about twenty miles apart, but the people in Freiburg and Libertyville have different values about family, property, and farm practice which are consistent with their ethnic origins. The German American farmers of Freiburg tend to value farming as a way of life, and they want at least one of their sons or daughters to carry on as farmers. According to one of Salamon's informants,

> The money's immaterial. I want a comfortable living for myself, the main thing is that it's something I've put together and I want to see it stay together. . . . I'd like to come back in 500 years and see if my great-great grandchildren still have it.[7]

These kinds of attitudes make the people of Freiburg very reluctant to sell land. Their wills specify that their farm will go to a child who will work the land and use farm proceeds to buy out any nonfarming siblings. Parents put considerable pressure on children to become farmers, but place relatively little importance on education. Salamon argues that these "yeoman" values are similar to those observed among peasant farmers in Europe and elsewhere. In contrast, the "Yankee" farmers of Libertyville regard their farms as profit-making businesses. They buy or rent land depending on economic conditions, and if the price is right, they sell. After a farmer sold out a good price, his neighbor commented approvingly, "[Y]ou don't make that money selling beans." Many farmers in Libertyville would like it if their children were to continue farming, but they see it as an individual decision. Some families help their children enter farming, but many do not, and they generally place a strong value on education.

The difference in values between Freiburg and Libertyville leads to different farming practices despite the nearness of the two towns and the similarity of their soils. Farms in Libertyville are about five hundred acres, nearly twice as large as those in Freiburg, because the Yankee farmers rent more land. Freiburg farmers are conservative, mainly farming the land they own, while Yankee farmers aggressively expand their operations by renting. The two communities also show striking differences in what they grow. In Libertyville as in most of southern Illinois, farmers specialize in grain production—it is the primary source of income for 77% of the farmers there. In Freiburg, farmers mix grain production with dairying or livestock raising, activities that are almost absent in Libertyville. Because these activities are labor intensive, they allow the German American farmers to accommodate larger families on more-limited acreage, consistent with the German farming goals. Yankee farmers don't go in for dairying and stock raising because "we could make more money from the land without all that work."[8]

The differing values of German American and Yankee farmers lead to differing patterns of land ownership in the two communities. In Freiburg, land rarely comes up for sale, and when it does, the price is higher than in neighboring areas. Salamon argues that the farmers there are willing to pay more for land because they are not solely maximizing profit—they want to provide land for their children. As a result, land is virtually never sold to non-Germans. In 1899, 90% of the land in Freiburg was owned by people of German ancestry, and by 1982 that figure rose to 97%. In Libertyville, land comes up for sale more often and at a lower price. The proportion of land owned by Yankee farmers has fluctuated considerably over the last one hundred years. Moreover, absentee landownership is more common in Libertyville—locals own 56% of the land in Libertyville, compared with 78% in Freiburg.

Similar patterns of ethnic variation exist elsewhere in Illinois. Salamon and her coworkers spent five years studying five ethnically distinct communities in east-central Illinois—German, Irish, Swedish, Yankee, and mixed German-Yankee.[9] As in the previous study, the five communities are near one another and have very similar soils. Their residents have many different beliefs and values, some of which are reflected in farming practices and patterns of land ownership. For example, the German and Yankee communities exhibit some of the same patterns of belief and behavior as in the southern Illinois study. Other groups, like the Irish and Swedes, differ in other ways.

The Nuer conquest of Dinka lands did not cause the Nuer to become like the Dinka

During the nineteenth and early twentieth centuries, two groups of people lived in the vast marshes of southern Sudan, the Nuer and the Dinka. Both groups lived a migratory existence, settling in villages and growing millet and maize in the wet season and then spreading out to graze their cattle on pastures uncovered by the subsiding flood in the dry season. The Nuer and the Dinka both numbered more than 100,000 people, and each was subdivided into many politically and militarily independent tribes numbering between three thousand and ten thousand people. Anthropologist Raymond Kelly provides a detailed account of the complex relationship between the Nuer and the Dinka over a period of half a century.[10] In about 1820, one of the Nuer tribes, the Jikany Nuer, migrated roughly three hundred kilometers to the east of their homeland, eventually invading land occupied by two Dinka tribes. Over the next sixty years, the Nuer expansion continued as tribes expanded south and west, conquering Dinka tribes and increasing their territory from a small area to more than half the swampland of the southern Sudan. Kelly estimates that more than 180,000 people, mostly Dinka, lived in the area conquered by the Nuer, and many were incorporated into Nuer society. There is every reason to believe that the Dinka eventually would have been eliminated had not the British intervened to suppress the conflict in the early 1900s.

Although they lived in the same environment, used the same technology, and were derived from the same common ancestors perhaps a thousand years ago, the Nuer and the Dinka differed in important ways. The

Nuer maintained larger herds, with about two cows for each bull, while the Dinka kept smaller herds, with about nine cows per bull. The Nuer rarely slaughtered cattle, subsisting mainly on milk, maize, and millet. In contrast, the Dinka frequently slaughtered and ate their cattle. As a result, Nuer population densities were about two-thirds those of the Dinka. The smaller human populations and larger cattle populations of the Nuer led to a number of differences between their yearly subsistence round and that of the Dinka. Most important, the dry-season settlements of the Nuer were much larger than those of the Dinka.

Another difference between the two peoples lay in their political systems. Among the Dinka, a tribe was the group of people who lived together in a wet-season encampment. In contrast, membership in Nuer tribes was based on kinship through the male line. As a result, the growth of Dinka tribes was constrained by geography, while Nuer tribes could in theory grow indefinitely. In fact, Nuer tribes seem to have been about three to four times larger than Dinka tribes. Kelly estimates that at the beginning of the expansion period, Nuer tribes averaged about ten thousand people, while Dinka tribes averaged only about three thousand.

Kelly argues that the differences in subsistence practices and political organization stemmed from the differences in "bride-price" customs. Among both the Nuer and the Dinka, the families of the bride and groom exchanged livestock at the time of a wedding. Custom specified the number of cows and goats that various classes of kin were expected to give and receive. Among both the Nuer and the Dinka, there was a net transfer of livestock from the groom's family to the bride's family, what anthropologists classify as bride price (rather than dowry). The details of such payments differed substantially between the Nuer and the Dinka. For the Nuer the minimum payment was about twenty head of cattle (the exact number varied); credit was not accepted. There was also an ideal payment of about thirty-six head. Between the minimum and the ideal payments, the groom's family had to pay all that it could, keeping only enough for subsistence. In contrast, the Dinka had no minimum payment and readily allowed credit. This meant that when times were tough, as during the rinderpest epidemic of the 1880s, Dinka weddings proceeded even though the bride's family might not receive any cows for an entire generation. The ideal and the minimum payments were substantially lower among the Dinka than among the Nuer, and Dinka payments often included goats. Kelly maintains that the Nuer kept larger herds to accommodate their larger and more inflexible bride wealth payments.

The distribution of livestock also varied. The Dinka gave livestock to the groom's paternal and maternal relatives, while the Nuer restricted bride-price payments to the groom's paternal relatives. This caused alliances to form among patrilateral kin in the Nuer and more-diffuse alliances to be established among the Dinka. Patrilateral alliances, in turn, caused the Nuer to develop a political system based on patrilineal clans, while the Dinka evolved one based on coresidence.

Distinctions between the Nuer and Dinka cannot be attributed solely to the environment. Both tribes lived in very similar habitats—seasonally flooded swamps. Of course, there are small environmental differences between the original Nuer homeland and the areas originally occupied by the Dinka, and people committed to strict environmental determination have argued that these are responsible for the behavioral differences between the two peoples. For example, anthropologist Maurice Glickman argued that the drier Nuer homeland allowed larger encampments during both the wet and dry seasons, giving rise to the other differences between the two groups.[11] But arguments of this kind all fail because the expansionist Nuer came to occupy exactly same environment as the departed and conquered Dinka. If environment determines culture, then the invading Nuer should have become like the Dinka, but the Nuer have continued to act like Nuer even after 100 years on former Dinka lands. Rather, tens of thousands of Dinka who remained in the conquered territories adopted the Nuer customs.

The social and economic variations between the Nuer and the Dinka had important consequences. Nuer military superiority allowed them to expand at the expense of the Dinka and was closely linked to other elements of their culture. Among both the Nuer and the Dinka, tribes were the units which conducted warfare. The Nuer did not conquer the Dinka; rather, various Nuer tribes conquered certain Dinka tribes. No Dinka tribe ever conquered a Nuer tribe, despite the fact that the military technology and tactics of the two groups were very similar. Nuer victories were routine because their tribes were larger. Nuer armies of fifteen hundred men easily defeated Dinka armies numbering about six hundred. The Nuer were able to recruit larger armies because their tribes were larger and because warfare typically occurred during the dry season, when Nuer encampments were larger. Notice that the Dinka did not adopt Nuer practices before they were conquered and assimilated, nor did they develop innovative military institutions to check the Nuer expansion. In chapter 6 we will consider some reasons we observe such cultural inertia.

A comparison of four East African groups shows
cultural variation is important

Anthropologist Robert Edgerton conducted a landmark study to investigate what happens when culturally similar peoples occupy quite different environments.[12] He focused on four East African tribes, the Sebei, Pokot, Kamba, and Hehe. Some communities of each of these tribes live in moist highlands, where they rely mainly on farming, while other communities of each group live in dry lowlands, where herding is more important. In each case, the highland and lowland groups had been in place for several generations, but there had been some contact between them over time.

Edgerton measured attitudes in each of these communities using a battery of psychological tests. For example, he asked people to respond to drawings which included scenes like a father confronting a misbehaving and disrespectful son, cattle damaging a maize farmer's field, and armed warriors raiding cattle protected by children. Respondents were asked to explain what was happening in the picture and what ought to happen in the scenes, as if they were taking place in the local village. Edgerton scored individual responses according to whether or not they included references to conflict avoidance, respect for authority, valuation of cattle, and self-control. Other measures involved more-structured questionnaires.

If culture played little role in shaping human behavior, the attitudes Edgerton measured should be associated with subsistence, not tribe. Migratory herding of cattle demands a much more fluid social organization than farming.[13] Farmers and herders should have different attitudes, but farmers from different tribes should be similar to each other, and so should herders. If culture is important, then tribe may be more important than subsistence. In this case, Kamba farmers and Kamba herders would be more similar than Kamba farmers and Sebei farmers or Kamba herders and Pokot herders.

Edgerton's results show the importance of culture. As he summarizes, "We . . . conclude that there can be no doubt that if we wished to know how someone in these four tribes would respond to the interview administered in this research, we would best predict that person's responses by knowing the tribe to which he belonged."[14] In a few cases, Edgerton did find evidence that ecological differences outweigh cultural ones: pastoralists, regardless of their tribal affiliation, have much more respect for authority than do farmers, which may result from the control over cattle maintained by senior men. However, an attempt to replicate this finding in southern

Tanzania by anthropologist Richard McElreath was only partly successful. McElreath found the same farmer/herder contrast in respect for authority among the Sangu, who, in different areas, pursue both subsistence systems. But among the Sukuma, a group of highly successful pastoralists, respect for authority is very low.[15] Instead, the Sukuma have a traditional system of collective social control and dispute resolution that commands great respect. This system requires that the leaders of the collective system be subject to sharp criticism for even minor infractions of rules.[16] Certainly, the cultural diversity of people living in the same environment should never be underestimated![17]

There are many similar cases

Many other examples tell the same story: people having different cultural and institutional histories behave differently in the same environment. Here are just a few more.

Sociologist Andrew Greeley used surveys to study the personality, political participation, respect for democracy, and family attitudes of Irish and Italian Americans.[18] He generated a series of hypotheses based on the assumption that resemblances to ancestral culture would persist for generations after immigration. For example, Irish immigrants disproportionately came from western Ireland, where rates of mass public participation in political activities were historically high; Italian immigrants mostly came from southern Italy, where political participation was low. Greeley hypothesized that rates of political participation of Irish and Italian Americans should mirror these historical differences. He found that immigrants do tend to converge toward the dominant Anglo norms in the United States, but slowly.

A study by political scientist Robert Putnam nicely complements Greeley's.[19] Putnam compared the performance of regional governments in Italy after widespread reforms in the 1970s devolved important powers on elected regional governments for the first time since the creation of the highly centralized Italian state in the 1870s. Responses to this change in the institutional "environment" differed dramatically among regions. To simplify a complex and quite interesting story, the northern Italian regions rapidly built powerful, competent, and relatively popular regional government organizations, as the reforms intended. The southern regions made much slower progress. Putnam provides historical evidence that this pattern is related to an old difference between north and south. From the late medieval period onward, northern Italy was a collection of self-governing city-states—Venice, Milan, Genoa, and Florence, among others—with a very lively tradition of large-scale community participation in governance. Southern Italy, in contrast, was governed by a succession of autocratic foreign imperial powers that ruled through appointed elites. Today, northern Italy has many more vibrant community institutions than the south; a century of common experience with centralized, nationally uniform political organizations has not erased different political traditions evolved over several centuries.

Geert Hofstede, an applied psychologist working in an IBM training center in Europe, collected a huge sample of questionnaire data about employees' work-related values.[20] He obtained samples of useful size from fifty countries and a few multinational regions. The data measured workplace values related to power, gender relations, uncertainty avoidance, and individualism. One might expect selection and training as an IBM employee to dampen cultural differences, but Hofstede found ample variation remaining. Culturally related societies tended to cluster together in his sample. British, American, and Australian employees reported similar values, as did Latin American and East Asian workers.

Sudden changes in the economic or institutional environment commonly elicit unique ethnic responses. The sudden change finds some groups accidentally preadapted to the change and others not, so the groups behave quite differently. In Nigeria, the experiences of the Ibo, Hausa, and Yoruba peoples provide a good example of this phenomenon. Ibo society before colonialism had social structures that emphasized individual achievement, whereas the Hausa and Yoruba emphasized hereditary statuses with less of an emphasis on individual ambition. The growth of market economies during colonial and postcolonial times gave the traditionally more-entrepreneurial Ibo a head start in adapting to the change.[21] A similar argument has been used to explain the striking entrepreneurial achievements of some rather simple Melanesian societies compared with seemingly more-sophisticated Polynesian societies in the same region.[22] Some Melanesian societies are so precociously private-entrepreneurial-capitalist that they seem to have been invented by Milton Friedman.

These examples indicate that many important differences between human groups result from conservative, transmissible determinants of behavior—either culture, genes, or persistent institutional differences. Shortly we will present evidence that institutions cannot be the whole story in explaining these differences, and that genes play little role. First, however, we need to briefly deal with the problem of technology.

Technology is culture, not environment

Natural experiments are not the only way to refute the argument that environmental differences are the main source of human variation. Some of the most extreme proponents of ecological and economic explanations of behavioral differences (for example, the late Marvin Harris)[23] take the tool kits used by various peoples to be part of the environment. This move is especially tempting in the case of the durable environmental modifications that technology is used to construct: road networks, impressive public buildings, rice terraces, and the like have profound effects on behavior. That people with different technologies behave differently in the same environment is not seen as a problem. For example, the introduction of steel tools may have changed the human ecology of tropical horticulturalists, because such tools reduced the cost of clearing new fields, which, in turn, increased population densities and reduced the reliance on hunting. Thus, the societies of steel-using people would be different in many ways from the societies of those people who had not obtained steel technology. Some argue that this is consistent with the all-environment position, because the tools are taken as part of the environment, but surely this is cheating. The knowledge necessary to extract iron ore, smelt it into steel, and work it into useful tools is not part of the environment, and people don't acquire this knowledge by themselves in a single generation. Rather, the necessary knowledge is accumulated slowly, transmitted from one generation to another by teaching and imitation. Of course, the development of this technology will also depend on environmental factors: Is the ore available? Are the tools worth the trouble? Are populations large enough to support specialists in metalworking? However, if people do not have the necessary knowledge, then none of these factors will be relevant.

Thus, even the strongest skeptics of culture's significance must make an exception for the culturally transmitted knowledge that produces technological differences in the same environment. Many might be comfortable with technological determinist explanations granting that aspect of culture important causal power. But cracking the door of dispute this far greatly weakens the environmental determinism argument, because there is no clear dividing line between technological knowledge and other forms of knowledge. Think about public health practices, such as boiling drinking water. People who believe in the germ theory of disease typically boil drinking water drawn from polluted sources. They believe that this practice is worthwhile, even though it is troublesome, because it reduces their chances of contracting cholera, diarrhea, and many other germ-borne diseases. However, as many public health workers have found, people who have other theories of disease do not readily adopt the practice of boiling drinking water.[24] To them, the beneficial effects of this practice are hard to observe, because people get sick for many reasons, and the costs, such as gathering extra fuel for cooking fires and purchasing containers for boiled water, are clearly evident. Thus, beliefs about the causes of disease must be considered part of a people's technological knowledge. But these beliefs are also typically tangled up with all sorts of beliefs about humanity, nature, and the supernatural.

Variation in the social environment is not enough to explain human variation

Many scholars, especially in sociology and social anthropology, would agree that human differences are not caused by differences in the natural environment, but they still reject the importance of culture. Instead, they argue that variation in the social environment, not culture, creates and maintains variation among societies. The idea here is that people's behavior depends on the behavior of others. To take a familiar example, driving on the right-hand side of the road makes sense if everyone else does the same. Once one form of behavior becomes common, it will be self-perpetuating, leading to a persistent pattern of behavior that we come to recognize as an institution. Social life, it is argued, is shot through with such institutions—marriage, familial obligations, career, and so on—and these institutions cause human societies to differ, even if they exist in the same environment.

It is important to distinguish two versions of this argument. In the strong version, everyday interactions perpetuate institutions. Driving on the right-hand side of the road is an institution in many countries, because the vast majority of people do so. The institution in question is a property of the society, not of individuals. Even if every one of us had total amnesia every time we stepped out of our car, we would rapidly relearn the proper rule once back behind the wheel. Of course, we do have the habit of driving on the correct side of the road for the country we're in, but it is a quite superficial thing. Americans and Continentals adapt quickly to driving on the left in Britain, and the Swedes switched from left to right overnight when they adapted to the Continental norm. Such "games of coordination" are self-policing. Everyone has a direct, though not necessarily quite so obvious, reason to conform to the prevailing rule, no matter what it is.

In the weaker version of the argument, people learn how to behave by observing the behavior of others. Americans do not form polygynous households because they believe that such behavior is morally reprehensible, and that polygamists will be scorned by their friends and neighbors. They acquire such beliefs through teaching, and occasionally they are reinforced when some would-be polygamist gets his (or her) just desserts.

The important point is that in the weak version, the social environment is just one form of cultural variation in the sense we define it here. People acquire and store information about how to behave by observing the behavior of others and by being taught local customs. In contrast, in the strong version, the information that perpetuates historical differences is not stored in human memory; rather, it is stored in the day-to-day behavior of individuals, enforced by the self-policing incentives of games of coordination. Perhaps such institutions are quite important compared to cultural information transmitted by imitation and stored in individuals' heads. Nevertheless, in this section we present arguments that the strong form of institutional variation cannot account for the bulk of human variation. Cultures can persist even when the chain of behavior linking the past to the present is broken, and institutional variation has difficulty accounting for persistent variation within cultures.

Cultures can "reappear" after long suppression

Ideas can be stubborn things. They often persist even when the overt behavior they prescribe is suppressed for long periods by a repressive social environment. You can test your own belief that differences are maintained by self-sustaining social interactions by conducting another thought experiment. Pick your favorite culture—say, the Mae Enga of the western highlands in Papua New Guinea. Now imagine that all of the practices that make the Mae Enga distinctive are interrupted. They are forbidden to practice their religion, their elaborate exchange rituals, and their habit of frequent violent conflict with their neighbors. Instead, a different pattern of behavior is imposed on them. However, they are not forbidden to teach their youngsters about the old Mae Enga ways. Further, imagine that this imposition persists for a generation or so, and then is removed. If you think that the Enga will continue with the patterns imposed on them, or evolve new patterns that are unrelated to their previous behavior, then you agree with the adherents of the strong institutional position that culture is unimportant. On the other hand, if you think that the new behavior of the Mae Enga will reflect their old culture in important ways, you believe that cultural continuity is not maintained solely by its daily performance. Rather, it rests in longer-lived memory. If culture, not self policing institutions, creates continuity, people of a culture might be compelled by circumstance to behave according to someone else's rules but still transmit some, much, or all of their culture to their children. If the force of circumstance disappears before the culture is readapted to the new environment, all or most of the old culture may still exist, and behavior may revert to the old ways if the compulsion is removed.

The posture of the Soviet state toward ethnic minorities provides a real, albeit brutally crude, version of this experiment. Anthropologist Anatoly Khazanov describes the history of ethnic differences and nationalism in the former Soviet Union. Between 1917 and 1979, the Soviet empire quite strenuously and ruthlessly attempted to impose the idea of a new soviet citizenship upon all of the very diverse peoples of that vast system. Moreover, for centuries the southern republics, the Ukraine, and many ethnic enclaves within the Russian Federation had been subject to Russian cultural influence and political control under the czars. According to Khazanov, the ultimate goal of Soviet national policy from Lenin down to Gorbachev's reforms in 1985 was the complete Russification of non-Russian nationalities under the slogan "merging the nations."

True, constitutional fictions portrayed ethnic non-Russians as having well-protected rights, and ethnic figureheads existed in the republics. Realities were different. The Russian language was gradually imposed upon other nationalities through the educational system, starting with higher education and working downward over time. By the 1960s study of minority languages in the Russian Federation had nearly disappeared. Similar policies were pursued in the non-Russian republics as well. Mass-media programming, book publication, street signs, maps, and official and semi-official meetings were dominated by the Russian language by the 1970s. In addition, emigration by Russians to the non-Russian republics was encouraged. Estonia went from 92% Estonian in 1940 to 61% in 1988. In Kazakhstan and Kyrgyzstan, indigenous people became a minority. By 1980, a majority of the population in most republics were fluent Russian speakers. For non-Russian members of the Soviet elite, conspicuous Russification was a prerequisite. In many republics, the Russification of the elite caused considerable grumbling among ordinary citizens, and language issues sparked strong resistance in some republics, such as Azerbaijan and Armenia. Many important institutions were effectively suppressed by the Soviets, including Islamic mosques and schools. The Soviet government kept Islamic institutions very small and servile, much like the Orthodox Church.

The Bolshevik Revolution was undoubtedly a social revolution that aspired to be a cultural revolution molding all the Soviet peoples into a new society in which ethnicity was limited to a few quaint customs. Despite the change in social environment and rigorous attempts at Russification, the end of the Soviet empire in 1989 led to an immediate, and to some a surprising, outbreak of nationalism. According to Khazanov, Russian chauvinism itself substantially counteracted Soviet communist ideals of intercultural unity, obstructing the effort to create an international Soviet socialist culture. The subject nationalities of the Soviet system maintained a strong, if necessarily covert, resistance to attempts at assimilation on Russo-Soviet terms. Ethnic sentiments remained (or reemerged as) a strong force after decades of Soviet rule. In the Central Asian republics, the mass of citizens still considered themselves Muslims, and by the 1960s underground clergy were conducting religious rituals and maintaining Islamic schools. Even those who were not able to participate regularly in Islamic religious life maintained identification with Islam. Other hints, such as the high birthrates in the Muslim south, suggest that a large suite of values were being maintained. Changes that did occur were substantially independent of those desired by Soviet policy. Outside the Soviet Union, the durability of Catholicism and nationalism in Poland, the spirit of private economic enterprise in China, and ethnic enmities in the Balkans impress us as examples of cultural continuity over generations in the face of severe institutional repression.

The exact means by which cultures were preserved during the Soviet period and the degree to which they remained intact is an untold story. Journalist Stephen Handelman chronicled some of these for an unusual quasi-ethnic group, the traditional Russian "Mafia."[25] The so-called Thieves' World subculture of organized crime has deep roots in czarist Russia. The Bolsheviks of the revolutionary period had a tendency to romanticize the Thieves' World as primitive revolutionaries, and expected its members to embrace the revolution after 1917. Instead, it persisted straight through Stalin's terror, operating as similar organizations do in the United States and Italy, as much from within prison as on the outside. Incredibly, in a state that tried to control its inhabitants' lives with a large and ruthless police bureaucracy, the iron rule of the Thieves' World meant that members may never take an official job. Even a powerful police state could not destroy such an organization. The Thieves' World's crisis came in the aftermath of World War II. During the war, large numbers of participants became sufficiently caught up in the patriotic fervor of resistance to the Nazis to become soldiers. This provoked a civil war within the Thieves' World that pitted returning soldiers against traditionalists who maintained that even service under such extreme circumstances violated the norm of no participation in legitimate organizations.

There are a number of other examples of this sort. In the United States, we have thus far utterly failed to win the war on drugs. Despite very high incarceration rates for drug offenses, and much official anti–drug use propaganda, drug subcultures are proving extremely durable in the face of repressive social environments. Another example can be found among Orthodox Christian communities that survived Ottoman repression throughout Anatolia and the Balkans.[26] The ability of heretical ideas to persist in Europe in the medieval and early modern periods, despite persecution by Catholic and Protestant authorities, kept a yeasty brew of beliefs and practices alive for centuries, and contributed to movements such as Masonry and Mormonism on the nineteenth-century American frontier.[27]

A mere disruption of the overt expression of culture will often fail to erase it. This does not mean that cultures are immutable; situations exist in which the desire to assimilate exceeds loyalty to tradition. However, socialization by parents and the willingness of priests and patriots to maintain underground organizations even at considerable risk to themselves can perpetuate substantial portions of a traditional culture in an extremely hostile and radically altered social environment. Culturally transmitted ideas do seem sufficient to reconstruct functioning social systems, even after long periods of suppression, which clearly falsifies the strong version of the institutional argument.

Social environment explanations have difficulty accounting for variation within groups

Not all people who live together are the same, and evidence suggests that culture plays a role in the differences. For example, the patterns of ethnic variation within the farm communities studied by Salamon are similar to those *among* communities.[28] Salamon studied the community of "Prairie Gem," which was settled by a mixture of Yankees and German immigrants. In 1890, Germans owned about 20% of the land; by 1978 they owned about 60%. In 1978, 66% of the absentee owners were Yankees, and only 43% of the resident owners were Yankees. Thus, Yankees living side by side with Germans in the same community behave much as they behave when they live in separate communities. A similar contrast exists in the predominantly Swedish community "Svedburg." The Swedes share with the Germans a strong commitment to keeping their farms in the family, and they are more likely to help their sons get started in farming than are Yankees. For example, 62% of the Swedes who were renters or part owners obtained their land with their father's help, while less than a quarter of the Yankee renters received parental assistance.

This kind of variation is difficult to explain in purely social-structural terms. In the cases where Germans or Yankees dominate a community, one might imagine that some institutional hypothesis could explain behavioral variation. But the Yankees and Germans of Prairie Gem interact every day —be it for business or social reasons. They farm the same soils in the same economic climate using the same technology. The only thing that distinguishes them is their ethnic heritage. How could day-to-day interaction in Prairie Gem motivate Germans to farm one way and Yankees to farm another way unless they had different culturally transmitted ideas, beliefs, and values?

Little behavioral variation between groups is genetic

Most people we know are rather immoderate on the question of whether behavioral differences among humans have a genetic basis. Many of our colleagues consider the question to be settled: there is no important genetic variation affecting behavior, and anybody who says that there is must have odious motives. At the same time, many of our friends and relatives seem to be thoroughgoing hereditarians. They say that their children get their good nature and quick wit from their parents, and they also say, particularly in unguarded moments, that the members of other ethnic groups are "born" different.[29] People are also usually confused, despite their passion, by the nature/nurture dichotomy.

We think that typical academics' beliefs about the heredity issue are barely better informed than folk psychology. Recent research in behavior genetics suggests that some behavioral variation among individuals has a substantial genetic component and a substantial environmental component. However, these results provide no evidence that variation among *groups* has any genetic component. Moreover, compelling natural experiments suggest that virtually none of the behavioral differences we see among the peoples of the world have a genetic basis.

Behavior genetics suggests that some differences among individuals are partially genetic

Most people think that children get basic values from their parents. Little Phyllis learns to condemn abortion from her conservative parents, while little Tom learns to favor a woman's right to choose from his liberal ones. This common view has long been endorsed by social science; innumerable studies show the similarity in attitudes of parents and offspring, and almost everyone [30] has assumed that this is because children learn social attitudes at home.

However, research by behavior geneticists casts doubt on the common view. The social attitudes of parents and offspring are correlated, all right, but these correlations result from genes that the children inherit.[31] These investigators administer questionnaires to large numbers of people, including identical twins, fraternal twins, related and unrelated people who lived in the same household during their childhood, and relatives who live in other households. There have been a number of different studies, but in each case all subjects were white middle-class citizens of a single country, either Australia, the United Kingdom, or the United States. The questions elicit attitudes toward topics such as modern art, capital punishment, and pajama parties. Statistical methods are used to cluster the answers into personality dimensions that psychologists label introversion-extroversion, neuroticism, psychoticism, religiosity, and conservatism.[32] Much work in psychology suggests that these dimensions tap fundamental aspects of personality. The importance of genetic and cultural transmission within the family is measured by statistically comparing the social attitudes of people who have the same family experience but different degrees of genetic similarity. For example, if learning from parents predominates, then pairs of adopted children, siblings, fraternal twins, and identical twins ought to be equally similar. If genetic transmission is most important, then identical twins should be most similar. Fraternal twins and siblings should be somewhat similar, while adoptees and their adopted relatives ought to be no more similar than any two people in the sample.

Results from several independent studies suggest that cultural transmission within the family is not very important; the similarity between parents and offspring is mainly due to genes. If these results stand up and generalize to other sorts of characters, then it would tell us that parents are less important in cultural transmission than many people suppose. Little Phyllis apparently abhors Democrats partly because she inherited genes from her parents that predispose her to adopt conservative views, and in part because of what she learns or observes or acquires by chance outside the family. While these studies have been criticized on a number of grounds,[33] the claim that there are heritable genetic differences among people is quite plausible. It is a truism among evolutionary biologists that all kinds of continuously varying traits show substantial genetic variation, including behavioral traits like the tendency of rodents to explore a cage, pigeons to return home, or dogs to "point." Given that the propensity of people to adopt one social attitude over another is likely affected by many aspects of brain chemistry and organization, and given that such aspects of the brain are likely affected by many different genes, it is certainly plausible that some of the variation in people's responses on a questionnaire, as well as perhaps their behavior, is affected by genetic variation. Indeed, if humans had *no* genetic variation at the individual level, we would be something new under the sun.

However, the existence of genetic variation does not mean that cultural transmission is unimportant. In most of the studies more than half of the variation in children's personality is associated with what behavior geneticists call nonfamily environment, which they interpret as the effects of the idiosyncratic events of individual lives. In this scenario, Joe had conservative parents, but is pro-choice because a good friend died as a result of an illegal abortion. But this is not the only sensible interpretation; the nonfamily environment could equally well be due to the effects of learning from other individuals: friends, clergy, fraternity brothers or sorority sisters, colleagues, and perhaps even professors. Since the behavior geneticists know only the attitudes of parents, they cannot exclude this interpretation of their results. Joe may have learned his views about abortion from a charismatic teacher. Moreover, this interpretation is consistent with the fact that the effect of family environment on some traits, most notably IQ, is fairly high for small children and then decreases as subjects approach adulthood. As the number of different individuals influencing a child's attitudes increases, the parental effect decreases until it drops below the level of resolution of the methods used in these studies.

Dialect variation is one example of a cultural system that is strongly influenced by nonfamily environment. Sociolinguists know a lot about the genesis of small-scale variations in dialect.[34] Children almost always learn their native language from their parents at home. However, as youngsters leave the household to interact with peers, they almost always switch their dialect from that of their parents to that of their peers. This is true of language evolution, which is led by younger people, whose dialect is detectably different from that of the older generation. It is also true of people who migrate across linguistic boundaries or gradients. Adults often struggle to conform to the norms of a new region, whereas younger children adjust completely. In terms of dialect *variation,* parents have almost no effect on children even if primary language socialization is, as it seems to be, *overwhelmingly* familial! If it happened to turn out (studies are lacking as far as we know) that innate vocal tract anatomy has a modest effect on dialect performance, then a dialect variable would have the same pattern as personality variables. There would be a genetic effect of parents acting through the heritability of anatomical features and a nonfamily environment effect due to dialect learning. The parents' large role in socialization disappears from view in this case even if most early language skills are learned from parents. In essence, parents normally transmit basic language traits to children, but the kids in turn acquire from peers the nuances that make up the variation.

High heritability within groups says nothing about variation between groups

Let's suppose that after much careful research, every sensible person was convinced that variation in social attitudes among white, middle-class Americans was largely due to genetic differences. For many people, this would imply that social attitudes are genetically transmitted. Obviously, social attitudes differ substantially among different populations—Scandinavians differ from Americans, who differ from Germans, and so on. If social attitudes were genetically transmitted within each society, wouldn't it follow that the variation in social attitudes that exist among groups are also genetic?

Our answer is a very testy NO!! That much of the variation in social attitudes among white, middle-class Virginians is genetic does not mean that social attitudes are genetically transmitted. It means that there is genetic variation which affects social attitudes, and that these effects are large compared with the effects of cultural and environmental differences among white, middle-class Virginians. It does not say that the differences in social attitudes between white, middle-class Virginians and, say, white, middle-class Danes are the result of genetic differences between these two groups. That would be true only if two quite different conditions held: first, a genetic difference must exist between Virginians and Danes *on the average,* and second, this average genetic difference must be large compared with the average difference in culture and environment between the two groups. That there is genetic variation *among* Virginians does not tell us whether they are genetically *on average* different from Danes. Nor does the relative

lack of environmental or cultural variation among Virginians tell us anything about the average difference in environment or culture between Virginians and Danes.

This is not rocket science; it is just common sense. Behavioral geneticists themselves are usually careful to underline the distinction between
heritable differences within populations and those between populations.[35]
Nonetheless, year after year undergraduates—and, alas, sometimes scientists who should know better—leap to the conclusion that differences
among groups are genetic even though they all are familiar with evidence
that ought to convince them of the opposite. It is to this evidence that we
now turn.

Little behavioral variation among groups is genetic

Two kinds of evidence show that much of the behavioral differences
among groups are not genetic. First, individual cross-cultural adoptees behave like members of their adopted culture, not the culture of their biological parents. Second, groups of people often change behavior much more
rapidly than natural selection could change gene frequencies. These data
are far too coarse to prove that there are *no* genetic differences between human groups, but we believe the evidence is sufficient to conclude that the
cultural differences between groups are much larger than any genetic variation that might exist.

Cross-cultural adoption

In recent years, cross-cultural adoption has become fairly common. Japanese, Korean, and Vietnamese children have been adopted into American
families, Navaho children have been adopted into Mormon families, and
Latino children have been adopted into Anglo families. If the differences between, for example, Korean society and American society were caused by
genetic differences between the two groups, adopted children would grow
up with the beliefs, values, and attitudes of their biological parents. But, of
course, this is not what happens. Adopted kids grow up with beliefs, values, and attitudes of the culture in which they are raised.

Only a few good studies of transcultural, especially transracial, adoptions exist, with[36] developmental psychologist Lois Lydens's study of 101
Korean children adopted by white American families being one of the best.
Her sample included 62 children adopted before the age of one year and
39 adopted after the age of six, most of whom became wholly acculturated,

successful "white" Americans. Adoptees develop perfectly healthy self-concepts, for example, differing little from the normal calibration sample employed in constructing the clinical test used. Children adopted later in life showed some significant deficits on subscales of the test that reflect self-certainty, global self concept, and adjustment in adolescence, but most of these effects had disappeared at the time of a retest in early adulthood. Even in adulthood, older adoptees had measurably, but only slightly, poorer feelings about their families than children adopted at younger ages. Lydens's sample clearly shows that growing up as a racial minority in a society with a significant amount of racial prejudice has some effects. For example, young adult adoptees had slightly below-normal satisfaction with their physical appearance. In free-form questions, both children and parents cited prejudice as a significant problem in the lives of the transracial adoptees.

The most striking thing is how little effect such prejudice had on the overall self- and even ethnic concepts of transracial adoptees. Many parents took care to be supportive of kids learning about their birth ethnic group, but few adoptees showed much sign of interest. Those that did were predominantly older adoptees. The adopted children studied were raised in mostly conservative religious homes with a strong commitment to making the adoptions work. As young adults, the adoptees were quite successful, with only four not graduating from high school and two unemployed. If there were big population-level genetic effects on behavior, one would predict that populations as distantly related as those from far western and far eastern Eurasia would encompass a fair fraction of the total human variation, and some detectable departures from Euro-American norms would turn up in Korean adoptees in the United States. Instead, adopted Koreans make perfectly assimilated Americans, except for the surprisingly minor hitch introduced by racism.

The ideal transcultural adoption "experiment" would include reciprocal adoptions. Would Anglo American kids adopted by Koreans make well-assimilated Koreans? Koreans, as it turns out, generated a one-way flow of adoptees to the United States because they oppose adoption outside the family. However, Anglo Americans did historically contribute a number of involuntary adoptees to American Indian parents, who are historically derived from northeast Asian populations. The aggressive frontier settlement of Anglo Americans generated the well-documented conflicts between these peoples, and as everyone knows, the Europeans often lost. This was particularly true during the long preindustrial period before 1776 when the frontier was only slowly moving westward. Victorious Indians often took captives; adult captives were normally killed, but children and adolescents were often adopted. Most often, Indian couples who had lost children took captives, mostly between five and twelve years of age, to replace them— swift-moving warriors were seldom able to manage infants and toddlers. Very strenuous efforts were made by whites to retake or ransom captives, even many years after the event, and French and British Canadians often helped American families recover captives from their tribal allies. Sometimes individuals who had been adopted at an early age and had lived decades as adoptees were recovered by their natal families following a serious, often final, defeat by the invading Anglo Americans. The pathos of the captives' stories led to a well-developed nonfictional (and fictional) literature detailing the experience, from which a fair sample of well-documented cases can be reconstructed.[37]

Historian Norman Heard assembled a sample of fifty-two captive accounts, weighted toward those in which adoptions took place, and in which information about the age, national origin, duration of captivity, and outcome of captivity were reasonably reliably recorded. The story of Cynthia Ann Parker is typical. She was taken captive at age nine in 1836 when a large party of Comanche and allies seized her father's trading fort in Texas. She had been taken along with three others, but they were redeemed fairly promptly. Eventually, Cynthia Ann was adopted by a Comanche family and lived twenty-four years with them. She married a chief and had three children, one of whom, Quanah, became an important chief in his own right. By Heard's estimate, Cynthia Ann became 100% Indian. In 1860 she was "redeemed" by a Texas Ranger and sent to live with an uncle, from whom she tried to escape several times. Although she regained the use of English and adapted to Anglo life, she retained her emotional attachment to the Comanche. Her "redemption" amounted to a second kidnapping, one to which she was too old to adapt. After the death of her little daughter, who was "redeemed" with her, Cynthia Ann fell into a depression and died herself.

In Heard's sample, age at capture, duration of capture, and type of treatment influenced whether assimilation to Indian life occurred. Young captives treated well for any length of time assimilated. Living with Indians into adulthood, especially forming an Indian family, generally resulted in individuals whose entire ethnic identification lay permanently with their adoptive group, as with Cynthia Ann Parker. Older children, treated badly and recovered shortly, generally remained essentially white, though a few teenage boys found the free and easy life of the Indians preferable to the straight-laced, hardworking Calvinism of their birth communities. "Good treatment" almost invariably meant formal adoption by an Indian family. Adopted individuals were treated with the same love and affection as Indian children and acquired the same rights and duties as any other member of the community. Western Indians sometimes kept child captives as domestic menials rather than adopting them, and the degree of assimilation of such captives, when they survived, was substantially reduced. Adopted children might lead a hard life for some period before they chanced to be adopted, and dated their real integration into the Indian community to adoption, not capture per se. Indian communities were only mildly racist, so physical difference between adoptees and birthright Indians was not a major handicap.[38] Among adoptees of Indians, the reminiscences of the strong emotional bonds to adoptive families primarily and their adoptive culture secondarily are remarkably parallel to those quoted by Lydens from the questionnaire responses of her Korean adoptee subjects.

In short, most children adopted into another culture before the age of ten or so, even with a history of traumatic capture or indifferent orphanage upbringing, will fully assimilate emotionally into another culture and become fully functional members of it. This result is not surprising to most people. Nonetheless, it is an extremely strong test of the theories under consideration here. If the behavioral differences between groups were substantially due to genetic differences, adoptees should show significant departures from norms of behavior of their foster culture.

Rapid cultural change

Many people erroneously think that natural selection always takes millions of years to do its work, but several lines of evidence suggest that it can act much more quickly. First, biologists have actually observed rapid evolutionary change in short periods of time. For example, a drought in the Galapagos reduced the availability of small, soft seeds preferred by one species of Darwin's finches. Careful studies by biologists Peter and Rosemary Grant[39] showed that those birds with thicker beaks were better able to process the larger, harder seeds that were available, and as a result were more likely to survive, and that beak thickness was heritable. Beak depth changed 4% in two years, a rate sufficient create a new species in less than forty years.[40] Artificial selection demonstrates that such changes can go on long enough to result in major changes in behavior and morphology. For example, all breeds of dogs are probably descended from wolves during the last fifteen thousand years. This means that artificial selection can change a wolf into a Pekinese in a few hundred generations. Finally, the fossil record indicates that substantial morphological change sometimes occurs on the timescale of a few thousand generations. At the beginning of the last interglacial period, about 120,000 years ago, rising sea levels caused the island of Jersey to be isolated from the European mainland. Fossil evidence shows that within six thousand years, the size of red deer (or in American nomenclature, elk) on the island had decreased by a factor of 2—in about one thousand generations, natural selection shrank red deer to the size of a large dog.

Human cultures can change even more quickly than the most rapid examples of genetic evolution by natural selection. We are all familiar with the frantic pace of cultural change during this century, and while this pace is unusual, it is not unique. For example, the complex artifacts, institutions, and behaviors we associate with the Plains Indians arose *after* the introduction of horses to the southern Great Plains by Spanish frontiersmen in northern Mexico in about 1650.[41] Before that time, the Great Plains were sparsely populated, because nomadic buffalo hunting was not a very productive subsistence strategy for foot hunters. Mounted hunters could match the mobility of the buffalo and reliably slaughter them in numbers. With the arrival of horses, people poured out onto the plains. From the East came people like the Crow, Cheyenne, and Sioux, who abandoned sedentary farming in river valleys where they had lived in large villages with kin-based clans and complex, large-scale political organization. From the West came nomadic hunter-gatherers such as the Comanche, and from the North came forest foragers like the Cree. These hunter-gatherers had lived in small family groups without permanent villages, complex kinship systems, or substantial political organization. During the late eighteenth and early nineteenth centuries, Great Plains tribes from the East, West, and North evolved a quite new way of life. During the summer, people who had spent the winter in small family groups gathered together in large groups for hunts and ceremonies. There, most tribes were governed by "police societies," a kind of political institution without close parallel in either the farmers of the East or the foragers of the West.

To be sure, different tribes carried many traces of the past—the Crow were matrilineal like their ancestors, while the Comanche had the flexible kinship system characteristic of their ancestors, but to a remarkable degree a wholly new economic and social system arose in less than twelve generations. Natural selection could not act so quickly, and so the original differences could not have been genetic. The possibility of diffusion of cultural innovations across group boundaries means that whole societies can, under favorable circumstances, acquire these innovations very rapidly. Once introduced into a group, obviously useful innovations will be imitated by everyone within a generation, more or less. Horses and riding spread rapidly beyond the Spanish frontier, and the various horse tribes traded innovations in social organization back and forth. We could use many other examples to illustrate the point. Behavioral change in human populations is very often too rapid to be easily explained by natural selection, and the intersocietal pattern of spread of the innovations is in any case inconsistent with a genetic explanation for the spread of a favored new behavior.

Much culture is not evoked

In their critique of what they characterize as the culture-saturated "Standard Social Science Model,"[42] evolutionary psychologists Leda Cosmides and John Tooby introduced the distinction between "epidemiological" and "evoked" culture. Epidemiological culture refers to what we simply call culture—differences between people that result from different ideas or values acquired from the people around them. Evoked "culture" refers to differences that are not transmitted at all, but rather are evoked by the local environment. Cosmides and Tooby argue that much of what social scientists call culture is, instead, evoked. They ask their readers to imagine a jukebox with a large repertoire of records and a program that causes a certain record to be played under particular local conditions. Thus, all the jukeboxes in Brazil will play one tune and all those in England will play another tune, because the same gene-based program orders up different tunes in different places. Tooby and Cosmides believe that anthropologists and historians overestimate the importance of epidemiological culture, and emphasize that much human variation results from genetically transmitted information that is evoked by environmental cues.

They are led to this conclusion by their belief that learning requires a modular, information-rich psychology. Cosmides, Tooby, and some other evolutionary psychologists[43] think that general-purpose learning mechanisms (like classical conditioning) are inefficient. When the environment confronts generation after generation of individuals with the same range of adaptive problems, selection will favor special-purpose cognitive modules that focus on particular environmental cues and then map these cues onto a menu of adaptive behaviors. Evidence from developmental cognitive psychology provides support for this picture of learning—small children seem to come equipped with a variety of preconceptions about how the physical, biological, and social world works, and these preconceptions shape how they use experience to learn about their environments.[44] Evolutionary psychologists think the same kind of modular psychology shapes social learning. They argue that culture is not "transmitted"—children make *inferences* by observing the behavior of others, and the kind of inferences that they make are strongly constrained by their evolved psychology. Linguist Noam Chomsky's argument that human languages are shaped by an innate universal grammar is the best-known version of this argument, but evolutionary psychologists think virtually all cultural domains are similarly structured.

For example, cognitive anthropologist Pascal Boyer argues that much religious belief derives from human psychology, not cultural transmission.[45] The Fang, a group in Cameroon studied by Boyer, have elaborate beliefs about ghosts. For the Fang, ghosts are malevolent beings that want to harm the living; they are invisible, they can pass through solid objects, and so on. Boyer argues that most of what the Fang believe about ghosts is not transmitted; rather, it is based on the innate, epistemological assumptions that underlie all cognition. Once young Fang children learn that ghosts are sentient beings, they don't need to learn that ghosts can see or that they have beliefs and desires—these components are provided by a sentient-being cognitive module that reliably develops in every environment. Like Cosmides and Tooby, Boyer thinks that many putatively cultural religious beliefs arise because different environmental cues evoke different innate information. Your neighbor believes in angels instead of ghosts because he grew up in an environment in which people talked about angels. However, most of what he knows about angels comes from the same sentient-being cognitive module that gives rise to Fang beliefs about ghosts, and the information that controls the development of this machinery is stored in the genome, an organism's genetic material. Cognitive anthropologist Scott Atran makes a similar argument for ecological knowledge.[46]

This picture of culture is a useful antidote to the simplistic view that culture is simply poured from one head into another. These scholars are surely right in stating that every form of learning, including social learning, requires an information-rich innate psychology, and that much of the adaptive complexity we see in cultures around the world stems from this information. However, ignoring transmitted culture completely is a big mistake. As we will see in chapter 4, the single most important adaptive feature of culture is that it allows the gradual, cumulative assembly of adaptations over many generations, adaptations that no single individual could evoke on his or her own. Cumulative cultural adaptation cannot be based directly, or in detail, on innate, genetically encoded information.

Evolutionary psychologists argue that our psychology is built of complex, information-rich, evolved modules that are adapted for the hunting and gathering life that almost all humans pursued up to a few thousand years ago. On this argument, humans can easily and naturally do the things we're adapted to do, like learn a language. Learning subjects such as differential calculus is much harder, and evolutionary psychologists are probably willing to make exceptions for modern societies and admit that cumulative evolved culture matters there. But what about hunting and gathering? Couldn't we learn that as easily as we learn language? Doesn't our brain contain the information necessary to follow hunting and gathering ways? Our lineage has lived as hunter-gatherers of some kind or another for the last two million or three million years. If we had to do so, couldn't we reinvent the things it takes to survive as a hunter-gatherer, in the same way that children reared in a multilingual community of immigrants are supposed to be able to invent a new language in a single generation?[47]

Good questions, but we think the answer is almost certainly "Are you nuts?!" Consider another thought experiment. Suppose we are stranded in some not-too-extreme desert environment (not the central Sahara or the Empty Quarter of Arabia). Our task is to survive and raise our kids. Deserts are fairly harsh environments, but harsh environments were the Pleistocene norm, and we know that hunting-gathering societies have adapted well to all but the harshest. We have spent considerable time in deserts. Like successful hunter-gatherers, we know a lot about their natural history compared to the average person, and have a good generic knowledge of how hunter-gathers exploit them. We're used to camping out and are fairly fit (in consideration of middle-age infirmities, allow us to begin this experiment twenty-five years ago). However, we certainly don't command any practiced hunter-gather skills. If such skills are needed to survive as hunter-gatherers in deserts, they had better be lying quietly, heretofore little used, in innate modules in our heads. Give us the resources to survive a few months in our new home before you take away our last steel tool and last can of beans—a little time to see what comes naturally.

Would we make it? Consider a typical desert subsistence task—crossing a long dry stretch of desert from one water source where resources are exhausted to another where they may be better. We have a particular trek in mind, from Sonoita in northwestern Mexico to Yuma, Arizona, on the Colorado River. The distance is about one hundred miles, and there are several fairly reliable "tanks" along the route where water can usually be had. We have a pretty good idea where these are, but have not actually taken the trouble to fix their locations precisely on past trips. Desert peoples have a number of tricks to find stored water so that they can survive these treks. In the American Southwest, they included using barrel cacti as emergency water sources, finding small "perched" aquifers in sandy wash bottoms, killing animals that have blood for drinking and wet flesh for eating, and so forth. Knowing this, we set out.

What are our chances of getting to Yuma alive? We guess only fair. Desert water holes are not easy to find unless you know exactly where to look. The locally adapted hunter-gatherer would know which birds need open water and could use them as clues for the distance and direction to water. Ditto for mammals that create a web of tracks centered on the tanks. We could use this kind of information if we had the skill to interpret the signs. Some plants that grow near water are visible at great distances—but only if you know what to look for. In our experience, a year is precious little time to come to know much about the habits of even one species of animal by personal observation, let alone many. We have read about all of the things we describe here, but it is only book learning—it tells us it's possible to do these things, but doesn't really provide much help in acquiring the skills we'd need to do them. We may find a way to craft some sort of canteen or water skin to transport water between tanks, but figuring out how to make such implements would take some time, and we will have many things to learn in our months of grace. The famous barrel cacti sound promising and are moderately abundant. But are all species useful? In all seasons? After a year of below-normal rain? Is this a year of normal rain or not? Lacking steel tools, how do you get past those pesky spines? Or is the barrel cactus idea mainly a legend of little or no practical utility? Even though we know where to start and we've read a lot of books and had months to practice, this trip is going to be an adventure to say the least.

In fact, the trip we describe is along the Camino del Diablo, "Devil's Road"—a bad stretch of the main land route from Old Mexico to California, used until the arrival of the railroad. For more than a century, Spanish, Mexican, and American travelers used El Camino del Diablo routinely. To get that far, every traveler had to be an experienced frontiersperson already, and no doubt most were hard-bitten, desert wise, and well equipped with familiar technology. It was the best of several bad routes and was comparatively well known and well marked. Still, it was an infamous leg of the journey, with more than its share of hasty graves dug alongside.

The Camino del Diablo area was also home to Tohono O'odam Indians, who not only traveled across the region but made a living there. If we were to do the same, we'd have to confront a succession of challenges, each of which is the same magnitude as our simple trek. Mastering them all, even starting with a goodly bit of relevant theory and some desert experience, doesn't seem to us a likely thing at all. Ethnographers remark on the subtlety of desert hunting and the complexity of hunting knowledge, belying the relative simplicity and paucity of the tools desert hunters use. A few pounds of wood, stone, and bone equipment is all you need, but you have to command a rather impressive amount of hard-won practical knowledge about natural history and have a system of supporting social institutions to make a go of it. We know from archaeology that the refinement of hunting and gathering technology to harsh environments of the high Arctic by the Eskimo and their predecessors took about eight thousand years. The same timescales obtain in provident environments like California, where the productive salmon- and acorn-based economy took about the same amount of time to evolve.[48] We think it very likely easier to acquire the skills required for hunting and gathering than to learn calculus, and this suggests that we may have some innate propensities for this lifestyle. Ethnographic accounts (and a bit of introspection) lead us to believe that most kids would rather spend time fiddling about with bows and arrows than practicing multiplication tables or mastering long division. But we'd trade a few hours of tutoring by a traditional Tohono O'odam for any number of months of trying to summon an innate knowledge of the desert if our task were to get to Yuma via the Camino del Diablo. (Untutored, it is an interesting junket if you have an SUV, five gallons of water, a full tank of gas, and permission from Barry Goldwater Bombing Range.)

Cultural adaptations evolve by the accumulation of small variations

There is yet another way that some evolutionary psychologists downplay the role of culture. For example, psycholinguist Steven Pinker writes,

> A complex meme does not arise from the retention of copying errors. It arises because some person knuckles down, racks his brain, musters his ingenuity, and composes or writes or paints or invents something. Granted the fabricator is influenced by ideas in the air, and may polish draft after draft, but neither projection is like natural selection.[49]

The idea here is that complex cultural adaptations do not arise gradually and blindly as they do in genetic evolution. New symphonies don't appear bit by bit as a consequence of the differential spread and elaboration of slightly better and better melodies. Rather, they emerge from people's minds, and their functional complexity arises from the action of those minds. The same goes for novels, paintings, and inventions, or so Pinker thinks. Culture is useful and adaptive because populations of human minds store the best efforts of previous generations of minds.[50]

On this view, culture is like a library. Libraries preserve knowledge created in the past. Librarians shape the contents of libraries as they decide which books are bought and which are discarded. But knowing about libraries and librarians does not help us understand the complex details of plot, character, and style that distinguish a masterpiece from a potboiler. To understand these things, you have to learn about the authors who wrote these books. How does universal human psychology shape the nature of storytelling? And how was the psychology of particular authors affected by their environments? In the same way, cultures store ideas and inventions, and people's "decisions" (often unconscious) about which ideas to adopt and which to reject shape the content of a culture. However, to understand a new complex, adaptive cultural practice, a new tool or institution, you have to understand the evolved psychology of the mind that gave rise to that complexity, and how that psychology interacts with its environment.

Students of the history of biology will recognize this picture of cultural evolution as similar to a frequently popular but incorrect theory of genetic evolution. Very few of Darwin's contemporaries accepted (or even understood) his idea that adaptations arose through the gradual accumulation of small variations. Some of his most ardent supporters, like T. H. Huxley, thought that new adaptations arose in big jumps, and then natural selection accepted or rejected these "hopeful monsters." In this century, biologist Richard Goldschmidt and paleontologist Stephen Jay Gould among others championed this theory of evolution.[51] It is wrong because the likelihood that a complex adaptation will arise by chance is vanishingly small. Of course, this objection does not have the same force for cultural evolution, because innovations are not random; and thus cultural evolution could conceivably mainly involve the culling of complex innovations, innovations that have to be understood only in terms of human psychology.

If complex culturally transmitted adaptations were mainly hopeful monsters, then the study of the population dynamics of ideas would be of some interest because it would help us understand why some hopeful monsters spread and others fail. However, population-based theory is much more important if most complex cultural adaptations were assembled by the gradual accumulation of small variations like organic adaptations. And, the evidence convinces us that this is exactly the way most cultural change occurs.

Culture usually evolves by the accumulation of small variations

Isaac Newton famously remarked that he stood on the shoulders of giants. For most innovators in most places at most times in human history, a different metaphor is closer to the truth. Even the greatest human innovators are, in the great scheme of things, midgets standing on the shoulders of a vast pyramid of other midgets. The evolution of languages, artifacts, and institutions can be divided up into many small steps, and during each step the changes are relatively modest. No single innovator contributes more than a small portion of the total, as any single gene substitution contributes only marginally to a complex organic adaptation. The limited imitative capacities of other animals seem to prevent the cumulative evolution of complex cultural features. At best, some chimpanzee innovations such as the use of hammers and anvils for cracking nuts may represent a two-step accumulation.[52]

The case of language illustrates the general principle that the cumulative effect of many small changes can be a powerful source of cultural change. In some cases, only a few differences of phonology, syntax, and lexicon separate closely related dialects. Careful dialect descriptions conducted in the United States in the 1930s allow contemporary linguists to describe in some detail the generation-to-generation change in language.[53] In one generation some dialect changes are rapid enough to be detectable to the trained ear. For example, New Yorkers are gradually tending to pronounce *r* at the end of words like *car* more often. Over time, these small changes accumulate. Without the benefit of an expert's notes, most of us miss many subtleties in Shakespeare's plays, and Chaucer is nearly impossible to follow. Still, to a comparative philologist, Middle English is closely related to Modern English. Modern English is even appreciably related to ancient Indo-European via a collection of words such as *agras* = field, from which the English *agrarian* is derived, which have cognates scattered across central and western Eurasia.

Most readers, we are sure, come to this book with the intuition that individual humans are pretty smart, and that this is mainly what is responsible for most of the spectacular accomplishments of our societies. However, there is much evidence that suggests that this view is wrong.[54] Psychological studies of human decision making indicate that human rationality is narrowly bounded. Human decisions and the psychological reasons that underlie those decisions are a fundamental part of cultural evolution.[55] We don't mean to denigrate individual human agency at all, merely scale it against the complexity of cultural adaptations arrived at by the cultural evolutionary process operating over considerable reaches of time and space.

The history of technology[56] shows that complex artifacts such as watches are not hopeful monsters created by single inventors. The watchmakers' skills have been built up piecemeal by the cumulative improvement of technologies at the hands of many innovators, each contributing a small improvement to the ultimately amazing instrument. Many competing innovations have been tried out at each step, most now forgotten except by historians of technology. A little too loosely, we think, historians of technology liken invention to mutation because both create variation, and compare the rise to prominence of the successful technology with the action of natural selection.[57] Forget watches for a moment. The historian of technology Henry Petroski documents how even simple modern artifacts like forks, pins, paper clips, and zippers evolve haltingly through many trials, some variants to capture the market's attention and others to fall by the wayside. No one knows how many failed designs languished on inventors' workbenches.[58] Most of the rest of this book is about how things are more complicated than bare-bones random variation and selective retention. To anticipate our argument, the decisions, choices, and preferences of individuals act at the population level as forces that shape cultural evolution, along with other processes like natural selection. We urge great care with loose analogies to mutation and selection because several distinct processes rooted in human decision making lead to the accumulation of beneficial cultural variations, each with a distinctive twist of its own and none exactly like natural selection.

While human innovations are not like random mutations, they have been small, incremental steps until recently. The design of a watch is not the work of an individual inventor but the product of a watch-making tradition from which the individual watchmaker derives most, but not quite all, of his design. This is not to take anything away from the real heroes of watch-making innovation, such as John Harrison. Harrison delivered a marine chronometer accurate enough to calculate longitude at sea to the British Board of Longitude in 1759. He used every device of the contemporary clockmaker's art and a number of clever tricks borrowed from other technologies of the time, such as using bimetallic strips (you have seen them coiled behind the needle of oven thermometers and thermostats) for compensating the critical temperature-sensitive timekeeping elements of his chronometers. His achievement is notable for the sheer number of clever innovations he made—the bimetallic temperature compensators, a superb escapement, jewel bearings requiring no lubrication, substitutes for the pendulum. It is also notable for his extraordinary personal dedication to the task. By dint of thirty-seven years of unremitting effort and a first-rate mechanical mind, sustained by incremental payments against a British Admiralty prize he was a good candidate to win, Harrison made a series of ever smaller, better, more-rugged seagoing clocks. Eventually he delivered "Number 4," with an accuracy of better than 1/40th of a second per day, a significant improvement over one minute per day for the best watches of his time.[59] Only the rarest of inventors makes an individual contribution of this magnitude. Yet, like every great inventor's machine, Number 4 is a beautiful homage to the art and craft of Harrison's predecessors and colleagues as much as to his own genius. Without a history of hundreds or thousands of ancient and mostly anonymous inventors, he would not even have conceived the idea of building a marine chronometer, much less succeeded in building one. The eighteenth-century theologian William Paley's famous Argument from Design would better support a polytheistic pantheon than his solitary Christian Creator; it takes many designers to make a watch.

Consider a much simpler nautical innovation, the mariners' magnetic compass. Its nameless innovators must have been as clever as Watt, Edison, Tesla, and the other icons of the Industrial Revolution whose life stories we know so much better.[60] First, someone had to notice the tendency of small magnetite objects to orient in the earth's weak magnetic field in nearly frictionless environments. The first known use of this effect was by Chinese geomancers, who placed polished magnetite spoons on smooth surfaces for purposes of divination. Later, Chinese mariners built small magnetite objects or magnetized needles that could be floated on water to indicate direction at sea. Ultimately, Chinese seamen developed a dry compass with the needle mounted on a vertical pin bearing, like a modern toy compass. Europeans acquired this form of compass in the late medieval period. European seamen developed the card compass, in which a large disk was attached to a pair of magnets and marked with thirty-two points. This compass was not merely used to indicate direction but was rigidly mounted at the helmsman's station, with a mark on the case indicating the bow of the ship. Now the helmsman could steer a course as accurate as 1/64th of a circle by aligning the bow mark on the case with the appropriate compass point. Compass makers learned to adjust iron balls near the compass to zero out the magnetic influence from the ship, an innovation that was critical after steel hulls were introduced. The first such step was a small one: replacing the iron nails of the compass box with brass screws. Later, the compass was filled with a viscous liquid and gimbaled to damp the ship's motion, making the helmsman's tracking of the correct heading still more accurate. Thus, even such a relatively simple tool as the mariner's compass was the product of numerous innovations over centuries and in space by the breadth of Eurasia.[61]

Other aspects of culture are similar. Take churches. Modern American churches are sophisticated organizations that supply social services to their parishioners.[62] The successful ones derive from a long tradition of incorporating good ideas and abandoning bad ones. Surprisingly, one of the unsuccessful ideas turns out to be hiring educated clergy. College-educated clergymen are good intellectuals, but too frequently deadly dull preachers, consumed with complex doubts about the traditional verities of Christian faith. In the United States, successful religious innovation is handsomely rewarded due to the free-market character of certain Protestant religious institutions. Many ambitious religious entrepreneurs organize small sects, mostly drawing upon a set of stock themes called fundamentalism. Only a tiny fraction of sects expand beyond the original cohort recruited by the initial innovator. The famous celibate Shakers are an example of a sect that failed to recruit followers, but there have been many others. A much smaller number are successful and have grown to become major religious institutions, largely replacing traditional denominations. The Methodists and the Mormons are examples of very successful sects that became major denominations.

Religious innovators build in small steps. Mormon theology is very different from that of most of American Protestantism. Nevertheless, historian John Brooke shows how founder Joseph Smith's cosmology mixes frontier Protestantism with hermetic ideas, Masonry, divination schemes for finding treasure, and spiritual wifery (polygamy).[63] He traces the spread of these ideas from Europe to specific families in Vermont and New York, where Smith and his family resided. Smith invented little and borrowed much, although we properly credit him with being a great religious innovator. His innovations were, like Harrison's, large compared to those introduced by most ambitious preachers.

Individuals are smart, but most of the cultural artifacts that we use, the social institutions that shape our lives, the languages that we speak, and so on are far too complex for even the most gifted innovator to create from scratch. Religious innovations are a lot like mutations, and successful religions are adapted in sophisticated ways beyond the ken of individual innovators. The small frequency of successful innovations suggests that most innovations degrade the adaptation of a religious tradition, and only a lucky few improve it. We don't mean to say that complex cultural institutions can't ever be improved by the application of rational thought. Human innovations are not *completely* blind, and if we understood cultural evolutionary processes better they would be less blind. But human cultural institutions are very complex and rarely have been improved in large steps by individual innovators.

It would be instructive to analyze a sample of complex bits of culture, like a fifteenth-century ship, and estimate the minimum number of innovations involved in their manufacture and the spatio-temporal distribution of the component innovations. For most, the number is surely very large, and the times and distances that separate the components great. The same technique could be applied to religions, artistic endeavors, and social institutions. The qualitative impression imparted by the few historians who have paid attention to the large-scale patterns of cultural evolution is that the compass is a good exemplar. Many people spread over a wide area and prolonged period contribute to human adaptations. True, a given musical composition, ship, or watch does have an individual designer, but if the work is at all complex, the designer taps a rich tradition of design in addition to whatever element of creativity he or she can muster.

Biologist Jared Diamond describes a major macroevolutionary pattern that is consistent with the hypothesis that culture evolves gradually by many small steps.[64] Europeans were strikingly successful at conquering and dominating the Americas, Australia, New Zealand, and many other smaller islands after the voyages of discovery. In contrast, though Europeans dominated and colonized Asia, the degree of domination was much less complete and much less enduring. China successfully resisted colonization, and India and Muslim Central Asia have shaken off the Europeans. On the other hand, the European possession of the Americas, New Zealand, and Australia is permanent. What is the secret of Eurasian success? Diamond argues that the greater size of the Eurasian continent, coupled with its east−west orientation, meant that it had more total innovations per unit of time than smaller land masses, and that these innovations could easily spread throughout long east−west bands of ecologically similar territory. The Americas are not only smaller but are oriented north−south, making it difficult to diffuse useful cultivars, like maize from (say) temperate North America to temperate South America, or domesticated animals like llamas in the opposite direction. As a result, the set of adaptations necessary to support complex urbanized societies was assembled more slowly in the Americas.

The magnitude of human variation is explained by culture

In this chapter we have focused on what biologists would call the *proximate* causes of human variation—that is, we have been talking about its immediate causes rather than its long-run evolutionary causes. If you came to this chapter doubting the proximal role of culture in human behavior, we hope that we have convinced you that many of the differences between people are cultural—people are different, at least in part, because they acquired different beliefs, attitudes, and values from others.

For those who came to the chapter already convinced of the importance of culture, our message is *almost* the opposite. We hope to have shaken your faith that the role of culture is truly well described. There are very few well-designed studies that critically address competing hypotheses about the source of human behavioral variation. Edgerton's pioneering study of the relative roles of environment and cultural history is unique. Reasonably well-controlled studies of change and persistence in immigrant communities are few. We are aware that some—perhaps all—of the studies we have cited here have skeptics and critics. In the end, the only way to finally silence the doubters of the role of culture is to multiply the number of good studies until we can chart the proximal roles of genes, culture, and environment in explaining human behavioral variation with real quantitative precision. Frankly, we think that the defenders of culture have grown complacent and lazy. Secure in the moral conviction that only people with evil intentions subscribe to racist notions like genetic explanations for human behavioral differences, or capitalist ones like rational choice, anthropologists, sociologists, and historians have neglected their knitting.

As it is, we think that even the most cautious, fair-minded reader will be sufficiently persuaded by the evidence to at least admit that the hypothesis that most behavioral variation between human groups is the product of culture is persuasive and worth pursuing. Such readers should be able to admit to any amount of skepticism concerning details and the significance of particular studies without being called names by defenders of cultural explanations. As proponents of strong cultural hypotheses, we have pushed the evidence about as hard as we believe it warrants. Students of culture owe their own subject, if not their critics, the hard work needed to get it right.

Understanding the ultimate causes of human variation is also important, particularly because humans are much more variable than any other species of animal. Other animals do vary. Consider baboons as an example.

Many biologists classify most baboons in a single species, *Papio cynoce-phalus*. These animals occupy a range that includes many different habitats: hot lowland forest, cool highland forest, savannah, scrub, and true desert. Within this range, baboons vary physically, especially in size and color. All baboons feed mainly on plant materials, and supplement their diet with insects, eggs, and small animal prey. However, across their range, the exact composition of their diets varies. The baboons in Amboseli, Kenya, dig up grass corms and crack open acacia pods, while the baboons of the Okavango delta eat figs and water-lily bulbs. Most savanna baboons live in multimale, multifemale groups of about thirty to seventy individuals. Females remain in these groups throughout their lives. However, in the highlands of southern Africa, baboons form much smaller, one-male groups, and females sometimes disperse between groups; in the forests of West Africa, baboons aggregate in enormous hordes that may number several hundred individuals. Social behavior also varies to some extent. In East Africa, males form coalitions with other males to compete for access to receptive females; these kinds of coalitions are never seen in southern Africa.

Now compare the amount of human variation that we see among people who occupy the same range of African environments. Like baboons, humans vary physically, mainly in size and color. Unlike baboons, the people in these regions get their daily bread and organize their social lives in very different ways. Until about ten thousand years ago, all people were foragers who lived by gathering plants and hunting mammals. However, even among hunter-gatherers there was great variation. !Kung bushmen have a simple system of kinship in which male and female relations are treated the same, while their neighbors, the !Xo, who live a few hundred miles to the south, have an elaborate system of clans based on relationship through the male line. The !Kung and the !Xo both hunt the game of the Kalahari with small bows, while the Kxoe bushmen live mainly by fishing in the nearby swamps of the Okavango. Some pygmies of the central African forest rely on large-scale cooperative hunting using nets, while the Hadza of the East African savannah hunt big game with great bows.

Of course, today most people in Africa are not hunter-gatherers. There are nomadic pastoralists like the Maasai of East Africa who live on the products of their cattle, moving from place to place in search of good grazing. Maasai political organization is based on cooperation and loyalty among age sets, groups of men who were circumcised the same time. Among other nomadic pastoralists loyalties are based on kinship—male kinship in the case of the Somalis and female kinship for the Himba of Namibia. Farming peoples grow a wide range of crops: millet and sorghum in the seasonally parched Sahel, peanuts, maize, and cassava in the forests of the Congo. They exhibit an equally wide range of social and political organizations: small family groups without any ranks or offices, elaborate kin-based clans, and great cities with full-time soldiers, priests, and rulers.

The behavioral variation within human groups is also much greater than the behavioral variation within groups of other animals. Again compare humans with baboons. The baboons living in a group do vary in their behavior. Male baboons are more likely to hunt than females; dominant females eat more of the most preferred foods, have the safest sleeping sites, and are harassed less than subordinate females; juveniles play more than adults; some females are more sociable than others; and so on. But all baboons must find their own food, keep a lookout for predators, and take care of their own infants. By comparison, even hunter-gather societies have part-time specialists in tool production, ritual activity, and food gathering. In complex farming societies the amount of variation explodes—there are butchers, bakers, candlestick makers, serfs, soldiers, sheriffs, kings, and clergy, who all have different knowledge, behavior, obligations, and subsistence tasks.

The difference between the range of human variation and that of other animals like baboons demands an evolutionary explanation. Ten million years ago (or thereabouts), our ancestors were an apelike species living in the forests and (perhaps) the savannahs of Africa whose range of variation was comparable with that of present-day baboons. Over the next ten million years, the processes of Darwinian evolution transformed that lineage into modern humans. Any theory that hopes to explain the behavior of contemporary humans *must* tell us what it is that causes humans to be so much more variable than any other species and why this peculiar capacity for variation was favored by natural selection. This burden falls particularly hard on models that try to account for human behavior invoking only individual learning mechanisms that also apply to other animals.

We think that the answer to the ultimate question about the magnitude of human variation is the same as the answer to the proximate question about its causes—culture. Our plan for the succeeding chapters is to assume that culture exists and ask if we can use this assumption to explain human peculiarities. In chapter 3 we begin by trying to explain why culture causes humans to be so variable, and in chapter 4 why culture was favored by natural selection.

Chapter Three

Culture Evolves

"When a dog bites a man, that is not news," goes the journalistic aphorism, "but when a man bites a dog, that is news."[1] To many anthropologists, the claim that culture evolves will seem more like "Dog bites man" than "Man bites dog"—it may or may not be true, but it certainly is not news. In fact, the idea that culture evolves is as old as the discipline of anthropology itself. The nineteenth-century founders of anthropology, Lewis Henry Morgan and Edward Tylor,[2] thought that all societies evolved from less complex to more complex through the (in)-famous stages of savagery, barbarism, and civilization. Such progressive evolutionary theories continued to be important throughout most of the twentieth century in the work of noted anthropologists like Leslie White, Marshall Sahlins, Julian Steward, and Marvin Harris. During this period, evolutionary theories became less ethnocentric and more realistic. Evolutionary stages were given less-loaded terms such as *bands, tribes, chiefdoms,* and *states,*[3] and models were developed that allowed for the effect of local ecology on the trajectories of cultural evolution.[4] Though evolutionary theories no longer dominate contemporary anthropology, they continue to have important defenders like Robert Carneiro, Allen Johnson, and Timothy Earle.[5] The attraction of such progressive evolutionary theories is plain to see. The archaeological and historical records leave no doubt that the average human society has become larger, more productive, and more complex over the last ten thousand years. Although unilineal theories of human progress have fallen out of favor, the general trend toward greater complexity is not in doubt.[6]

However, we mean something quite different when we say culture evolves. Remember that the essential feature of Darwin's theory of evolution is population thinking. Species are populations of individuals that carry a pool of genetically acquired information through time. All of the large-scale features of life—its beautiful adaptations and its intricate historical patterns—can be explained by the events in individual lives that cause some genetic variants to spread and others to diminish. The progressive evolutionary theories debated by generations of anthropologists have almost nothing in common with this Darwinian notion of evolution. Very little of this work focuses on the processes that shape cultural variation; it is mainly descriptive. Those accounts of cultural evolution that do provide mechanisms typically focus on external causes of change. People's choices change their environment, and these changes lead to different choices. For example, a common argument is that the evolution of political and social complexity is driven by population growth—denser populations require economic intensification and facilitate political complexity, division of labor, and so on.[7] Such processes are more akin to ecological succession than evolution. In the same way that lichen colonizing a glacial moraine change the environment, making the soil suitable for grasses which in turn further change the soil, making way for shrubs, simpler societies change their environments in ways that make more-complex societies necessary.

There is little doubt that such successional processes have played a role in human history. However, they are far from the whole story;[8] culture evolves. Human populations carry a pool of culturally acquired information, and in order to explain why particular cultures are as they are, we need to keep track of the processes that cause some cultural variants to spread and persist while others disappear. The key is to focus on the details of individual lives. Kids imitate one another, their parents, and other adults, and both children and adults are taught by others. As children grow up they acquire cultural influences, skills, beliefs, and values, which affect the way that they lead their lives, and the extent to which others imitate them in turn. Some people may marry and raise many children, while others may remain childless but achieve prestigious social positions. As these events go on year after year and generation after generation, some cultural variants thrive while others do not. Some ideas are easier to learn or remember, some values are more likely to lead to influential social roles. The Darwinian theory of cultural evolution is an account of how such processes cause populations to come to have the culture they have.

The Darwinian theory of culture presented here emphasizes the generic properties of different types of processes. For example, some cultural variants may be easier to learn and remember than others, and this will, all other things being equal, cause such variants to spread, a process we call biased transmission. The basic kinds of processes are the *forces* of cultural evolution, analogous to the forces of genetic evolution, selection, mutation, and drift. In any particular situation, the concrete events in the lives of real people are what really goes on. However, by collecting similar processes together, and working out their generic properties, we build a handy conceptual tool kit that makes it easier to compare and generalize across cases. While we make no pretense that our scheme is a finished and final account, we do think that the tools in hand are useful for understanding how culture evolves.

A Darwinian account of culture does not imply that culture must be divisible into tiny, independent genelike bits that are faithfully replicated. Rather, the best evidence suggests that cultural variants are only loosely analogous to genes. Cultural transmission often does not involve high-fidelity replication; nor are cultural variants always tiny snippets of information. Nonetheless, cultural evolution is fundamentally Darwinian in its basic structure. Analogies to ordinary biological evolution are useful, but only because they provide us with a handy, ready-made tool kit to use in building a theory rooted in the best social science.

Skeptics who distrust Darwinism are common, particularly in the social sciences. But Darwinism is not inherently an individualist, adaptationist footpad sneaking into the social sciences to explain everything by genetic reductionism. Nor does it signal a return to the progressive, Eurocentric ideas of the past. A great variety of substantive theories arise when the all-important details are specified. Some models end up looking a lot like rational choice; and in others, arbitrary cultural differences can arise from the dynamics of interacting cultural elements. Some models lead to long-term directional change in which artifacts or institutions become more efficient, while others lack such trends.

Culture is (mostly) information in brains

The first step in applying population thinking to human culture is to specify the nature of the information that is being transmitted. Culture is (mostly) information stored in human brains, and gets transmitted from brain to brain by way of a variety of social learning processes.

Every human culture contains an enormous amount of information. Think about how much information must be transmitted just to maintain a spoken language. A lexicon requires something like sixty thousand associations between words and their meanings. Grammar entails a complex set of rules regulating how words are combined into sentences; and although some of these rules may arise from innate, genetically transmitted structures, clearly the rules that underlie the grammatical differences separating languages are culturally transmitted. Subsistence techniques also entail large amounts of information. For example, the !Kung San of southern Africa have a very detailed knowledge of the natural history of the Kalahari Desert—so detailed in fact that the researchers who studied them were unable to judge the accuracy of much of !Kung knowledge, because it exceeded the expertise of Western biology.[9] As anyone who has ever tried to make a decent stone tool can attest, the manufacture of even the simplest implement requires lots of knowledge; more-complex technology requires even more. Imagine the instruction manual for constructing a seaworthy kayak from materials available on the north slope of Alaska. The institutions that regulate social interactions incorporate still more information. Property rights, religious custom, roles, and obligations all require a considerable amount of detailed knowledge to make them work.

The vast store of information that exists in every culture must be encoded in some material object. In societies without widespread literacy, the main objects in the environment capable of storing this information are human brains and human genes. Undoubtedly some cultural information is stored in artifacts. The designs that are used to decorate pots are stored on the pots themselves, and when young potters learn how to make pots they use old pots, not old potters, as models. In the same way, the architecture of the church may help store information about the rituals performed within. Without writing, however, artifacts can't store much information. The young potter cannot learn how to fire a pot simply by studying existing ones. Without written language, how can an artifact store the notion that Kalahari porcupines are monogamous, or the rules that govern bride-price transactions? With the advent of literacy, some important cultural information could be encoded on the pages of books.[10] Even now, however, the most important aspects of culture still tend to be those stored in our heads.

Behavior depends on skills, beliefs, values, and attitudes

Unfortunately, there is little scientific agreement about *how* information is stored by human brains. In some parts of the social sciences, especially history, people's behavior is often understood in terms of their values, desires, and beliefs. In other parts of the social sciences, the notions of values and beliefs are formalized under the "rational actor" model, in which values are represented by a "utility function," a mathematical rule that assigns a number to every state of the world that an individual might experience. Beliefs are represented as a Bayesian probability distribution that specifies the individual's subjective probability that each state of the world will occur. Individuals make choices that maximize the expected value of their utility. Many find the rational actor account of human psychology to be convincing because of its theoretical elegance; mathematicians have shown that only by maximizing expected utility can people avoid grossly irrational behavior—preferring ice cream to pickles, pickles to pizza, and pizza to ice cream, for example.

Psychologists of all stripes caution us that values and beliefs are folk psychology, culture-bound folk psychology at that,[11] and most care nothing for formal elegance and everything for empirical verisimilitude. Psychologists also believe that the brain is crucial for understanding all aspects of human behavior, from "low-level" functions such as processing visual information to "higher-level" functions such as reasoning or speech production. Since the real world of the human mind is complex and poorly understood, deep disagreements exist within psychology about how such information is stored and how it shapes behavior. Behaviorists concentrate on observable behavior and cognitive scientists speak of mental rules and representations,[12] while others deny the relevance of such entities and argue that only neurophysiological descriptions are useful.[13] It is unclear whether these pictures of the human mind can be integrated. To quote the eminent psycholinguist Ray Jackendoff,

> What is pretty much a mystery at this point is how linguistic rules and representations are neurally instantiated—that is how [the] physical structure of the brain could make possible the combinatorial regularities discovered by linguistic research. In fact, other than certain aspects of

low-level vision, I know of no success at relating systematicities of mental representations to the details of neural architecture.[14]

A lot of progress can be made without solving these problems. However, we need some expedient agreement about what to call the information stored in people's brains. This problem is not trivial, because psychologists have deep disagreements about the nature of cognition and social learning. Adopting a terminology may mean taking sides in these controversies, something that is neither necessary nor desirable. But, we can't go on saying "information stored in people's heads"—it's just too awkward. Some authors use the term *meme* coined by the evolutionary biologist Richard Dawkins, but this connotes a discrete, faithfully transmitted genelike entity, and we have good reasons to believe that a lot of culturally transmitted information is neither discrete nor faithfully transmitted. So we will use the term *cultural variant*. We will also sometimes use the ordinary English words *idea, skill, belief, attitude,* and *value* without meaning to imply that introspection is necessarily a reliable guide to what is stored in your own brain, or that what people tell you is necessarily a reliable guide to what is stored in *their* brains. Psychologists will one day exchange the terms of folk psychology for clearly defined, scientifically reliable concepts; in the meantime we use these terms in the interests of producing readable prose.

Cultural variants are acquired by social learning

Many of the beliefs, ideas, and values that influence people's behavior are acquired from other people through social learning.[15] We will loosely say that people imitate other people, but in fact ideas get from one head to another by a variety of complex processes. Consider how you learn to tie a knot, say, a bowline. As simple as it is, almost no one invents such a clever knot; they learn it from others, but they do so in many different ways. Some learn by verbal instruction. Someone tells you that a bowline is a strong knot that can nonetheless be easily untied. Someone else teaches you the algorithm "The rabbit comes out the hole, up the tree, around the tree, and back down the hole." You can learn by watching somebody tying a bowline, or you might come upon an example of a bowline in a book and learn how to tie it by yourself. You can learn from us by studying the picture in figure 3.1. (Try it! It's much better than an overhand knot for many everyday tasks.) What these forms of social learning have in common is that information in one person's brain generates some behavior—some words, the act of tying a knot, or the knot itself—that gives rise to information in

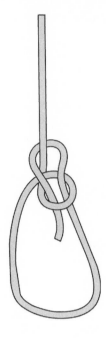

Figure 3.1. Although the bowline is strong and easy to untie, it can accidentally come untied.

a second person's brain that generates a similar behavior. If we could look inside people's heads, we might find out that different individuals have different mental representations of a bowline, even when they tie it exactly the same way.

Cultural evolution is Darwinian

Now, let's see how we can use population thinking to link these facts about how culture is stored and transmitted by individuals to the two central facts about cultural variation: traditions exist, and traditions change.

Consider a simple, hypothetical example inspired by Salamon's account of German and Yankee farmers. This is not a real model of cultural evolution in Illinois; rather, it is a way of illustrating the logic of Darwinian methods.[16] The standard way to modularize an evolutionary problem is to think about the main events in the life cycle of an individual, divide that life cycle into stages in which only one process operates, specify the processes, develop the statistical machinery to scale up from individuals to the population, and then use this machinery to keep track of the distribution of cultural variants as the population marches through history, one generation at a time.

First, we must define the problem. What are the boundaries of the population? And, what cultural variants are present in the population? Assume that basic values about farm and family are only acquired from members of the local community, which means that we can take the community as our population. If we were interested in the evolution of some other trait, say, preferences for recorded music, the population would be different, because these preferences are strongly influenced by people outside the community. Let us also assume that there are only two variants: people have either yeoman values or entrepreneurial ones. Of course, reality is much more complicated, and we will consider how to deal with such complications later on; but for now it helps to keep things simple. We also need to decide how to represent the distribution of cultural variants in the population at any one time. Because there are only two variants, it is convenient for this purpose to keep track of the fraction of the population who hold each belief. In other situations we use other statistics to describe the distribution of beliefs.

Next, we consider what happens at each stage of the cultural "life cycle" (fig. 3.2). Here we assume that children initially acquire the beliefs of their biological parents. Children growing up in families with two parents having yeoman values acquire yeoman values; children with two entrepreneurial parents acquire entrepreneurial values; and children whose parents differ are equally likely to acquire yeoman values and entrepreneurial values. This means that transmission from parents to offspring leaves the population unchanged from one generation to the next. This model assumes accurate replication of cultural variants, although social learning in practice will probably introduce frequent errors.[17] The basic framework can easily be modified to allow for this possibility.

As children grow older they are exposed to people other than their parents, some of whom may cause them to modify their beliefs. Suppose that young adults get experience with other farm operations (perhaps as the result of participating in young-farmer groups like 4-H). They observe that farmers with yeoman values work longer hours and make less money, but have closer family ties than do their entrepreneurial counterparts. These observations cause some young adults to adopt new values—some switch from yeoman values to entrepreneurial ones, and some do the opposite. For most young adults a close family doesn't compensate for long days and low wages, so more of them switch from yeoman to entrepreneurial values than

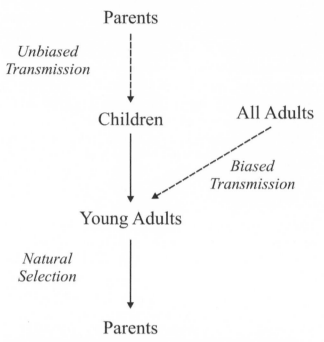

Figure 3.2. A diagram of the life cycle described in the text. Children acquire beliefs and values about farming from their parents. Then, as they grow older, their beliefs and values may also be affected by other adults. Next, as adults, they marry and choose a career. Those who abandon farming and leave the community have no further effect on the values in the community.

the reverse. This is an example of *biased* cultural transmission, which occurs when people tend to acquire some cultural variants rather than others. Biases may be innate preferences, or they may be cultural preferences acquired in an earlier episode of social learning.

Eventually, young adults grow up. Some obtain a farm and remain in the community, while others abandon farming to become mechanics, salespeople, lawyers, or academics. Salamon's data suggest that people who hold yeoman values are more likely to remain in the community. Since only adults who remain in the community influence the values of community members of the next generation, selective emigration, a type of natural selection of cultural variants, has the effect of increasing the proportion of the community holding yeoman values.

Finally, people get married and have children. According to Salamon, the descendants of German immigrants have about 3.3 children per family, while those descended from Yankees have only 2.6.[18] Suppose that this difference in family size results from the same belief system that causes differences in farm management and inheritance patterns. Since children initially acquire their values from their parents, this means that differential reproduction also leads to the spread of yeoman values in the community. This process is another form of natural selection, and rather strong selection at that.

Now let's use this model to explain the why cultural differences persist. So far we have seen how various processes lead to cultural continuity and change within a single generation. To explain long-term persistence, we iterate the model from generation to generation to determine what happens over time.

The ancestors of the Yankees and Germans of Salamon's study came to Illinois with different values that led to significant differences in behavior even though they farmed on similar soils and faced the same technical and economic constraints. In the simplified world of our model, this means that the net effect of all the social learning processes operating in each population is to leave each population more or less unchanged. If yeoman values are common in one generation, then they will be common in the next. If entrepreneurial values are common, they will remain common.

"Cultural inertia" can arise in two ways. It can arise from a tendency to conform to the beliefs of the majority. However, in the current model, the most natural explanation is a combination of unbiased sampling and faithful copying. You can think of children as being exposed to a sample of two of the cultural variants of the previous generation. Sometimes both parents hold entrepreneurial values, sometimes both hold yeoman values, and sometimes parents differ. As long as holding yeoman values doesn't have too big an effect on family size, these samples will be representative of the population from which they were drawn, meaning that the probability that a parent holds yeoman values is approximately the same as the frequency of yeoman values in the population. Then, as long as the cultural learning process is accurate and unbiased, the probability that a child acquires yeoman values will also be approximately the same as their frequency in the population of parents—transmission from parents to offspring won't change the cultural composition of the population. The same goes for social learning by young adults. Again, they are exposed to a sample of adults from the previous generation. If the sample is representative of the population, and if young adults are not strongly predisposed to acquire entrepreneurial values, transmission will lead to little change.

We also want to explain how cultures change. In the present case, there are three possibilities. One possibility is that the effect of biased transmission is very strong—almost everyone who starts out with yeoman values switches to entrepreneurial ones, and almost everyone who starts with entrepreneurial values retains them. Then entrepreneurial values will spread in the community, because people are predisposed to choose such values. Second, biased transmission could be relatively weak—some people switch from one set of values to another, but most people retain the values that they learned from their parents. Then yeoman values will spread, because people with such values are more likely to stay in the community and because they have larger families. This is what actually seems to be happening in the communities that Salamon studied. Third, the community might settle down to a stable mix of the two types.

The forces of cultural evolution

We call the processes that cause the culture to change *forces* of cultural evolution. We divide the evolving system into two parts. One is the "inertial" part—the processes that tend to keep the population the same from one time period to the next. In this model cultural inertia comes from unbiased sampling and faithful copying of models. The other part consists of the forces—the processes that cause changes in the numbers of different types of cultural variants in the population. These processes overcome the inertia and generate evolutionary change.[19]

In our stripped-down version of the lives of German and Yankee farmers, two forces are at work. Biased transmission causes entrepreneurial values to increase, and natural selection causes yeoman values to increase. These two processes exemplify two distinct classes of forces. Transmission biases are forces that arise because people's psychology makes them more likely to adopt some beliefs rather than others. Natural selection is a force that results from what happens to people who hold different cultural variants. We focus on biased transmission and natural selection here as a device to introduce the logic underlying our models of cultural evolution, and in subsequent chapters we extend our analysis to include the other forces introduced in table 3.1.

Biased transmission

Biased cultural transmission occurs when people preferentially adopt some cultural variants rather than others. Think of it as comparison shop-

Table 3.1 A list of cultural evolutionary forces discussed in this book

Random forces

Cultural mutation. Effects due to random individual-level processes, such as misremembering an item of culture.

Cultural drift. Effects caused by statistical anomalies in small populations. For example, in simple societies some skills, such as boat-building, may be practiced by a few specialists. If all the specialists in a particular generation happen, by chance, to die young or to have personalities that discourage apprentices, boat-building will die out.

Decision-making forces

Guided variation. Nonrandom changes in cultural variants by individuals that are subsequently transmitted. This force results from transformations during social learning, or the learning, invention, or adaptive modification of cultural variants.

Biased transmission

Content-based (or direct) bias. Individuals are more likely to learn or remember some cultural variants based on the their content. Content-based bias can result from calculation of costs and benefits associated with alternative variants, or because the structure of cognition makes some variants easier to learn or remember.

Frequency-based bias. The use of the commonness or rarity of a cultural variant as a basis for choice. For example, the most advantageous variant is often likely to be the commonest. If so, a conformity bias is an easy way to acquire the correct variant.

Model-based bias. Choice of trait based on the observable attributes of the individuals who exhibit the trait. Plausible model-based biases include a predisposition to imitate successful or prestigious individuals, and a predisposition to imitate individuals similar to oneself.

Natural selection.

Changes in the cultural composition of a population caused by the effects of holding one cultural variant rather than others. The natural selection of cultural variants can occur at individual or group levels.

ping. People are exposed to alternative ideas or values and then choose among them (although the choice may not be a conscious one).[20] The diffusion of innovations provides a fund of well-studied examples of how biased transmission works. This body of work was pioneered by a landmark study by sociologists Bryce Ryan and Neal Gross of the spread of hybrid corn (maize) in two Iowa farm communities in the early 1940s. Following their lead, thousands of case studies of the diffusion of innovations have been published.[21] These studies indicate that in both traditional and contemporary societies, innovations often spread as the result of personal contact. People adopt an innovation like hybrid maize after observing the behavior of friends and neighbors who have already adopted the innovation. Once they have observed the innovation firsthand, their decision about whether to adopt the innovation is strongly affected by the perceived utilitarian advantage of the new crop. Is the hybrid seed more resistant to disease? Is there a ready market for the new crop? If so, people will tend to adopt the new crop and the innovation will spread.[22] The decision to adopt a new idea, crop, or any other cultural variant may also be affected by the number or prestige of the people who have already adopted it, leading to varieties of biased transmission that we will consider in detail in chapter 4.

Because biased transmission results from the (not necessarily conscious) comparison of alternative variants, the resulting rate of cultural change depends on the variability in the population. Initially, innovations spread slowly because few people practice them, and so few other people are in a position to observe the innovation and compare it with their existing behavior. As the innovation becomes more common, more people are exposed to it and can compare it with other behaviors, and the rate of adoption of the innovation accelerates. As the old behavior becomes rare, there are fewer people still practicing it and fewer opportunities to make the comparison, so the rate of spread of the new behavior slows. This process, which has been documented in many different cases, generates a characteristic S-shaped trajectory.

The rate at which a population changes by biased transmission also depends on how hard it is to evaluate alternative behaviors. If a new crop variety has substantially higher yields than existing crops, then farmers will easily detect the difference. Hybrid corn had about a 20% yield advantage over traditional varieties, so its use spread rapidly. Similarly, after sweet potatoes were introduced to coastal New Guinea from the New World sometime in the 1700s, they swiftly replaced other crops in the cool highlands because they performed much better than typical tropical plants. This happened even though the Europeans who brought the sweet potatoes to New Guinea went no further than the coast and didn't even know that people lived in the highlands until the 1930s.[23] However, the benefits of many other desirable traits may be much harder to detect. The practice of boiling drinking water substantially reduces infant mortality from diarrhea. Nonetheless, the practice may fail to spread, because the effects of boiling water are difficult to discern. There are other ways of getting diarrhea, and people can't see the microbes in the water. People who believe that disease is caused by magic may find it hard to believe that boiling drinking water is useful. Figuring out which variant is best is often hard even if they have very different payoffs. Traits whose beneficial effects only become apparent over time are especially difficult to evaluate.

Biased transmission doesn't always result from an attempt to evaluate alternative cultural variants according to cultural standards or rules. Biases are often caused by universal characteristics of human cognition or perception. For instance, many linguists believe that some linguistic features are "marked," meaning that they are harder to produce and perceive than alternative unmarked features. Languages that denote the subject and object of sentences with word order are less marked that languages accomplishing this function by changing the form of the noun. Such unmarked features are simpler, and accordingly appear earlier in first language acquisition. Many linguists also believe that "internal" language change (as opposed to change that results from contact between languages) typically proceeds from marked to unmarked. Such changes tend to make the language easier to produce and understand. Thus, language learners confronted with two slightly different grammatical variants will tend to adopt the less marked of the two, and in this way biased transmission can drive language change.[24] This hypothesis is somewhat controversial, but if it turns out to be true, it will provide a good example of how biases may arise from the workings of human psychology.

Biased transmission depends on learning rules

The strength and direction of biased transmission always depend on what is going on in the minds of imitators. The explanation for the increase in the frequency of entrepreneurial values in rural Illinois lies in the values of young adults. Why do they value cash and comfort over family? In some cases, values may result from universal human propensities—desires for wealth, comfort, and control over your life are likely built into human psychology. In other cases, values may stem from other cultural variants—cash and comfort might win in contemporary Illinois, but family loyalty wins in rural China.

Anthropologist William Durham distinguishes between genetically acquired learning rules, which he calls "primary value selection," and culturally acquired learning rules, which he refers to as "secondary value selection."[25] The rules that underlie change in the way words are pronounced (phonology, in linguistic jargon) provide a good example of this distinction. To a first approximation, the pronunciation of vowels can be represented in a two-dimensional space which represents the vertical and horizontal position of the tongue. Ample evidence from many different languages shows that pronunciation evolves so that the distance between vowels in this space is maximized. Presumably, people subconsciously prefer widely spaced vowels because they facilitate both pronunciation and understanding.[26] Young people who are establishing their dialect listen to the pronunciation of others and tend to adopt speech variants of people whose vowels are most evenly spaced. That this process has been documented in a wide range of different languages suggests that the preference for evenly spaced vowels is what Durham would call a primary value.

Language change also provides examples of secondary values. When people speaking different languages come into contact, all kinds of linguistic variants can diffuse from one language to the other. The rate at which this occurs depends on how similar the languages are. When languages are similar, people hear a new form, find it understandable, and then can incorporate it into their own language. If languages are very different, it's harder to learn foreign words or grammatical forms, and borrowing is inhibited. Thus, the attractiveness of a new form depends on the language that you and your community already speak, which is an example of what Durham labels secondary values.

The relative importance of primary and secondary values selection is controversial. Some evolutionary biologists, such as Richard Alexander, Charles Lumsden, and Edward Wilson, advocate a dominant role for primary values.[27] Durham makes a case for the importance of secondary values, although his terminology implies that secondary values derive from the primary ones. Our hunch is that primary and secondary values virtually always interact. Consider the effects of contact-induced language change. The usefulness and intelligibility of new forms is governed by the similarity of the two languages in contact. But why do people want to communicate effectively? Why don't people choose the less- rather than more-intelligible forms? Sometimes they do: think of lawyers, politicians, or sometimes, alas, scientists.[28] People may prefer gratuitously complex linguistic forms to signal that they occupy a particular social role or for similar culture-specific reasons. The reason people often do prefer less marked forms must lie in the basic nature of human psychology—people (usually) want to be understood. The difficulty of convincing people to boil their drinking water illustrates the same point. The desire to avoid unnecessary work, like gathering extra fuel to boil water, and the desire for children to thrive are likely to be primary values that have deep, genetically influenced psychological roots. Belief in the germ theory of disease creates a secondary value. The decision about whether or not to boil drinking water depends on both these primary and secondary values.

How cultural variants compete

So far we have tacitly assumed that cultural variants compete with each other:[29] that farmers either hold yeoman values *or* entrepreneurial values, that people use one dialect *or* another, that they either adopt innovation *or* retain their present behavior. This either/or dichotomy is appropriate for genes, but it may not be for culture. The competition between different versions of the same gene results from the machinery of genetic replication. Every gene sits at a particular site, or locus, on a particular chromosome. For example, in a population of one thousand individuals there are two thousand chromosomes that can carry any given gene. If the number of chromosomes that carries one version of a gene increases from one generation to the next, the number of chromosomes that carries alternative versions of the same gene must decrease. Cultural replication need not have the same dichotomous character. People can learn and remember more than one variant. For example, they could know how to speak two different dialects, so a new dialect can spread through a population without other dialects declining.

We think that cultural variants compete in two related ways. First, they compete for the cognitive resources of the learner, both during the process of social learning and afterward, when the learner must expend some effort maintaining the variant in memory. Learning things takes time and energy that could be devoted to other valued activities, and may compete with remembering old ones. This constraint may not be very important for knowledge that is easy to acquire. For example, a bowline, a fisherman's knot, and a figure-eight knot (fig. 3.3) can all be used to tie a loop at the end of a rope, and you easily can learn them all. The amount of time that it takes you to learn a bowline doesn't prevent you from learning the others. Learning a new knot takes only a few minutes.

But, for knowledge that is more difficult to acquire, the cost of learning leads to sharp competition between variants. Mastering a new academic discipline or learning a new language requires a substantial investment of time and energy, and this may require us to choose among alternatives. Some years ago we spent a year at a German university, and we both thought it would be a good thing to learn German, but we both chose to spend the time working on this book instead. Competition between cultural variants for time and energy is diffuse compared to competition between genes at a locus. It does not necessarily lead to competition between variants that affect the same behavior; rather, it causes competition between all the variants that a person might acquire at a given time. German did not

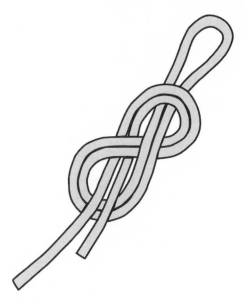

Figure 3.3. The figure-eight knot is strong, unlikely to become accidentally untied, and easy to untie if loaded, but it is a bit slow to tie.

compete with French for our all-too-limited time and attention. German competed with learning historical linguistics and studying the history of technology. The diffuse competition for our time and energy seems to limit our willingness to build up big repertoires of even such simple, useful skills as tying knots.

The second and, we think, more-stringent form of competition between cultural variants is for control of behavior. People learn a great deal by observing others, and if a cultural variant doesn't affect behavior, it won't be transmitted. Unlike genes, culture is a system of inheriting acquired variation. It has no analog of recessive or silent genes that do not influence phenotype—an organism's observable properties produced by the interaction of its genetic material with the environment—but are transmitted anyway. If you believe that a figure-eight knot is the best knot for making a loop in the end of a rope, and you always use this knot, then the people cannot learn other knots from you, even if you know how to tie other knots. The competition between cultural variants will be particularly acute when they affect many aspects of a person's life. An Illinois farmer who holds yeoman values will behave differently almost every day of his life than one who holds entrepreneurial values. Protestant converts to Catholicism or Buddhism may remember all the Protestant doctrine they learned, yet they will cease being models for Protestantism.

A long-unused variant may also be forgotten. We have all experienced the distressing loss of some hard-earned skill like differentiation, playing the clarinet, or carving a parallel turn. Use it or lose it.

People also learn ideas and values through overt teaching.[30] Here the effect is more subtle. The same sorts of things that cause a cultural variant to be used will also cause it to be taught, and to be used by those who acquire it. If you believe that the figure-eight knot is the best knot because it is strong, unlikely to accidentally untie, but easily untied after being under tension, then it is likely that this will be the knot you teach to others. Even if you teach people to tie other knots, they will be more likely to use the figure-eight knot if they accept your argument about why it is best.

Competition for control of behavior is much less diffuse than competition for attention. If two variants specify different behavior in the same context, typically only one of them can control behavior. We can drive on the right or the left, but only drunks and foolish teens try both. In bilingual environments people may switch rapidly from one language to the other, even in midsentence, but word by word, or at least word fragment by word fragment, they can be speaking only one. This example also illustrates the interaction between the two forms of competition. If a trait can easily be learned, it will not matter so much that it rarely affects behavior—occasional demonstrations will allow it to persist. One of us learned a rare but very useful knot, the trucker's hitch (fig. 3.4), from a single demonstration

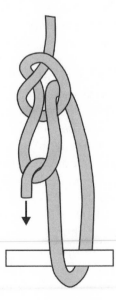

Figure 3.4. The trucker's hitch is useful for securing loads because of its mechanical advantage.

many years ago, the first and only time he has ever seen anyone else tie it. On the other hand, skills and knowledge that are only acquired over long periods of observation will be strongly affected by the amount of time that they can be observed.

Natural selection of cultural variations

The logic of natural selection applies to culturally transmitted variation every bit as much as it applies to genetic variation. For natural selection on culture to occur,

- people must vary because they have acquired different beliefs or values through social learning,
- this variation must affect people's behavior in ways that affect the probability that they transmit their beliefs to others, and
- the total number of cultural variants that can exist in the population must be limited in some way.

Or, in other words, cultural variants must compete.

You can substitute the appropriate genetic terms in this list to recover the standard textbook account of how genes evolve by natural selection. The basic logic is identical. All other things being equal, beliefs that cause people to behave in ways that make their beliefs more likely to be transmitted will increase in frequency. If the behaviors that are shaped by the beliefs acquired by imitation are important ones, they may affect many aspects of individuals' lives: who they meet, how long they live, how many children they have, or whether they earn tenure. All of these factors could affect the probability that an individual becomes available as a model for others to imitate or a teacher with the opportunity to instruct the naive.

To the extent that people acquire beliefs from their parents, natural selection acts on culture in almost exactly the same way it does on genes. For example, religious beliefs affect both the survival and the reproduction of people who practice them. Sociologists Susan Janssen and Robert Hauser compared the fertility of a large sample of people living in Wisconsin.[31] Catholics (both men and women) had 20% more children, on average, than did non-Catholics. Similarly, epidemiologists L. McEvoy and G. Land report that members of the Reformed Latter-Day Saints Church of Missouri have age-adjusted mortalities about 20% lower than control populations belonging to other religions.[32] Behavior genetic studies indicate that religious affiliation (whether you are a Mormon or a Catholic) is culturally transmitted.[33] In Janssen and Hauser's case, people's religious beliefs are strongly correlated with the beliefs of their parents. Thus, beliefs that lead to high fertility and low mortality will increase, because people holding such beliefs are more likely to survive to adulthood and have larger families if they do, and because the children in these families will tend to have the same beliefs as their parents.

Whenever individuals are culturally influenced by teachers, peers, celebrities, and so on, natural selection acting on cultural variation can favor the increase of behaviors that increase the chance of attaining such non-parental roles. In this same scenario, when the traits that maximize success in becoming a parent are different from those that maximize success as a teacher, priest, or celebrity, natural selection acting on cultural variation can cause genetically maladaptive traits to spread.

Consider one of the most bizarre traditions in the whole ethnographic record: the existence of a subculture of people who devote more time to, and are prouder of, the length of their publication list than the number of their children. The phenomenon is potentially explicable by the effect of selection on cultural variation. We, of course, are members of this odd group and can testify to the evolutionary pressures from firsthand experience. Some of our readers will have observed university faculty at close range and may well share our experiences. To see how the selection valuing long CVs can overwhelm the complex, powerful mixture of primary and secondary urges favoring having children, consider the young assistant professor just beginning her career. Entering a new university, she needs to acquire many new beliefs or modify old ones acquired as a graduate student. She needs to know how hard to work on teaching, what the standards are by which committee work is judged, and how much time should be devoted to graduate students. And, most critical of all, how much effort should she to devote to her research? Is career advancement possible if time is also devoted to family and recreation?

In making their choices, many assistant professors decide to follow the example of older and more-experienced faculty. These senior faculty represent a biased sample of the original population of assistant professors hired, because those who did not work hard and publish lots of papers were not promoted to tenure and hence aren't available to pass on their experience. Imitating tenured faculty will cause our new assistant professor to aspire to high standards in research and likely enough to postpone starting a family and limit the number of her children. This force operating on many assistant professors over several generations has produced a population that puts very high value on publications and substantially curtails childbearing. Note that we have simplified the whole story here. Throughout the educational career of our aspiring professor, she has been exposed to teachers who have faced similar career/family dilemmas, and the most successful and most influential will have been mainly those who favored career. She is liable to have fallen in love with one of her ambitious graduate school peers who shares the same background socialization and career ambitions. A successful midcareer anthropologist of our acquaintance describes the sympathetic concern of her African friends. So proud of their big families, they could not comprehend that a healthy woman would "freely" choose to have but one child.[34]

Selection for successful research faculty is driving behavior in a quite different direction from what we would predict if it were acting on genes. The role of tenured faculty member is a kind of cultural parent and social selection agent rolled into one. Potentially, natural selection on cultural variation can select for success in any role that is active in cultural transmission—biological parent, friend, leader, teacher, grandparent, and so on. The biological system is much simpler in this regard, as long as we stick with conventional organisms. There are only two roles, male and female, to worry about, and both parents make equal contributions of genes to the offspring. There are many patterns of genetic transmission that lead to the same general sorts of complexities as culture, such as Y chromosomes (transmitted from fathers to sons) and mitochondrial DNA (transmitted only by mothers),[35] but nothing quite like human culture.

Of course our young assistant professor will also take her own preferences into account as she makes decisions. If she is ambivalent about having children, she may readily adopt the publish-or-perish mentality of her most ambitious colleagues. If she is very eager to have children, she will hope that her tenure committee is more impressed by quality than quantity, and think about starting a family soon. The effect of preferences that bias decision making will lead to biased transmission. If the bias is strong, the effect of selection on the pool of models will have little effect. Plausibly, however, the bias will be weak in this case. In deciding how much time to devote to their families, young professionals must estimate not only the immediate effect on their careers and home lives but also the long-run effects on the development of their children. Biological urges to have children may be satisfied by having one or two, and the urge to achieve professional success seems to tap deeply felt biases as well. In such cases the information available to individuals may be very poor, and sentiments conflicted. Plausibly, aspiring academics will rely almost entirely on traditional beliefs, and if they do, the selective process that winnows tenured faculty will have an important effect on how faculty behave.

Why distinguish selection and biased transmission?

Biased transmission occurs because people preferentially adopt some cultural variants rather than others, while selection occurs because some cultural variants affect the lives of their bearers in ways that make those bearers more likely to be imitated. Almost every other author who has written about this topic, including biologists Luigi Cavalli-Sforza, Marcus Feldman, and Richard Dawkins, and anthropologist William Durham,[36] describes biased transmission as a form of selection, often using the term *cultural selection*. This is not unreasonable—biased transmission is a process of selective retention. Human populations are culturally variable. Some variants are more likely to be imitated than others, and thus some variants have higher relative "cultural fitness."

Nonetheless, we think that distinguishing between biased transmission and *natural* selection is very important. Biased transmission depends on what is going on in the brains of imitators, but in most forms of natural selection, the fitness of different genes depends on their effect on survival and reproduction, independent of human desires, choices, and preferences. We can understand the evolution of beak morphology in birds by asking how beaks of different size and shape affect the bird's ability to acquire food. True, we need to know something about other aspects of the bird's phenotype, so the fitness of genes affecting beak size does depend on other genes, but the dependence is much weaker than for biased transmission. Biased transmission is more like a genetic evolutionary process called meiotic drive, in which "driver" genes cause the chromosomes carrying them to be disproportionately likely to be incorporated in eggs and sperm. Meiotic drive is clearly a form of selection, but most biologists think that it is useful to distinguish it from plain vanilla natural selection.

We think that the same kind of distinction should be made in the case of cultural transmission. Consider something such as acquiring an aversion to addictive drugs. If this bias is common, it will tend to suppress the spread of addiction. But even people with biases against drugs may sometimes be tempted and succumb to an addiction that could land them behind bars, or otherwise remove them from the pool of people who exercise strong cultural influence on others. Both effects may be quite important in keeping rates of drug addiction down. The aversion to addictive substances is an example of biased transmission, while the processes that influence the number of addicts available as models exemplify selection. Although distinguishing the effects of biased transmission and selection in specific empirical cases is not always easy, the distinction is important, because these processes often lead to very different evolutionary outcomes.

In our experience, most people's intuition is that psychological forces like biased transmission are much more important than natural selection in cultural evolution. They feel in control of their culture and believe they came by most of it by choice. But the truth is, we often have much less choice than we think. As Mark Twain put it,

> We know why Catholics are Catholics; why Presbyterians are Presbyterians; why Baptists are Baptists; why Mormons are Mormons; why thieves are thieves; why monarchists are monarchists; why Republicans are Republicans and Democrats, Democrats. We know that it is a matter of association and sympathy, not reasoning and examination; that hardly a man in the world has an opinion on morals, politics, and religion that he got otherwise than through his associations and sympathies.[37]

Crucial questions hang on the relative importance of biased transmission and natural selection. If the psychological forces are much more important, then the causes of cultural evolution will ultimately trace back to innate primary values—all complex, adaptive behavior will ultimately be explained in terms of how natural selection shaped the innate aspects of psychology—and culture will have only a proximate role. However, if natural selection acting on cultural variation is important, then it is also an ultimate cause. Perhaps Durham's culturally transmitted secondary values are not always secondary at all. And so we will argue!

Population thinking is useful even if cultural variants aren't much like genes

Adopting a Darwinian approach to culture does not mean that you have to also believe that culture is made of miniscule, genelike particles that are faithfully replicated during cultural transmission. The evidence suggests that sometimes cultural variants *are* somewhat genelike, while at other times they are decidedly not. But—and this is a big but—in either case, the Darwinian approach remains useful.

You are forgiven if you find this assertion surprising. Over the last decade or so, a lot of ink has been spilled in discussions of whether cultural variants are genelike particles. On one side of this debate are "universal Darwinists" like evolutionary biologist Richard Dawkins, philosopher Daniel Dennett, and psychologist Susan Blackmore. These authors sometimes seem to be arguing that genelike replicators are necessary for adaptive evolution, and they also think that cultural variants, which they refer to as memes, are discrete, faithfully replicating genelike particles. Because cultural variants are genelike, Darwinian theory can be applied to cultural evolution, more or less unchanged.[38] On the other side are a diverse group of critics like the anthropologists Dan Sperber and Christopher Hallpike, who argue that cultural variants are not particulate and are not faithfully replicated, so Darwinian ideas of variation and selective retention cannot be used to understand cultural evolution.

We don't agree with either side in this argument. We heartily endorse the argument that cultural evolution will proceed according to Darwinian principles, but at the same time we think that cultural evolution may be based on "units" that are quite unlike genes. We encourage you not to think of cultural variants as close analogs to genes but as different entities entirely, about which we know distressingly little. They must be genelike to the extent that they carry the cultural information necessary to create cultural continuity. But, as you will see, this can be accomplished in most un-genelike ways.

The modest requirements for the properties of cultural variants are a potent rejoinder to those who believe that we can't theorize about cultural evolution until we understand exactly what cultural variants are like. If it were true that adaptive evolution depended critically on the units of transmission, Darwin and all his followers would still be marking time, waiting for the developmental work definitively showing how genes give rise to the properties of organisms. Understanding how complexes of genes interact in development to create the traits upon which selection falls is a current hot topic in biology, if not *the* hot topic. Darwin had a very un-genelike picture of how organic inheritance worked, complete with the inheritance of acquired variation. He nonetheless did remarkably well, because the essential Darwinian processes are tolerant of how heritable variation is maintained. For the same reason, we can black-box the problem of how culture is stored in brains by using plausible models based on observable features that we do understand, and forge ahead.

Cultural variants are not replicators

In his book, *The Extended Phenotype,* Richard Dawkins eloquently argues that cumulative, adaptive evolution depends on the existence of what he calls "replicators"—entities that reproduce faithfully, that are long enough lived to affect the world, and that can increase in number. Replicators give rise to cumulative, adaptive evolution because they are targets of natural selection. Genes are replicators—they are copied with astounding accuracy, they can spread rapidly, and they persist throughout the lifetime of an organism, directing its machinery of life. Dawkins thinks that beliefs and ideas are also replicators, and coined the term *meme* to describe a cultural replicator. Memes, Dawkins thinks, can be reproduced, copied from one mind to another, thereby spreading through a population, controlling the behavior of people who hold them.[39]

We doubt that beliefs and skills are replicators, at least in the same sense that genes are. As has been forcefully argued by the cognitive anthropologist Dan Sperber,[40] ideas are not transmitted intact from one brain to another. Instead, the cultural variant in one brain generates some behavior, somebody else observes this behavior, and then (somehow) creates a cultural variant that generates more or less similar behavior. The problem is that the cultural variant in the second brain is quite likely to be different from that in the first. For any phenotypic performance there is a potentially infinite number of rules that could generate that performance. Information will be *replicated* as it is transmitted from brain to brain only if most people induce a unique rule from a given phenotypic performance. While this may often be the case, genetic, cultural, or developmental differences among people may cause them to infer different cultural variants from the same observation. Language no doubt helps get many ideas from one person to another accurately, but words are subject to multiple interpretations. As teachers, we struggle mightily to be correctly understood by our students, but in many cases we fail. To the extent that these differences shape future cultural change, the replicator model captures only part of cultural evolution.

The generativist model of phonological change illustrates the problem. According to the generativist school of linguistics, pronunciation is governed by a complex set of rules that takes as input the desired sequence of words and produces as output the sequence of sounds.[41] Generativists also believe that adults can modify their pronunciation only by adding new rules that act at the *end* of the chain of existing rules. Children, on the other hand, are not so constrained and instead induce the simplest set of grammatical rules that will account for the performances they hear. Although the children's rules produce the same performance, they can have a different structure, and therefore allow further changes by rule addition that would not have been possible under the old rules.[42]

The following example[43] illustrates how this phenomenon might work. In some English dialects, people pronounce words that begin with *wh* (*whether*) using what linguists call an "unvoiced" sound, while they pronounce words beginning with *w* using a voiced sound (*weather*). (Unvoiced sounds are produced with the glottis open, resulting in a breathy sound, while voiced sounds are produced with the glottis closed, causing a resonant tone.) People who speak these dialects must have mental representations of the two sounds and rules to assign them to appropriate words. Now suppose that people in such a population come into contact with other people who only use the voiced *w* sound. Further suppose that this second group of people is more prestigious, and people in the first group modify their speech so that they, too, use only voiced *w*'s. According to the generativists, they will accomplish this change by adding a new rule that says, "Voice all unvoiced *w*'s." So, when Larry wants to say "Whether it is better to endure . . . ," the part of his brain that takes care of such things looks up the mental representations for each of the words in this sentence, including *whether* with an unvoiced *w* (because that is the way Larry learned to speak as a child). Then, after any other processing for stress or tone, the new rule changes the *w* in *whether* to a voiced *w*. In the next generation, children never hear an unvoiced *w* and adopt the same underlying representation for *whether* and *weather*. Thus, even though there is no perceptible difference in the speech of parents and children, their cultural variants differ. This difference may be important, because it will affect further changes. For example, if linguistic rules were truly replicated, future generations might recover unvoiced pronunciation of the *wh* words, whereas if they are copied from behavior, all distinctions between *wh* and *w* words will have been lost.

Replicators are not necessary for cumulative evolution

Dan Sperber and his colleagues cognitive anthropologists Pascal Boyer and Scott Atran have argued that because cultural variants do not replicate, cumulative cultural evolution is unlikely to result from the selective retention of cultural variants. They believe that the transformations that arise during cultural transmission are usually so large as to swamp the relatively weak evolutionary forces like biased transmission and natural selection.

This argument comes in two different flavors: Sometimes, Sperber and his colleagues maintain, social learning leads to systematic transformation, so that people observing a variety of different behaviors tend to infer the same underlying cultural variant. Sperber refers to such preferred variants as "attractors," because systematic transformations create a new nonselective force that moves the population toward nearby attractors. He thinks that this process is usually so strong that selective processes can be ignored.[44] In other situations, Sperber argues that the transformations that occur during social learning are unsystematic, so that people observing the same behavior infer very different cultural variants; consequently, cultural replication is so noisy and inaccurate that weak selective forces would be swamped.[45] Let's consider each of these arguments in turn.

Weak bias and selection can be important even when guided variation is strong

In many parts of the world, agricultural landowners receive a share of the crops raised on their land in lieu of rent, a practice called sharecropping. Economic theory predicts that the landowner's share will depend on the quality of the land. Owners of high-quality land should get a larger share, because they provide a more-valuable input. Since land quality varies continuously, there should be all kinds of sharecrop contracts—62.3% for the landowner, 36.8% for the landowner, and so on and so on. However, typically sharecrop contracts fall into a few simple ratios. In Illinois, for example, the vast majority of contracts are of two types: 1:1 and 2:1 for the farmer.[46] Now suppose that there is a cultural variant that is the farmer's mental representation of the optimal sharecrop contract. This could take on any share between zero and one. However, further suppose that there are attractors at simple integer ratios, perhaps because such shares are easier to learn and remember. In a particular county, the optimal share might be 1.16:1. Farmers who used this contract might be more attractive as models because they make more money, and thus biased transmission would favor a 1.16:1 contract. However, the attractor would tend to increase the frequency of 1:1 contracts, and if this force were strong compared to bias, most farmers would end up believing that the 1:1 contract is best, even though they could make more money by demanding the larger share.

This example also shows that if there are multiple attractors, weak selective forces can be important even if attractors are overwhelmingly strong. Suppose that there are two equally strong attractors for sharecrop contracts, 1:1 and 2:1, and that a population of farmers starts out with a range of contracts. After a short while, everybody will think one of the two simple ratios is the best contract—some 1 : 1 and others 2 : 1. Because these are strong attractors, they will be transmitted extremely faithfully. People who observe somebody using a 1 : 1 contract will correctly infer that that person thinks even shares are the best contract. Similarly, people observing a 2 : 1 contract in action will correctly infer the underlying belief. If the 2 : 1 contract is a little more profitable for landlords, 2 : 1 contracts will gradually replace the 1 : 1 contract, because other landlords are more likely to imitate the successful. In effect, *multiple* strong attractors lead to discrete, genelike cultural variants. Only if one attractor is stronger than the sum of all the other forces acting on other attractors will they completely determine the evolutionary outcome.

Adaptive evolution can occur even when transmission is very noisy

When cultural transmission is noisy, it cannot produce cultural inertia for exactly the same reasons that genetic transmission does. To see this, suppose there are only two cultural variants in some domain, labeled A and B. Each generates different but overlapping distributions of observable behavior. When cultural learning occurs, naive individuals, perhaps children, observe a sample of individuals from these distributions, make inferences, and then adopt their own mental representation. This process is very sloppy—a naive individual who observes an A infers that the individual is an A 80% of the time and a B 20% of the time. Similarly, a naive individual who observes a B infers B 80% of the time and A 20% of the time. It is clear that this kind of social learning will not lead to replication at the population level. Suppose that 100% of the people initially have cultural variant A. After one generation 80% will be A, after two generations it will be 68%, and by generation 5 or so, the population will have converged to a random distribution of cultural variants. Only very strong selection or bias could generate cumulative adaptation.

However, just because cultural transmission is inaccurate, it does not necessarily follow that there can be no cultural inertia or cumulative evolution of adaptations. Transmission processes can lead to accurate replication at the level of the population even when individual social learning is loaded with errors. As before, suppose that every naive individual observes the behavior of a number of models and makes inferences about the beliefs that gave rise to each person's behavior, and that people make the wrong inference 20% of the time. Now, suppose that individuals adopt the cultural variant that they believe is *most common* among their models. This is a form of biased transmission, because some variants are more likely to be adopted than others. However, unlike the biases discussed above, the nature of the bias is independent of content. It depends only on which variant is more common, and represents a "conformist" bias in social learning. In the next chapter you will see that there is good evidence that people do have a conformist bias, and that there are good evolutionary reasons why this should be the case. A conformist bias at the individual level leads to reasonably accurate replication at the population level even when individual inference about underlying mental representations is inaccurate. For example, if everyone is A, 20% of the As are mistaken for Bs, but the chances are that most naive individuals will observe samples in which A is the most common variant as long as these samples are large. Conformist bias corrects for the effect of errors because it increases the chance that individuals will acquire the more common of the two variants.

Yet the combination of high error rates and a conformist bias does not result in the same kind of "frictionless" adaptation as genetic replication. Highly accurate, unbiased genetic replication allows minute selective forces to generate and preserve adaptations over millions of years. Error-prone cultural replication, even when corrected by a conformist bias, imposes modest, but still significant forces on the cultural composition of the population. This means that only selective forces of similar magnitude will lead to cumulative adaptation. We do not think this is a problem: the forces of bias and natural selection acting on cultural variation are probably much stronger than those that shape genetic variation because they work on shorter timescales, and are often driven by psychological processes, not demographic events. The empirical record supports this somewhat, providing examples of innovations that spread over decades, not millennia.

Cultural replication can be quite accurate

Cultural transmission does not *have* to be biased and inaccurate. In fact, sometimes arbitrary cultural variants are transmitted with considerable fidelity. Take word learning, for example. The average high school graduate has mastered about sixty thousand words—an astounding feat. Learning words is a difficult inferential problem for the reasons already mentioned. The child on the nursery floor hears the word *ball* and surveys the scene. Perhaps the adult is referring to the red ball rolling across the floor, but many other inferences are possible. It could be that the adult is referring to moving red objects, the fact that it is warm, or the fact that the ball is rolling north. Despite seemingly endless opportunities for confusion, children acquire about ten new associations between a range of sounds and a meaning every day.

According to developmental linguist Paul Bloom, children use a variety of strategies to acquire their immense vocabularies.[47] They behave as if they start with the assumption that words refer to objects, and even very young children have innate presumptions about what objects are. Our hypothetical child will interpret the red ball as an object because it is connected, bounded, and moves as a unit unless some further evidence proves otherwise.[48] "Joint attention" provides another important mechanism for learning language.[49] Children follow the gaze of adults, who can often be induced to pay attention to what a child is paying attention to. In the course of these games, the adult often names the object of joint attention, usually as a part of a more-complex utterance: "A red ball! I'll roll you the red ball!" To extract *ball* and *red* out of such a language stream as names of a certain kind of round object and a color that applies to many objects is quite a feat, but the potential ambiguity is sharply limited by the assumption that the utterance is only relevant to the object of joint attention, the red ball. Another strategy children use is what psychologists call "fast mapping." Suppose a three-year-old is presented with two balls, one red and one turquoise. An experimenter asks, "Toss me the chromium ball, not the red one, the chromium one!" The child knows the color term *red* very well but not *chromium* or *turquoise*. Typically the child simply assumes that *chromium* means "turquoise" and many retain this false hypothesis for at least a week. In many cases, further experience confirms hypotheses formed by fast mapping and they go on to become a durable part of the vocabulary. Grammatical cues also play a role in language learning. For example, the child knows that *red ball* is not an action from its role in the sentence. These are only a few of the mechanisms that allow kids to accurately acquire a huge vocabulary without any innate predispositions about what words mean.

Historical linguistics suggests that these mechanisms can maintain detectable similarities in languages over hundreds of generations. Sir William Jones, the Chief Justice of India, launched the discipline of historical linguistics at the end of the eighteenth century by demonstrating that Sanskrit has certain remarkable resemblances to European languages such as Greek and Latin, resemblances too numerous to be explained by chance. Instead, these languages and a variety of others belonging to the Indo-European language family are all descendants of a single language known as Proto-Indo-European. As the people speaking this language spread out across

Eurasia, linguistic communities became isolated and the languages gradually diverged. Exactly how long ago this occurred is controversial. Some think that the speakers of Proto-Indo-European were the earliest farmers who dispersed from their agrarian homeland in southwestern Asia beginning about ten thousand years ago. Others think that they were horse-mounted nomadic herders who emerged from Central Asia or southeastern Europe about six thousand years ago.[50] To be conservative, let's suppose that Proto-Indo-European was spoken six thousand years ago, or roughly 240 human generations in the past. Contemporary Indo-European languages are connected to the speakers of Proto-Indo-European by a chain of cultural transmission 240 generations long. Each generation, children learned the sound-meaning associations from adults, and then served as models for the next generation. Thus the similarities that historical linguists use to link these languages have survived 480 generations of cultural transmission, indicating that cultural transmission can be quite accurate indeed.

Cultural variants need not be particulate

Many people believe that cultural inheritance must be particulate if it is to undergo Darwinian evolution because, the story goes, only particulate inheritance conserves the variation necessary for the action of natural selection. Biology textbooks often illustrate this idea by explaining how the discovery of Mendelian genetics rescued Darwin from the problem posed by a British engineer named Fleeming Jenkin. Jenkin was nobody's fool—a longtime associate of the great but antievolutionist physicist Lord Kelvin, he played a key role in the design and construction of the first transatlantic cable and made important contributions to economics, including inventing the supply and demand curve. Nowadays, however, he is mainly known for pointing out that if inheritance works by taking the average of the parental genetic contributions, as Darwin proposed, then the amount of variation would be reduced by half each generation. Therefore, the variation necessary for natural selection to be effective would rapidly disappear. This critique vexed Darwin greatly, but it wasn't resolved until geneticists like R. A. Fisher showed that variation persists because genes don't mix; each parent's genes remain separate particles in offspring.

This story is true but misleading. Because mutation rates are very low, the particulate nature of genetic inheritance is crucial for maintaining genetic variation. However, perhaps the analog of mutation in cultural transmission is not so low.[51] We can even imagine that cultural transmission is sufficiently noisy and error prone that blending inheritance would be an *advantage* in keeping cultural variation from growing disastrously large. In a noisy world, taking the average of many models may be necessary to uncover a reasonable approximation of the true value of a particular trait. For example, when you speak, the sounds that come out of your mouth depend on the geometry of your vocal tract. For example, the consonant *p* in *spit* is created by momentarily bringing your lips together with the glottis open. Narrowing the glottis converts this consonant to *b*, as in *bib*. Leaving the glottis open and slightly opening the lips produces *pf*, as in the German word *apfel*. Linguists have shown that even within a single speech community, individuals vary in the exact geometry of the vocal tract used to produce any given word. Thus, quite plausibly, individuals vary in the culturally acquired rule about how to arrange the inside of the mouth when they are saying any particular word. Languages vary in the sounds used, and this variation can be very long lived. For example, in dialects spoken in the northwest of Germany, *p* is substituted for *pf* in *apfel* and many similar words. This difference arose about AD 500 and has persisted ever since.[52]

Now suppose that children are exposed to the speech of a number of adults who vary in the way that they pronounce *pf*. Each child unconsciously computes the average of all the pronunciations that she hears and adopts the tongue position that produces approximately the average. There is no doubt that this act of averaging would tend to decrease the amount of variation in the population each generation. However, phenotypic performances also will vary as a result of age, social context, vocal tract anatomy, and so on. Moreover, learners will often misperceive a performance. These sorts of errors in transmission will keep pumping variation into a population as blending bleeds it away. Further note that the errors one makes will affect one's performance and will thus affect what learners use as the basis for constructing their own way of saying *pf*. Some variation will always remain if any heritable errors occur in the cultural transmission process, as surely they do.

With this sort of averaging mechanism, mental rules are not particulate, nor do they replicate. A child may well adopt a rule that is unlike any of the rules in the brains of its models. The phonological system can nonetheless evolve in a quite Darwinian way. More-attractive forms of pronunciation can increase if they have a disproportionate effect on the average. Rules affecting different aspects of pronunciation can recombine and thus lead to the cumulative evolution of complex phonological rules. In fact, this model faithfully mimics all the usual properties of ordinary genetic evolution. We are confident of this claim, because models exactly like it have been used in population genetics to represent the evolution of characters such as height that are affected by many genes, each with a small effect. They provide a good approximation to genetically more-realistic models and are much easier to analyze.[53]

Cultural variants need not be small, independent bits

Many people believe that a Darwinian approach to cultural evolution requires breaking culture into little, independent bits, an anathema to many anthropologists who believe that cultures are tightly integrated systems of shared meanings. Just as the syntax of a language is made up of a system of interdependent rules, so are the cultural meanings embedded in systems of kinship, cosmology, law, and ritual. Since Darwinian models require that cultures be decomposed into independent, atomistic traits, the argument runs, Darwinian models must be wrong. For example, Christopher Hallpike complains:

> The absence of any . . . structural concepts inevitably reduces the examples of memes and culturgens to ridiculous laundry lists of odds and ends—Dawkins's tunes, catch-phrases and ways of making pots, and Lumsden and Wilson's food items, colour classifications, 6000 attributes of camels among the Arabs, and the ten-second-slow-downs by which drivers cause traffic jams.
>
> In fact, such theories of basic units of culture do not rest on any evidence, or any sociological theory at all, but are simply proposed because if one is trying to explain culture on the basis of a neo-Darwinian theory of natural selection, it is highly inconvenient *not* to have a "unit" like a meme or culturgen, quantifications of which can be treated as continuously variable over time like the gene.[54]

This criticism misses the mark. Perhaps we (and others of our persuasion) have fostered this view by choosing very simple examples to illustrate our ideas, but there is absolutely nothing in the theory that requires that cultural variants be little bits of culture. People may choose between great, linked cultural complexes—between speaking Spanish or Guarani, or between remaining a Catholic or becoming a Seventh-Day Adventist; or they may choose between smaller, more loosely linked items of knowledge—

between pronouncing *r* at the end of a word or not, or between different views about the morality of contraception. At a *formal* level, Darwinian methods will apply equally well in either case. We keep track of the different variants, independent little bits or big complexes as the case may be, present in a population, and try to understand what processes cause some variants to increase and others to decline. The same logic applies whether the variants are individual phonological rules or entire grammars.

Cultures are not tightly structured wholes

Whether cultures actually *are* tightly integrated wholes is an important empirical question. While there has been surprisingly little systematic attention paid to this problem, a great mass of observational data bear on it. We believe that these data suggest that culture is a complex mixture of structures. Some cultural variants are linked into coherent wholes, while others float promiscuously from culture to culture.

The data from linguistics suggest that even the tightly interlinked rules underlying language sometimes diffuse and recombine. Words, phonological rules, and syntax all can diffuse and recombine independently, and as a result, different components of a single language often have a different evolutionary history. You can see this in the history of English. Some words in the English lexicon are derived from French, while others come from German. In German, the object sometimes comes before the verb in a sentence, but in French the object always follows the verb. English adopts the French syntax, although the majority of spoken English vocabulary is derived from German. Most English phonology is descended from a Germanic language; but unlike German speakers, English speakers distinguish [v], as in *veal,* from [f], as in *feel,* apparently as a result of the influence of Norman "loan words." Linguists Sarah Thomason and Terrence Kaufman[55] provide many examples from other languages, including the Ma'a language spoken in northern Tanzania that has a basic lexicon related to Cushitic languages and a grammar related to Bantu languages. They summarize by saying that "any linguistic feature can be transferred from any language to any other language."[56] They go on to argue that it is the actual pattern of social, political, and cultural interaction that determines the extent and kinds of diffusion among languages.

While the linguistic data suggest that any linguistic feature can diffuse from one language to another, they also suggest that the rate at which different features diffuse depends on a number of linguistic and social factors. What linguists call "typological distance" seems to be the most important linguistic factor. Typological distance measures the extent to which two languages have similar structure. All other things being equal, the more similar two languages are the higher the rate of borrowing. In turn, more highly structured subsystems of language diffuse and recombine at a slower rate than less structured systems. Individual words are more or less independent of each other, and as a result, they are the first items to diffuse when two languages come into contact. Inflectional morphology (for example, different verb forms that depend on the person, timing, or type of action) is linked in a complex, multidimensional system and therefore will diffuse very slowly unless the inflectional morphology of neighboring languages shares a similar structure.[57] For example, Norse had a substantial impact on English grammar even though only a small number of Danes occupied a small part of England for a relatively short time, because the typological distance between Norse and Old English was small. The rate and direction of diffusion is also strongly influenced by many social factors, the extent of bilingualism, the context in which bilingual speakers use each language, and the relative prestige of groups speaking different languages.[58]

Good evidence also suggests that language is not a good predictor of material culture—anthropological jargon for the kinds of tools, containers, dwellings, and clothing that people use. One recent study compared the artifacts collected at a number of villages on the northern coast of New Guinea during the early 1900s with the languages now spoken in those villages.[59] There was no association between language spoken and the kinds of artifacts used when the distance between villages was held constant. This means that the material cultures of two villages thirty kilometers apart with closely related languages are no more similar than the material culture of two villages thirty kilometers apart in which completely unrelated languages are spoken. Studies in Africa and North America come to the same general conclusion.[60]

A vast amount of anecdotal data provides circumstantial evidence that other components of cultures are a mix of loosely and more tightly linked elements. There are obviously many examples of important cultural similarities and differences that do not map onto linguistic differences. For example, male and female genital mutilation are common customs throughout central and eastern Africa and are practiced by people who speak very distantly related languages. California acorn-salmon hunter-gatherers and maize farmers of the Southwest both encompassed diverse language groups. The spread of religious practices, including the Sun Dance on the Great Plains, Islam across central Asia, and millenarian movements in

Melanesia, along with the contemporary spread of Protestantism in Latin America, provide additional examples of cultural practices diffusing across many different cultures/languages. On the other hand, that ritual practices and systems of religious belief can be identified as they diffuse among widely different cultures suggests that the many beliefs that make them up *are* reasonably tightly integrated and as a result *do* cohere. Some scholars, such as philologist Georges Dumézil,[61] argue that cultures have a set of core beliefs, and these core beliefs create cultural continuity over thousands of years.

Population thinking helps explain variation in cultural coherence

That cultures are not made up of independently evolving bits but composed of at least partly integrated complexes of beliefs and values is not an embarrassment for the Darwinian approach. Quite to the contrary, population-based evolutionary theory has tools to help us think clearly about the degree, pattern, and process of integration. What we mean by integration here is that the various components of a particular aspect of culture covary in space or time for particular reasons. Because a population-based theory of culture focuses on patterns of variation, it also provides a natural framework to describe patterns of integration.

Sometimes the existence of one variant doesn't create any bias for or against other variants. Such is often the case for lexicon. You can use the Spanish loanword *arroyo* for a dry gully without also having to adopt *gato* for cat. In this case, the mixing of individuals from different populations has a powerful tendency to erase differences between populations, destroying any structure that previously existed. On the other hand, the effect of mixing is limited if you learn one set of things from one person and other sets of things from others. This may produce independent subcultures within a population, subcultures that can even coexist within a single individual. For example, the subculture of science is reasonably coherent and coexists with the subculture of rock climbers, and in English-speaking countries both groups share the same language. There are even a few scientists who climb rocks and speak English, but they certainly don't form a subculture of rock-climbing, English-speaking scientists—especially if scientists who climb rocks make no special effort to recruit their students to become rock climbers or to persuade their rock-climbing buddies to become scientists. Being a scientist may have no impact on your success as a rock climber and no more impact on your social status than having any one of a number of other middle-class occupations. In this case, evolutionary processes will have independent effects on each of the three trait complexes. The evolution of some traits can be substantially decoupled from the evolution of others.

When the interaction between elements is strong, biased transmission can build coherence even in the face of substantial mixing pressure. Suppose that rock climbing has the effect of enhancing cognitive skills that are particularly useful for physical environmental scientists (geologists, meteorologists, and the like), although they detract from one's ability to be a good biologist and teach exactly the wrong lessons for social scientists. Rock climbers would then tend to be especially successful environmental scientists, but very poor social scientists. If successful environmental scientists tend to attract more students who learn both science and rock climbing from their mentors, a correlation between rock climbing and environmental science would arise. On the other hand, few successful social scientists would be rock climbers, and wouldn't encourage this hobby among their students. Successful social scientists might be prone to, say, play soccer. Eventually, a complex of coherent traits may arise which separates physical and social scientists. The gulf between the physical and social sciences is real, although we have no reason to think that rock climbing or soccer played any role in their estrangement!

Why bother with evolutionary models?

Evolutionary models aren't the only way to study how human behavior and human societies change through time. Historians, and historically minded scholars in other disciplines, have long studied social change without any reference to evolution, evolutionary forces, or anything of the like. Instead, historians seek to generate a reliable narrative account of particular sequences of historical events, and have developed rigorous methods for answering questions like What motivations led the Continental Congress to declare American independence in 1776?[62] The goal is a true historical narrative of events. Historians typically eschew simple abstract models that can be applied to a variety of cases. Instead, they focus their efforts on developing a rich explanation of events within a particular historical frame. This approach is without doubt successful in accounting for temporal change in human societies, so historians could reasonably ask, why should we abandon it in favor of simple, process-based models?

The answer is that you don't have to choose between simple abstract models and rich historical explanation—these modes of explanation are complementary, not competing.[63] Historians are certainly right: every concrete problem in cultural evolution is embedded in a complex, historically contingent frame, and all causes of events are local to that frame. However, the same is true for genetic evolution—the evolutionary biologist knows complexity and diversity as intimately as the historian. Biologists are responsible for millions of species with a huge range of characteristics and complex histories, and for the interactions of many species in complex communities. Successful field biologists typically have steeped themselves in natural history from their teenage years onward.[64] If they followed the practice of many historians and anthropologists, they would give up the concept of natural selection and speak simply in terms of the concrete events in the lives of particular organisms living in particular places and particular times that caused some genes to spread and others to diminish. After all, these local causes are all that natural selection can ever amount to in concrete terms.[65]

Instead, these very same biologists typically have a love of simple explanatory models. What gives? The answer is that such explanatory models are not laws but tools to be taken up or not as the situation warrants. Good models are like good tools: they are known to do a certain job reasonably well. Simple models that work well for a wide variety of jobs are an especially valuable part of the biologist's tool kit.

Having a toolbox filled with such models brings three important benefits. First, it is economical. The complexity of any interesting problem is likely to demand more hard thinking than any given investigator can bring to bear by himself. Person-months, if not person-years, have gone into the development of existing models, and no single investigator is likely to develop anything half as good on the spot. A mechanic who insisted on building all his tools from scratch could not be nearly as productive as one who shops at the hardware store. When available models don't work, the reasons they don't provide clues about what to try next, usually a modification of an existing model.

Second, simple models provide islands of conceptual clarity in the midst of otherwise mind-numbing complexity and diversity. Although this is not a book about formal models of cultural evolution,[66] our thinking about the major issues in cultural evolution is schooled by mathematical formalism borrowed from population genetics, game theory, and economics. These three disciplines share an enthusiasm for *simple, general* models. And these models can prevent serious errors in reasoning—errors that are all too frequent in disciplines that eschew such models.[67]

Third, by using a standardized conceptual tool kit, we increase the chance that we will detect useful generalizations in spite of the complexity and diversity of human behavior. Evolutionary biology and ecology are not without encouraging results in this regard. Although historical contingency and local uniqueness clearly matter, we can detect some general patterns in the worlds we study.[68] From the theory-as-tool kit perspective, every study provides a bit of information about the circumstances in which specific tools succeed or fail. Your colleagues provide the tools to carry to the work, and you in turn provide what help you can to the investigator with a similar problem by explaining which tools worked for you and which did not. Science advances by developing better methods, and an expanding set of empirically useful theoretical models.[69]

Darwinian tools help get the right answer

We are advocating that social scientists change the way they do business, *supplementing* their usual tool kit with ideas imported from biology. Naturally enough, many of them resent unsolicited advice from outside their disciplines. The philosopher Elliot Sober has captured one common reaction in a paper in which he argues that population-based models of cultural change will be of little interest to social scientists, because cultural evolution depends on learning rules.[70] As he puts it,

> My main reason for skepticism is that these models concern themselves with the *consequences* of transmission systems and fitness differences, not with their *sources* [his emphasis].[71]

To understand why some ideas spread but others do not, you need to know people's learning rules, their transmission biases, and the like. Why did someone invent a given cultural variant in the first place? Why is it attractive to others? You have to know which ideas will be imitated and which will be ignored. This knowledge does not come from within the Darwinian model, Sober argues; rather, it has to come from some other theory. Given learning rules, Darwinian models can predict the trajectory of cultural change, but according to Sober, this is of much less interest to social scientists than people's preferences. In other words, Sober thinks that population-based theories take all the important stuff as given, and concentrate on the stuff that nobody really cares about. The hard parts of social science don't involve its population-level properties, and the population level, unlike the biological case, is trivial. This critique has in common with many others the idea that cultural evolution is somehow so different from organic evolution that population-level processes simply don't matter.

There are three things wrong with this argument. First, it assumes that content-driven biases are the only important process affecting cultural change, and this is simply false. Biases are important, but so are processes like natural selection, which can only be understood in terms of the population dynamics of alternative cultural variants. Second, it assumes that once you know people's learning rules, how they make choices about which culture to imitate and perform, it's easy predict the evolutionary outcome. Or, in other words, we are all good intuitive population thinkers. Much experience in the relatively simpler world of evolutionary biology suggests that this is not the case. Finally, the biases are themselves the result of interacting genetic and cultural evolutionary processes. Understanding the evolution of the rules requires a theory that can work out how rules influence the social environment, which in turn influences what social information is available.

Conclusion: We are ready to get to work

We have now introduced you to all of the essential components of the Darwinian analysis of cultural evolution.

The basic steps of Darwinian analysis are

- draw up a model of the life history of individuals;
- fit an individual-level model of the cultural (and genetic, if relevant) transmission processes to the life history;
- decide which cultural (and genetic) variants to consider;
- fit an individual-level model of the ecological effects to the life history and to the variants;
- scale up by embedding the individual-level processes in a population; and
- extend over time by iterating the one-generation model generation after generation.

In a theoretical model, the final product will contain mathematical terms and operations representing each of these steps. For a large set of models built on these principles, see our earlier book and works in the same genre.[72] In an empirical investigation, we want descriptions and measurements of as many of these components as we can manage.

In order to actually make progress with theoretical or empirical work, you have to be willing to simplify, simplify, and then simplify some more. The Darwinian tradition encourages us to modularize problems and deal with highly simplified bits of nature one at a time. We are fond of simple models that are deliberate caricatures of the real world. We are also fond of abstract experiments that admit only a tiny bit of the realism. We are fond of field data that clearly show the effects of one process and hate data where several processes interact to produce an unintelligible mishmash. We don't have these preferences because we think that the real world normally resembles these kinds of simple models, experiments, and field situations. No sensible scientist thinks that the complexity of the organic or cultural world can be subsumed under a few fundamental laws of nature or captured in a small range of experiments. The "reductionism" of evolutionary science is purely tactical. We do what we can do in the face of an awesome amount of diversity and complexity. Simple, deliberately unrealistic models and highly controlled experiments have great heuristic value, because they capture manageable bits of realism. We use them to school our intuitions. We undertake empirical studies looking at limited aspects of a phenomenon— technology, politics, or art, say—because we haven't the mental or physical resources to do more. We look for the simplest real cases we can find to develop some confidence that our models and experiments are at least sometimes true.[73]

We hope your mind is racing ahead, anticipating the modifications and extensions to this rudimentary map of cultural evolution. If so, you may well already be in uncharted territory. The possible avenues of exploration are large relative to those traveled thus far. In what follows, we will repeat the exercise of this chapter for several more forces of cultural evolution, examine the results of the models in light of the current evidence, and sketch what we believe is a basic picture of the cultural evolutionary process in humans. If we don't do justice to your favorite regions, we aim to leave you with the tools for doing so at home. You can't hurt yourself.

Chapter Four

Culture Is an Adaptation

In this chapter we are going to spill a lot of ink talking about why culture is an adaptation. Experience discussing this with students, friends, and colleagues leads us to expect that many readers will think that this is a ridiculous waste of time and effort. The advantages of social learning seem obvious. Individual learning is costly, and without social learning everybody would have to learn everything for themselves. Teaching, imitation, and other forms of social learning allow us to inherit a vast store of useful knowledge while avoiding the costs of learning. In fact, we have made exactly this argument ourselves, and so have many other authors whose work we admire.[1]

Unfortunately, this reasoning, though intuitive, is wrong. As we will see, if the only benefit of social learning is that it allows most individuals to avoid the cost of individual learning, social learning can evolve all right, but—and this is a big but—at evolutionary equilibrium social learning does not increase the fitness of the imitators, or the population. The reason is that imitators are parasites who free ride on the learning of others. They contribute nothing to the capacity of the population to adapt to the local environment. To see this, imagine a population in which people acquire some behavior only by imitation, so that everyone copies someone who copied someone else, who in turn copied someone else, and so on ad infinitum. Since no one learns, there is no connection to the state of the environment, and no reason that behavior should be adaptive.

Thus we are left with a puzzle: It seems clear that culture is highly adaptive. It allows human populations to accumulate complex, highly adaptive tools and institutions that in turn have allowed people to expand their range to every corner of the globe. The puzzle is, how?

The exceptional nature of the human species deepens the puzzle—if culture is so great, why don't lots of other species have it? One of Charles Darwin's rare blunders was his conviction that the ability to imitate was a common animal adaptation. Many other complex adaptations like camera-style eyes evolved long ago, evolved independently in distantly related lineages, and are retained in most of their descendants. While many vertebrates do have simple forms of culture, only a few other species are even tolerably sophisticated social learners compared to humans. Why can't natural selection scale these protocultural systems up to the human level the way it scaled up simple eyes to complex ones? Why not long ago and in lots of species? If the presence of advanced culture in humans is not puzzling, then surely its rarity in others species is. Imagine that only humans had advanced eyes and the rest of the vertebrates were blind or nearly so. We call this complex of vexing issues the adaptationist's dilemma. The harder you think about humans the stranger we seem, not least in culture's adaptive properties.

In this chapter we try to ferret out how imitating others can increase individual fitness, and when this advantage will be great. We begin by presenting data that strongly suggest that even monkeys and our fellow apes acquire relatively little of their behavior by social learning. This fact suggests that human social learning is not a byproduct of sociality and individual learning capabilities, but requires special-purpose mental mechanisms. We then assume that these mechanisms might have been shaped by natural selection, and ask how and when culture is adaptive. Then we address the problem of why culture on the human scale is so rare. Finally, we test the hypotheses that emerge from our models of social learning with macroevolutionary data on human origins and parallel events in other lineages. We contrive an explanation for how to solve the adaptationist's dilemma; you can see what you think. We are under no illusions that it is the last word!

Figure 4.1. A mystery gadget.

Why study adaptations?

We know a woman who plays an inventive game with her daughter. In every high-end cooking store is a gizmo department—a wall covered with inexpensive little gadgets, each of which is supposed to help with a specific kitchen task, like pitting cherries, making radish rosettes, or stripping asparagus. Occasionally, when one of the women happens to be in one of these stores, she goes to the gizmo department, buys the strangest and most obscure gizmo that she can find, removes the instructions and any other indication of what the gizmo is for, and sends it to the other. The object of the game is for the recipient to figure out what the gizmo is supposed to do.[2] Sometimes this turns out to be really hard. Figure 4.1 shows one of these gadgets. It is complicated and clearly designed for something, but what? Study it for a while, and if you have to give up (we both did), turn to page 137, where its function is revealed. Amazing, isn't it? Until you know what the gadget does, you are hard put to figure out what its various parts are for and how they work; but once you know what it's supposed to do, how it works is obvious.

Biologists study adaptation for exactly this reason. Plants and animals are very complicated contraptions with many parts that interact in complicated ways. One of the most important goals in biology is to figure out how organisms work, and one of the most useful tools for solving this task is the working hypothesis that the parts are adaptive. For example, scientists studying the complicated feeding organs of bivalve mollusks assume that these organs are well-designed machines for the purpose of extracting small bits of food from the water, and the assumption provides a powerful tool for understanding how the various parts of these organs work. Behavior is studied in the same way. People studying great tits assume that the foraging strategies of these birds maximize their rate of energy intake. This facilitates understanding the details of foraging behavior: Which items should the birds take? How long should they stay in a patch? How are these decisions affected by handling time, travel time, and risk of being eaten by a predator?[3]

Surprisingly, the study of adaptations is controversial these days. The late paleontologist Stephen Jay Gould and evolutionary biologist Richard Lewontin have convinced many people, including many social scientists, that adaptive explanations are usually unjustified.[4] Their position is that many features of organisms are historical accidents or side effects of adaptive changes in other characters, and that one must be extremely cautious in invoking adaptive explanations.

We couldn't agree less. Of course, there *are* many reasons that organisms may not be well adapted to their present circumstances. Unknown trade-offs may cause the evolution of the characters of interest to be affected by changes in other characters. Genetic or developmental constraints may prevent natural selection from achieving the optimal morphology or behavior. Environments may be changing so rapidly that selection cannot keep up. However, the mere existence of such mechanisms does not justify Gould and Lewontin's extreme conservatism about adaptive explanations. Such skepticism would be justified only if, in addition, nonadaptive outcomes were much more common than adaptive ones, or if the cost of mistakenly invoking an adaptive explanation was very much higher than the cost of mistakenly invoking a nonadaptive explanation. We do not think that either of these two things is true.

Much of the variation we see in nature likely *is* adaptive. Functional studies demonstrate that organisms are well designed, and a vast body of evidence from every part of biology illustrates that all kinds of traits can be understood by asking how these parts function to promote reproductive success. In his book, *The Blind Watchmaker,* evolutionary biologist Richard Dawkins cites the human eye as an example of complex organic design. The eye has a myriad of complex parts, carefully arranged to permit sight. No mechanism other than natural selection can account for the existence of such adaptive complexity. Comparative studies show that the differences in the structure of eyes among species are adaptations to different environments. Consider, for example, fish eyes. Unlike the eyes of humans and other terrestrial mammals, fish eyes have a spherical lens. The index of refraction of the lens varies smoothly from the same value as water at the surface of the lens to much higher values at its center. This lens design allows the fish to keep one entire 180-degree hemisphere in focus without needing muscles to distort the shape of the lens. Terrestrial creatures cannot use this design. Both fish eyes and human eyes must have a cornea, a transparent cover that allows light to enter the eye but protects and contains the interior of the eye. Because air has a lower index of refraction than any tissue, human corneas can act as a lens, and this fact frees the design of the remaining lens elements. In contrast, fish corneas have an index of refraction very close to that of water and thus have no effect on the entering light.[5]

Nor is attempting an adaptive analysis of a neutral or maladaptive character particularly costly. Typically, adaptive analyses make many detailed predictions about the character in question—explanations that can often be tested by studying the structure and behavior of the organism in the field. In contrast, explanations based on random historical events or developmental constraints are usually difficult to test, because they involve events in the distant past or poorly understood tradeoffs. Gould and Lewontin are surely right that we should be cautious about casually accepting adaptive "just-so" stories about the function of traits that we observe. But we should be equally cautious, perhaps more cautious, about casually accepting nonadaptive just-so stories that invoke mysterious unspecified events or tradeoffs.

Culture is a derived trait in humans

Some animals have socially transmitted traditions that produce behavioral differences between populations of genetically similar individuals living in similar environments. Some observers are inclined to quarrel about whether such traditions qualify as culture in the sense that we apply that term to humans. People who are inclined to keep some distance between ourselves and the common run of beasts argue that traditions observed in other animals lack essential features of human culture: traditions that are symbolically encoded and are widely shared.[6] Others, who believe in the continuity between humans and other animals, argue that those who deny culture to nonhuman animals are applying a double standard—if the kind of behavioral variation observed among some other primate populations were observed among human populations, anthropologists would surely regard it as cultural.[7]

Despite having a lot of respect for the protagonists of these debates, we think this argument is a waste of time. Just as limbs evolved from fins, the machinery that allows people to learn by observing others must have evolved from homologous machinery in the brains of our ancestors. Moreover, the function of cultural transmission in humans could well be related to its function in other species, whether or not the psychological structures involved have evolved from a common ancestral structure. The study of the evolution of human culture must be based on categories that allow human cultural behavior to be compared to potentially homologous and functionally similar behavior in other organisms. At the same time, such categories should be able to recognize distinctions between human behavior and the behavior of other organisms, because the evidence strongly suggests that human culture differs in important ways from similar behavior in other species.

Social transmission of behavior is common

Many species of animals have socially transmitted behavioral differences that are analogous to human culture. In a review of social transmission of foraging behavior, comparative psychologists Louis Lefebvre and Boris Palameta give 97 examples of socially learned variation in animals as diverse as baboons, sparrows, lizards, and fish.[8] Some of the most detailed work on culture in other animals comes from studies of songbirds and the social transmission of their song dialects.

Three decades of fieldwork across Africa suggest that chimpanzees exhibit cultural variation in subsistence techniques, tool use, and social behavior.[9] For example, chimpanzees in the Mahale Mountains of Tanzania often adopt a grooming posture in which both partners extend one arm over their heads, clasp hands, and then groom each other's exposed armpits. These grooming handclasps occur often and are performed by all members of the group. Chimpanzees at Gombe Stream Reserve, who live less than one hundred kilometers away in a similar type of habitat, groom often but never perform this behavior. At Mt. Assirik in Senegal, chimpanzees strip the bark from twigs before using them to fish for termites, while Gombe chimps use the same plant for termite-extracting tools but discard the twig and use the bark. Chimpanzees from some populations living in the Taï Forest of the Ivory Coast crack open hard-shelled nuts with stone hammers that they pound against other stones and exposed tree roots, while chimpanzees from nearby populations don't, though they have access to both the same nuts and suitable stones. Primatologist William Mc-Grew has reviewed all of the field observations of chimpanzee tool use in wild populations,[10] and argues that the complexity of chimpanzee tool traditions rivals those of the simplest modern human tool kit known, that of the Aboriginal Tasmanians.[11]

Orangutans use tools, but not bonobos ("pygmy" chimpanzees) or gorillas, so far as is known. Orangutans in some areas of Sumatra use sticks to extract oily, energy-rich seeds from amid the irritant hairs that cover *Neesia* fruit.[12] Orangutans elsewhere in Sumatra and in Borneo often do not use tools even where *Neesia* are common. These geographical patterns do not seem to be the result of ecological differences, because *Neesia* seeds are the top-ranked food in terms of energy gained per unit time, and it is not likely that there is any environment in which orangutans would not eat them if they could.

In a few cases, scientists have observed the spread of a novel behavior. The most famous example occurred on Koshima Island in Japan in a group of Japanese macaques whose home range included a sandy beach. The monkeys were fed sweet potatoes, and one young female accidentally dropped her sweet potato into the sea as she was trying to rub sand off it. She must have liked the result, because she began to carry all of her potatoes to the sea to wash them. Other monkeys followed suit. However, other members of the group took quite some time to acquire the behavior, and many monkeys never washed their potatoes. Another example comes from the work of psychologist Marc Hauser, who saw an old female vervet monkey dip an *Acacia* pod into a pool of liquid that had collected in a cavity in a tree trunk. She soaked the pod for several minutes and then ate it. This behavior had never been seen before, though this group of monkeys had been observed regularly for many years. Within nine days, four other members of the old female's family had dipped their pods, and eventually seven of the ten group members learned the behavior.

Some of the most impressive field evidence for social learning in non-humans comes from species other than primates, such as whales. Zoologists Luke Rendell and Hal Whitehead have recently surveyed the whale data.[13] As with chimpanzees, studies of humpbacked whales, sperm whales, killer whales, and bottle-nosed dolphins show an impressive amount of geographical variation in behaviors ranging from vocalizations to feeding strategies that are plausibly culturally transmitted. The toothed whales (sperm whales, killer whales, and dolphins) live in stable matrilineal groups, and animals living in different matrilines often behave quite differently when the groups occupy the same environment. These behaviors can be quite complex. Some killer whale matrilines deliberately beach themselves to capture seals. Observations suggest that imitation and even teaching by mothers is a part of learning this risky behavior. Humpbacked whales cooperate to blow bubble curtains that form a sort of net to concentrate prey for subsequent capture. In the Gulf of Maine, observers noted the addition of an innovative fluke-slapping behavior at the end of the curtain-formation sequence, probably designed to stun or confuse their prey. This behavior spread to other whales in the vicinity in an exponential fashion consistent with cultural transmission. And, field observations suggest that other animals such as parrots[14] and elephants[15] have complex cultural repertoires.

The problem with field evidence is that it is very difficult to tell whether behavior really is acquired culturally. For example, it is hard to exclude the possibility that some obscure difference between the environments gives rise to the observed differences in tool use between neighboring groups of chimpanzees. But, social learning has also been studied in the laboratory, where researchers can control opportunities for individual and social learning. Experimental evidence indicates that a number of behaviors, including song dialects, novel food preferences, and other foraging strategies, are socially transmitted. The most famous case is the transmission of song dialects in birds like the white-crowned sparrow.[16] These birds have a specialized social learning system for imitating the song patterns of local adults. The song of this species varies from place to place—different local variants are called dialects. Experiments show that young birds who do not hear the conspecific song develop only a simplified version of the typical song of their species. However, if young birds are exposed to the adults singing the local song dialect, they acquire that dialect in all its complexity. Comparative psychologist Bennett Galef and his students have demonstrated that Norway rats learn about novel foods from the smell of nest mates' fur when they return from foraging trips.[17] Louis Lefebvre and his colleagues, working with pigeons and their relatives, have demonstrated the social transmission of food acquisition strategies.[18] Even humbler sorts of organisms, such as guppies,[19] show evidence of social learning under controlled conditions. These experiments provide convincing evidence that animals can learn new behaviors from one another.[20]

Cumulative cultural evolution is rare in nature

While researchers debate culture in nonhuman animals, one thing is fairly clear: only humans show much evidence of *cumulative* cultural evolution. By cumulative cultural evolution, we mean behaviors or artifacts that are transmitted and modified over many generations, leading to complex artifacts and behaviors. Humans can add one innovation after another to a tradition until the results resemble organs of extreme perfection, like the eye. Even an implement as simple as a hunter-gatherer's spear is composed of several elements: a carefully worked, aerodynamic wooden shaft, a knapped stone point, and a hafting system to fasten the point to the shaft. Several other tools have to be used to produce the parts of a spear: scrapers and wrenches to shape and straighten the shaft, knives to dissect sinew for the hafting system, hammers to knap the stone point. As we explained in chapter 2, complex artifacts like this are not invented by individuals; they evolve gradually over many generations. In nonhuman animals, the evidence for cumulative cultural evolution is scanty and controversial; social learning leads to the spread of behaviors that individuals could, and routinely do, learn on their own. In many cases, these traditions are short-lived. Norway rats, for example, constantly sample new foods on their own and eventually will come to eat most of the edible foods they find without social cues. They also forget foods that they have eaten only a few days before—their traditions don't last longer than a week or so unless they are reinforced by the continued presence of the food item.

A few nonhuman social traditions are durable and based on innovations that are difficult for individuals to learn on their own. In an Israeli pine plantation, black rats use a simple but difficult to invent technique to extract seeds from pinecones. The seeds are arranged in a spiral and are protected by tough scales. Sufficiently hungry naive rats will attempt to scale the cones, but their technique requires more energy than they gain from eating the seeds. Knowledgeable rats start by removing the unrewarding basal seedless scales, following the spiral around until they reach the second row and start uncovering seeds.[21] Zoologist Joseph Terkel and his coworkers demonstrated experimentally that young pups learn this "spiral" technique from their mothers. The trick is simple, but no rat tested learned the technique by individual trial and error. One unusually lucky, persistent, or smart rat must have invented this tradition. In black rats, unlike Norway rats, marked traditional differences between local populations might arise because such traits are hard to learn and are inherited by social learning.[22] The song dialects in birds such as the white-crowned sparrow have multiple elements. Each generation of birds learns the details of the local dialect by listening to others. However, errors and sampling variation introduce innovations that sometimes spread in local populations. As a consequence, song dialects can be traced over many generations and substantial geographic distances, much like human dialects.[23] Some of the field observations, such as the humpback whales' addition of fluke slapping to bubble curtains, and the hammer-plus-anvil nut-cracking technique of chimpanzees, may prove to be examples in which a few sequential innovations have created modestly complex culture. Hal Whitehead predicts that killer-whale hunting strategies will eventually be shown to resemble those of humans in their complexity and diversity.

Human culture requires derived psychological mechanisms

Considerable evidence suggests that the ability to acquire novel behaviors by observation is essential for cumulative cultural change. Students of animal social learning distinguish *observational learning* or *true imitation* (hereafter, plain *imitation*) from other kinds of social transmission. Imitation occurs when animals learn a novel behavior by observing the behavior of more-experienced animals.[24] Simpler kinds of social transmission are much more common.[25] For example, *local enhancement* occurs when the activity of older animals in a particular location increases the chance that younger animals will visit that spot and then learn the older animal's behavior on their own. Thus, young chimpanzees that frequently accompany their mothers to termite mounds are more likely to acquire termiting skills than individuals whose mothers never termite. A similar mechanism, *stimulus enhancement,* occurs when a social cue makes a given stimulus salient to the animal. For example, smelling food particles on nest mates makes Norway rats more likely to sample these foods when foraging. Young individuals do not acquire the information necessary to perform the behavior by observing older individuals in either of these cases. Instead, the activity of others causes them to be more likely to acquire this information through their own interaction with the environment.

Local and stimulus enhancement and imitation both can lead to persistent behavioral differences among populations, but only imitation gives rise to the *cumulative* cultural evolution of complex behaviors and artifacts.[26] To see why, consider the cultural transmission of stone tool use. Suppose that an early hominid learned, on its own, to strike rocks to make useful flake tools. Her companions, who spent time near her, would be exposed to the same kinds of conditions, and some of them might learn to make flakes, too, entirely on their own. This behavior could be preserved by local enhancement, because groups in which tools were used would spend more time in proximity to the appropriate stones. However, that would be as far as toolmaking would go. Even if an especially talented individual found a way to improve the flakes, say by blunting the back to protect the hand, this innovation would not spread to other members of the group because each individual has to learn the behavior independently, and individual learning is time consuming and chancy. Local and stimulus enhancement are limited by the learning capabilities of individuals, and by the fact that each new learner must start from scratch with only the barest clues from other animals to go by. Imitation allows each new innovation to be added to an individual's behavioral repertoire, because the information about how to perform the behavior is acquired by observing the behavior of others. To the extent that observers can rapidly and accurately use the behavior of models as a starting point, imitation leads to the cumulative evolution of behaviors that no single individual could invent on its own.

Several lines of evidence suggest that imitation is usually not responsible for protocultural traditions in other animals. First, as we have already said, many socially learned behaviors, like potato washing in Japanese macaques, are relatively simple and could be learned independently by individuals in each generation. Second, new behaviors like potato washing often take a long time to spread through the group, a pace more consistent with the idea that each individual had to learn the behavior on its own, aided only by weak clues of stimulus or local enhancement. Finally, sophisticated laboratory experiments capable of distinguishing imitation from other forms of social transmission like local enhancement have usually failed to demonstrate observational learning, except for the specialized song-learning system of some birds.[27]

Adaptation by cumulative cultural evolution is not a byproduct of intelligence and social life. We say "monkey see, monkey do," and use *ape* as a verb, but in fact monkeys and even apes do not seem to be especially clever imitators compared to humans. The best evidence comes from experiments in which the imitative capacities of children and apes have been compared.[28] Primatologists Andrew Whiten and Deborah Custance designed an artificial "fruit," a rugged, transparent plastic box that held treats inside. Experimental participants could open the box by manipulating a latch consisting of either bolts or a pin-and-handle arrangement. The participants were eight chimpanzees three to eight years of age and three groups of children with mean ages of 2.5, 3.5, and 4.5 years. They watched a familiar human demonstrate a specific technique for opening the fruit, and then were allowed to attempt open it themselves. The experimenters recorded whether the participants used the same technique that they had been shown. By most measures, chimpanzee imitative performances exceeded chance. However, 2.5-year-old children did even better, and older children were dramatically more proficient imitators than the chimpanzees.

Psychologist Michael Tomasello and his coworkers conducted similar experiments in which chimpanzees and children were shown how to use rakelike tools to obtain food that was out of reach. The chimps who watched expert demonstrators were more successful than untrained chimps in using the tool to obtain the food reward, but they did not imitate the precise method that their demonstrators had used. Children, on the other hand, followed the method they had been shown. Tomasello describes the ape technique as *emulation* rather than *imitation;* apes learn that a tool can be used to cause some desired effect by watching a demonstrator, but they don't pay close attention to the details of how the tool is used. Children imitate so faithfully that they persist in using an inefficient technique, one that the chimpanzees usually abandon in favor of the more-efficient alternative. Children aren't *smarter* than chimpanzees in general, just much more imitative.[29] Taken together, these experiments suggest that social learning in apes and humans is not the same. Children imitate very faithfully, while apes emulate or at least imitate less faithfully.

Although the evidence on hand suggests that most cultural traditions in other animals are not the product of imitation, some caution is in order. Negative results are always difficult to interpret; experiments can fail for many reasons. A recent clear demonstration of imitation by marmosets suggests that better experiments might detect imitation in a wider range of species.[30] Experimental data from bottle-nosed dolphins suggests that they are excellent vocal and motor imitators, consistent with the field evidence.[31] Thus, we don't claim that imitation is unique to humans. However, the current evidence suggests that (1) cumulative cultural evolution is rare, and perhaps absent, in other species; and (2) even our closest relatives, the chimpanzees, rely on different modes of social learning than humans.

So far, we know of no convincing evidence that any other species has a cultural item as complex as a stone-tipped spear. Rudimentary forms of observational learning are certainly present in chimpanzees, orangutans, whales, crows, various songbirds, and parrots,[32] but as Darwin put it, a "great gap" exists between humans and other animals. No other species seems to depend on culture to anywhere near the degree that humans do, and none seem adept at piling innovation atop innovation to create culturally evolved adaptations of extreme perfection. In fact, there is no evidence that *humans* made tools as complex as a stone-tipped spear until about four hundred thousand years ago.

As an aside, we are disappointed by the seeming lack of imitation and cumulative cultural evolution in other species, and we'd love it if future work showed more-sophisticated social learning in nonhuman animals. The more the great gap is closed up, the more we can put the comparative techniques familiar to both evolutionists and social scientists to work. The sober chore is to estimate the width of the gap as accurately as we can, and the trend of the best current evidence seems to us to favor a gap even larger than Darwin imagined.[33] This fact leaves the adaptationist on the horns of the puzzling dilemma with which we opened this chapter. In the remainder of this chapter, we will explore the conundrum of culture as an adaptive system.

Why is culture adaptive?

In 1988, anthropologist Alan Rogers published a theoretical model demonstrating that avoiding the costs of learning is an important benefit of imitation, but this alone is not sufficient to explain the evolutionary origin of human culture. To see why, let's consider Rogers's argument.

Reducing learning costs may allow culture to evolve, but that alone does not increase adaptability

Rogers's conclusions are based on a model of the evolution of imitation in a very simple hypothetical organism. These hypothetical creatures live in an environment that can be in either of two states; let us call them wet and dry. The environment has a constant random probability of switching from wet to dry each generation and the same probability of switching from dry to wet. Over the long run, the environment is equally likely to be in either state. The probability of switching is a measure of the predictability of the environment. When the environment switches often, knowing the state of the environment in one generation tells you little about the state of the environment in the next generation. In contrast, when the environment switches states less often, the environment of the past generation was likely to have been the same as the present generation. The organisms have one of two possible behaviors: one best in wet conditions and one best in dry conditions. They can be one of two genotypes: learners and copiers. Learners figure out whether the environment is wet or dry on their own and always adopt the appropriate behavior. However, the learning process is costly, because trial-and-error learning takes time and energy. Copiers simply pick a random individual and copy it. Copiers don't pay the cost of learning. Copying thus does not have any direct effect on survival or reproduction, but copiers may acquire the wrong behavior for their environment. Rogers then used some simple but clever mathematics to determine which genotype wins in the long run.[34]

The answer is surprising (at least it was to us). The long-run outcome of evolution is always a mixture of learners and copiers in which both types have the same fitness as purely individual learners in a population without copiers. In other words, natural selection favors culture, but culture provides no benefit at equilibrium. The organisms are no better off than they were without any imitation. To understand the logic of this counterintuitive result, think of the imitators in Rogers's model as *information scroungers* and the learners as *information producers*.[35] Information producers bear a cost to learn. When scroungers are rare and producers common, almost all scroungers will imitate a producer. Most scroungers will obtain the same benefits of good information as producers but will not bear the cost of production. However, when scroungers are common, they will often imitate one another. If the environment changes, any scroungers that imitate scroungers will get caught out with bad information, whereas producers will adapt. The system equilibrates when the cost of production by producers just equals the cost of being wrong to scroungers when environments change. At evolutionary equilibrium, scroungers gain no advantage over producers. Both types are exactly where all the producers were when the evolution of scrounging began. Moreover, the theoretical result is robust; you can change the model in lots of ways, but as long as the only benefit of imitation is avoiding the costs of learning, you get the same answer. Information scrounging is known to exist from experiments on humans and on pigeons.[36] Perhaps many cases of simple culture and even aspects of culture in humans approximate Rogers's model.

This result is disturbing to most people, because it conflicts with their intuitions about the role of culture in the human species. Since the first appearance of tools and other evidence of culture in the archaeological record, the human species has increased its range from part of Africa to the entire world, increased in numbers by many orders of magnitude, exterminated many competitors and prey species, and radically altered the earth's biota. Rogers's model must be incomplete. Culture *is* adaptive. However, figuring out *what* is wrong with the simple producer-scrounger model is an interesting exercise, because what is missing will help us isolate which features of culture are the ones crucial to our extraordinary success.

Culture is adaptive when it makes individual learning more effective

Thinking about imitation in terms of costs and benefits reveals the crucial missing element in Rogers's model. Social learning improves the average fitness of a population only if it increases the fitness of individual learners who produce information, not just those who imitate. In other words, increasing the frequency of imitators must make information production cheaper or more accurate. We have been able to think of two ways that this can happen.

Imitation allows selective learning

Imitation may increase the average fitness of learners by allowing organisms to learn more selectively. Learning opportunities often vary—sometimes the best behavior is easy to determine, other times not. Organisms that can't imitate must rely on learning, take the information that nature offers, for better or worse. For example, consider individuals trying to decide which of two foraging techniques is better. They try them both out, and choose the one that yields the highest return. Because yields will vary for many reasons, individuals' trials may often yield misleading results—the technique with the higher return during the trial may have a lower return over the long run. Without imitation, every individual must decide based on the information each has available. Even if trials suggest that both techniques have the same return, one must decide which to adopt.

In contrast, an organism capable of imitation can afford to be choosy, learning when learning is cheap and accurate, and imitating when learning is likely to be costly or inaccurate. For example, individuals could use a contingent rule such as "Try out the two techniques and if one yields twice as much as the other, adopt that technique; otherwise, use the technique that Mom used." The use of such a rule would cause those individuals who do rely on imitation to make fewer errors than those who always rely on individual learning. It would also cause them to imitate often, but not always.

A more-stringent rule, say, adopt the technique only if it yields four times as much the other, would further reduce the errors made by learners (and increase their fitness on that account), but would further increase the number of individuals who imitate (leading those who rely on imitation to be more susceptible to environmental change). In this model, everyone both produces and scrounges, depending upon circumstances. Now, increasing the frequency of imitating increases the average fitness of learning, because relying only on more-definitive information cuts the cost of learning. At the same time, a higher frequency of imitating steadily reduces the fitness benefits of imitating, because the population doesn't keep up with environmental changes as well as when learning is more common. Eventually, an equilibrium is reached in which individuals mix learning and imitation optimally, trading off the higher cost of learning when cues are less obvious against the risk of imitating outdated information. But now average fitness is higher than in an ancestral population entirely dependent on individual learning. By becoming a selective learner, an individual gains most of the advantages of both learning and imitation.

Imitation allows cumulative improvement

Imitation also raises the average fitness of cultural creatures by allowing learned improvements to accumulate from one generation to the next. So far we have only considered two alternative behaviors. Many kinds of behavior admit successive improvements toward some optimum, as in adding a sharp, hard stone tip to a spear instead of merely trying to sharpen the wood itself. Individuals acquire an initial "guess" about the best behavior by imitation, and then invest time and effort in improving their performance. For example, a spear maker might tinker with the taper on the shaft of his spears in order to get them to fly straighter. For a given amount of time and effort, the better an individual's initial traditional spear, the better on average his final performance. Now, imagine that the environment varies, so that different behaviors are optimal in different environments. Game populations fluctuate. Sometimes a spear stout enough to stab large, slow animals is best; other times a slim aerodynamic one to toss at fleeter, smaller animals is better. Still other times, some compromise design may be best. Organisms that cannot imitate must start with whatever initial guess is provided by their genotype. They can then learn and improve their behavior. However, when they die, these improvements die with them, and their offspring must begin again at the genetically inherited initial guess. In contrast, imitators can acquire their parents' behavior after it has been improved by learning. Therefore, imitators will start their search closer to the best prevailing design than purely individual learners, and can invest the information production efforts efficiently in further improvements. Then they can transmit *those* improvements to the grandkids, and so on down the generations until quite sophisticated artifacts evolve (and re-evolve to meet the needs of changing environments). Historians of technology have demonstrated quite nicely how this step-by-step improvement gradually diversifies and improves tools and other artifacts.[37] Even such seemingly simple items as spears, hammers, dinner forks, paper clips, and our mystery gadget are the product of many stepwise, cumulative improvements over a number of generations.

When is culture adaptive?

What kinds of environments favor a system of sophisticated imitation and teaching that in turn produces cumulative cultural evolution? When is such a cultural system liable to be worth any costs it may impose, such as the cost of having a big, expensive brain in order to imitate accurately? These are crucial questions, because the human species' extreme reliance on culture fundamentally transforms many aspects of the evolutionary process. The evolutionary potential of culture makes possible unprecedented adaptations like our modern complex societies based on cooperation with unrelated people, *and* some almost equally spectacular maladaptations, such as the collapse of fertility in these same modern societies. The conditions under which selection might favor a strong reliance on imitation are all-important for understanding what sort of animal we are.

The force of guided variation

In our elementary models of adaptive cultural transmission, individuals acquire beliefs and values by unbiased imitation or some other form of social transmission. They can modify their beliefs and values based on any effort they invest in learning for themselves as opposed to blindly sticking with tradition. People may modify existing beliefs, or even invent completely new ones, as a result of their experiences. When such people are subsequently imitated, they transmit the modified beliefs, and the next generation can engage in more individual learning and further hone the trait. When the beliefs of one generation are linked to the next by cultural transmission, learning can lead to cumulative, often adaptive, change. We say that such change results from the force of *guided variation*. The system is a little like an imaginary genetic system in which mutations tend to be fitness-enhancing rather than random.

Like biased transmission, guided variation depends on learning rules, and it's likely that many of the same psychological mechanisms underpin both processes. Because they both depend on decision-making rules, we will refer to them collectively as *decision-making forces*. However, there are also important differences between the two. Biased transmission results from the comparison of different cultural variants already present in the population. As a result, biased transmission is a culling process like natural selection. Some variants in the population are more likely to be transmitted than others, and those variants spread. Thus, like natural selection, the strength of biased transmission depends on the amount of variation in the population. When a favorable trait is very rare, only a few people will have the opportunity to benefit from a comparison with a less-favored trait. As the favored trait becomes more common, more people will have the advantage of the comparison, and the rate of increase of the favored trait will accelerate. As the favored trait becomes even more common, fewer and fewer people will have the disfavored trait and the rate of change will drop again.

Guided variation works quite differently, because it is *not* a culling process. Individuals modify their own behavior by some form of learning, and other people acquire their modified behavior by imitation. As a result, the strength of guided variation does not depend on the amount of variability in the population. A population in which every individual believed exactly the same thing can change by guided variation just as readily as a population in which people vary. This difference means that the time paths of cultural change that result from biased transmission and guided variation are quite different when a favored trait is rare. If the bias force must wait until a favorable variant is introduced by chance, then progress is slow until an appreciable number of individuals so acquire it. Individual learners, by contrast, have the most influence when the trait is rare, potentially getting the evolution of a newly favored trait off to a very fast start compared to a case with only random variation and bias (or bias and natural selection). While biased transmission has important analogies to natural selection, guided variation definitely does not. It is a source of cultural change that has no good analog in genetic evolution.[38]

Culture is adaptive when learning is difficult and environments are unpredictable

The strength of guided variation and biased transmission affects the heritability of cultural variants. When these decision-making forces are weak, most people end up with the same beliefs as their parents and their friends—cultural differences are heritable. For example, weak decision-making forces are one way to explain the slow change in beliefs and values that affect farming practices in the Illinois farming towns of Freiburg and Prairie Gem. German kids who grow up surrounded by people who believe that farming is a valuable way of life end up with the same yeoman values themselves, as do Yankee kids who grow up among people holding entrepreneurial values. Now compare this situation to beliefs subject to a strong decision-making force—say, about whether one should suppress weeds by mechanical cultivation or by using chemical herbicides. Suppose that almost everyone tries herbicides and decides that they are superior to mechanical cultivation. Now what people believe has little to do with the culture in which they were raised and everything to do with the decisions they have made based on their own experience—cultural differences are not very heritable.

When decision-making forces are weak, cultural variants are highly heritable, and this means that other evolutionary processes that depend on the existence of heritable variation can operate. When decision-making forces are strong, there will be little heritable variation, and other processes can have little effect. Remember that natural selection favored yeoman values because people who hold such values had larger families and were more likely to remain in farming, but selection can have an interesting effect only if decision-making forces are weak. Suppose that biased transmission is very strong—so that almost everyone who starts out with yeoman values switches to entrepreneurial ones and almost everyone who starts with entrepreneurial values stays that way. After a very short time, everyone will have entrepreneurial values and there will be no cultural variation for natural selection or further bias to act upon. The same goes for herbicide use. Suppose that organic-agriculture advocates are right in believing that using herbicides actually reduces profitability. Perhaps the sight of hated weeds dying a lingering death is so much more satisfying than their merciful end by mechanical cultivation that farmers systematically overestimate the value of herbicides. Now natural selection among farms will favor mechanical cultivation. Farmers who use herbicides will earn lower profit and therefore be more likely to go out of business. However, if biased transmission acts sufficiently strongly in a maladaptive direction, almost all farmers will erroneously use herbicides, and natural selection will have little effect.[39]

In the next chapter, you will see how cultural evolution can lead to outcomes not easily predicted by simple adaptive considerations; this is important because it enables a theory rooted in basic Darwinism to generate a rich enough variety of outcomes to explain the complexity and diversity of human behavior. However, these processes can only be important if there is sufficient heritable cultural variation. Are there circumstances in which natural selection will favor a sufficient reliance on accurate, unbiased cultural transmission to support heritable cultural variation? Or put very simply, when does natural selection favor doing something "just because" other people are doing it? You can think of this exercise as a basic account for the evolution of any system of social transmission. All organisms have means of adjusting their behavior and anatomy to local conditions. When can selection favor a costly system for transmitting these adjustments to offspring or other social learners?

We have analyzed this problem using several mathematical models of the evolution of imitation, and all of them tell the same story.[40] Selection favors a heavy reliance on imitation whenever individual learning is error prone or costly, and environments are neither too variable nor too stable. When these conditions are satisfied, our models suggest that natural selection can favor individuals who pay *almost* no attention to their own experience, and are *almost* totally bound to what Francis Bacon called the "dead hand of custom."

This result is quite intuitive. If people can accurately determine the best behavior, then there is no need to imitate; just do it. You don't need to observe your neighbors to duck into shelter when it rains or find shade when it is hot. If the environment changes rapidly, there is no sense in copying what has worked in the past, because what worked for Mom and Dad will be of little help today. No matter how error prone your best guess is about what to do, you are bound to do better than imitating someone whose behavior is surely out-of-date. For imitation to be beneficial, the environment must change slowly enough that the accumulation of imperfect, socially learned information over many generations is better than individual learning, but not so slowly that an innate instinct under the influence of natural selection alone is sufficient.

These models paint a consistent, intuitively pleasing picture of why capacities for culture evolve, but given that environments almost always vary, they seem to predict that culture ought to be much more common than it is. True, the culture we assume in the models is rather simple, and simple systems of social learning are common. Students of nonhuman social learning have reason to be happy with the theory. However, we remain stuck with the stubborn fact of humans' overwhelming success using an exceedingly rare form complex culture.

Are the models a true depiction of the adaptive properties of culture? Unfortunately, we don't know. Usually evolutionary biologists test models of this kind by applying the comparative method. But in this case, one would have to collect data on a range of species that vary in the extent to which they rely on social learning, and then look to see whether more social learning occurs in the circumstances predicted by the model. However, there is so little data on the costs and benefits of social learning in other animals that this kind of test is not currently feasible. Interestingly, the best-known animal social learning systems occur in Norway rats and feral pigeons. These are the animal equivalent of weeds, species that do well in a wide range of environments, especially in disturbed habitats associated with humans. If a broader comparative study of animal social learning showed a significant correlation between environmental variability and capacities for social learning, the models would be supported.

Two more adaptive cultural mechanisms

Before we try to dig our way out of the adaptive puzzle of human culture, let's heap some more material on the pile by introducing two variants of the bias force that further enhance the adaptive power of cultural evolution. So far we have considered why and when accurate imitation can be favored by natural selection. We have imagined that people have the ability, albeit limited, to judge the relative merit of alternative beliefs and values, and to choose between them.

Such imitation strategies can be thought of as *heuristics* for guessing the right thing to do in a complex and variable environment. Psychologists have studied how human decision makers cope given our limited cognitive abilities. For example, a group led by Gerd Gigerenzer has investigated "fast and frugal" heuristics that generate correct answers to a class of problems quickly with minimal demands for data or computational effort.[41] In one experiment, Gigerenzer's group gave a list of pairs of German cities to American college students and asked them to judge which was larger. In this case, the information that Americans have is poor, but a simple heuristic turns out to be quite accurate. A city Americans have heard of, such as Frankfurt, is almost always larger than the one they haven't, for instance Bielefeld. Many fast, frugal heuristics are very nearly as accurate as the best statistical procedures, and for some classes of problems they often do a little better. Social learning can also be thought of as a decision-making heuristic. When in doubt about what to do, stop fretting and copy Mom, Dad, or your best friend. Our models of guided variation suggest that this is a useful heuristic whenever your own experience is not very telling.

But why stop there? Very often, decision makers who detect that Dad's way of doing things is quite outdated will be ill advised to start a brute force trial-and-error search for a solution to their problem. A biased search for a better model is a relatively cheap alternative. But even what we call content bias—careful comparison shopping among existing ideas—is likely to involve a costly search for good data and be a demanding calculating chore if conducted by the methods taught in statistics and research methods courses. Given the size and complexity of our cultural repertoire, it defies imagination that we can use costly heuristics to bias many of our behavior-adoption decisions. Life is short, and rewards come from getting on with it. If fast and frugal heuristics exist that are less costly than guided variation and content bias, but are still better than merely blindly copying Dad, then natural selection will have favored incorporating them into our bag of tricks for managing our cultural repertoire. No doubt the fast and frugal heuristics that Gigerenzer and his colleagues study are often applied in the form of strategies to learn for oneself and to bias the acquisition of cultural variants. In addition, culture affords the opportunity to use other types of cute tricks. We have been able to think of two:

Imitate the common type

Recall the old saw "When in Rome, do as the Romans do." This strategy makes good evolutionary sense under a broad range of conditions. A number of processes, including guided variation, content bias, and natural selection, all tend to cause the adaptive behavior to become more common than maladaptive behavior. Thus, all other things being equal, imitating the most common behavior in the population is better than imitating at random. We label this general process *frequency-dependent bias,* because the bias depends on the commonness of the behavior, not its characteristics, as in a content bias. In the case of weighting the common type more heavily, we have a *conformist* bias. Conformity is not just simple cultural influence, but a differential weighting of one's models by the commonness of the trait.

If you regard your oddball friend Jane as a lovable eccentric and are as prone to imitate her as any of your more-conventional friends, you are not exercising conformist bias. If you treat her as a barely tolerable deviant and actively avoid imitating her, you are a conformist. If you admire her spunky independence and are especially *prone* to imitate her, then you are applying a *nonconformist* bias, another type of frequency-dependent bias we shan't discuss further, though it has some obvious domains of applicability, such as selecting an occupation in a world where faddish choices tend to drive down wages in overfavored lines of work.

A hypothetical example illustrates how a conformist bias might be favored by selection. Consider a population of early humans in the process of expanding their range from tropical savanna into temperate woodland, a habitat that favors quite different behaviors. This is easy to see for things related to subsistence—the foods that have the highest payoff, the habits of prey, shelter construction methods, and so on. However, different habitats may also favor different beliefs and values affecting social organization: What is the best group size? When should a woman accept being a man's second wife? What foods should be shared? Individuals will have difficulty making these decisions, and as a result, pioneering groups on the margin of the range will evolve slowly toward the most adaptive behavior. This improvement will be counteracted by the influx of beliefs and values brought by immigrants from the savanna that will often cause some people in woodland populations to hold beliefs more appropriate to life in the savanna than to life in the woodland. However, once a peripheral woodland population is isolated enough that adaptive processes cause the best variants to be most common, those who imitate the most common variant are less likely to acquire inappropriate beliefs than those who imitate at random. If this conformist tendency is genetically or culturally heritable, it will be favored by natural selection.

We have modeled the evolution of a conformist bias to see whether these intuitions are correct.[42] We assume that a population is subdivided into a number of partially isolated local populations that are linked by migration. The model has two environmental states, and each local population lives in a habitat that switches back and forth between these two states with a constant probability. The model has two cultural variants—one is better in one environment and the other better in the other environment. As before, individuals have imperfect information about which variants are best in the local environment. However, we now also assume that individuals observe the behavior of more than two models. Individuals vary in two dimensions: the extent to which they imitate the behavior of others (as opposed to rely on their own information about the state of the environments) and, given that they do imitate, the extent to which they are influenced by the more-common type among their models. Finally, we assume that variation in both dimensions has a heritable genetic basis. We then combine the effects of biased social learning, individual learning, and natural selection to estimate the net effect of these processes on the joint distribution of cultural and genetic variants in the population. To project the long-run consequences, we iterate this process over many generations. We then ask, what amount of conformist transmission will be favored by natural selection? If there were an office pool, what value of conformity would you guess is optimal?

And (the envelope please) the winner is . . . a strong conformist tendency. As before, a reliance on social learning is favored when environments change slowly and the information available to individuals is poor. Any combination of these two factors that leads to the evolution of a strong reliance on social learning also favors a strong conformist tendency. In fact, selection favors a strong conformist tendency even when there is only a modest reliance on social learning. Thus, the psychology of social learning should plausibly be arranged so that people have a strong tendency to adopt the views of the majority of those around them. Anyone who has raised (or been) a teenager knows that people have a strong urge to conform, and a great deal of evidence from social psychology confirms this impression. Classic studies by social psychologists Muzafer Sherif, Solomon Asch, and Stanley Milgram established that individuals adjust their behavior to that of others.[43] Sherif used an "autokinetic" procedure to demonstrate the effect of conformity. Subjects sit a dark room in which a point of light is shown on a screen for a few seconds. Although the point of light is stationary, it appears to move, a trick of visual perception. When subjects are asked how far the light moves, estimates vary considerably, but on average people estimate that it moves about four inches. Nevertheless, small *groups* of individuals that have different perceptions will cause deviant individuals to change their perceptions quite dramatically. For example, a person who initially estimates that the light moves eight inches can be induced to conform to an estimate of two inches, if the other two people in the group have initial estimates of half an inch and two inches.[44]

Most conformity studies do not distinguish between simple cultural transmission and the curvilinear effects of conformity. For example, many experiments have several confederates who behave in a certain, usually highly odd way, and just one real subject. Subjects markedly conform in such a case, but they would do so whether the cultural effect was conformist or not. Also, only a few studies have checked to see how durable conformity effects are. They are of little interest if conformity is mere polite agreement with the group that vanishes when individuals leave it.

A few studies do demonstrate durable influences.[45] Psychologist Robert Jacobs conducted one of the most informative experiments. He used the same autokinetic procedure as Sherif,[46] and set up microsocieties of two to four people. Each "generation," the subjects viewed the fixed dot and reported their estimates of its movement. Then the "oldest" experienced subject was removed from the society and a new naïve subject introduced. The experiments continued for ten generations. To create interesting initial conditions, some of the members of the initial generation were the experimenter's confederates. In one pair of experiments, Jacobs set up two three-person microsocieties. In both cases, confederates reported that the light moved sixteen inches, a highly deviant value. In one experiment, two of the three initial members of the society were confederates, and in the other experiment, only one of the three initial members was a confederate. When real subjects faced two confederates, estimates were more than twice as far from the "true" movement of four inches compared with real subjects who were in groups with just one confederate. In both societies, the effect of the initial deviant models was temporary. Both microsocieties evolved toward the average estimate of uninfluenced naive subjects, although the society with the initially largest deviation took considerably longer to reach equilibrium. In this experiment, guided variation was a powerful enough force to overbalance the conformist-bias effect in the long run.

Conformity does not stir much interest among contemporary social psychologists; the work conducted between 1950 and 1980 is still the main stuff of modern textbooks.[47] Conformist *transmission* remains very poorly studied, and we believe this illustrates a common phenomenon. Without Darwinian concepts and tools, the population-level consequences of individual behavior are not intuitive. Social psychologists following their noses did not discover the role of conformity in cultural evolution, whereas Jacobs, who worked on his project with the pioneering evolutionary psychologist Donald Campbell, asked an evolutionary question and devised the proper experiment to answer it. Darwinian analysis reveals a mass of largely unexplored questions surrounding the psychology of cultural transmission and the biases that affect what we learn from others. Small, dull effects at the individual level are the stuff of powerful forces of evolution at the level of populations.[48] Understanding rather precisely how *individuals* deploy their kit of imitation heuristics is necessary to understand the rates and direction of cultural evolution, and work on the problem has hardly begun.

Imitate the successful

People often imitate the successful—aspiring pop stars imitate Madonna's vocal style and sartorial panache, and aspiring NBA stars imitate Michael Jordan's slash to the hoop, his solution to male-pattern baldness, and, if the Sara Lee Corporation[49] has spent its money wisely, his taste in underwear. On the face of it, this strategy seems odd, but advertising executives earn handsome rewards for getting inside our heads. Mass-media celebrities notwithstanding, our attraction to the successful makes much adaptive sense. Determining *who* is a success is much easier than to determining *how* to be a success. By imitating the successful, you have a chance of acquiring the behaviors that cause success, even if you do not know anything about which characteristics of the successful are responsible for their success. If you can accurately imitate everything they do, you ought to be a success too, at least insofar as success is based on culturally transmissible characters. Even when the exact behaviors that contribute most to fitness are very hard to evaluate, there may be easily observable traits that are correlated with fitness, such as wealth, fame, and good health. If so, you can try to imitate everything that wealthy people do in an effort to acquire the traits that make them wealthy, but without actually trying to determine exactly how wealth is produced. We call this process *model-based bias,* because the bias depends not on the characteristics of the cultural variant itself, but instead depends on some other characteristic of individuals modeling the variant, such as indicators of prestige. Anthropologist Joe Henrich and psychologist Francisco Gil-White argue that we grant prestige, and the favors that go with it, to people we perceive as having superior cultural variants as a means of compensating them for the privilege of their company and the opportunity to imitate them. They contrast human prestige with the more-widespread phenomenon of dominance, where strong or guileful individuals usurp resources from the weaker.[50] We can think of other forms of model-based bias besides the prestige bias, but we'll stick with the more-evocative term in what follows.

To see how prestige bias might evolve, consider once again, the hypothetical population of early humans expanding their range from tropical savanna into temperate woodland. Assume that individuals living in the woodland have a hard time determining the best way to behave, and as a result peripheral populations contain a mix of behaviors, some good and some not so good. People who happen to acquire the best behavior will be, on average, more successful. They will be healthier and have larger families or more political power. Thus, people who imitate the successful will, all other things being equal, be more likely to acquire the locally adaptive behavior. If the tendency to imitate the successful is genetically (or culturally) variable, it will increase by natural selection.

Simple mathematical models show that the strength of prestige bias depends on the correlation between the traits that *indicate* success and the traits that *cause* success.[51] They also show that prestige bias can lead to an unstable, runaway process much like the one that may give rise to exaggerated characters such as peacock tails.

Many social psychological experiments suggest that we are predisposed to imitate successful, prestigious people, even in domains not obviously related to their success. In one study, for example, subjects were asked their opinions on "student activism" in one of three scenarios: after hearing the opinion of somebody identified as an expert on the topic, after hearing the opinion of an expert on the Ming dynasty, and after a control condition in which they didn't hear anybody's opinion. Subjects tended to voice opinions similar to either of the two experts, and they were equally likely to adopt the opinions of experts on activism and the Ming dynasty.[52] Other experiments are consistent with the prediction that the tendency to imitate the prestigious should be greater when individuals have difficulty figuring out the best alternative on their own. Field studies are also consistent with the idea that prestige plays an important role in social learning. For example, people often use prestige bias to acquire new traits, tending to adopt the practices of high-status "opinion leaders."[53] This is particularly true for the poor and less educated, whose ability to bear the costs of direct evaluation of innovations is limited. Interestingly, the poor and less educated typically imitate people of high *local* status, not socially distant elites whose life situation is far from potential adopters. A poor Turkoman herder is probably well advised to imitate the herd management practices of his wealthier neighbors and to ignore the advice of technical experts from Colorado, Switzerland, or New Zealand. Studies of dialect evolution also support this hypothesis; locally prestigious women tend to be the most advanced speakers of evolving dialects.[54] Indeed, the data suggest that popular preteen girls of the working or lower middle class are usually the most important leaders of language evolution in American cities. (We get perverse pleasure out of teasing our sometimes language-elitist academic colleagues with this fact.) The patterns of prestige in human societies are also consistent with the idea that information, not power, gets you prestige. For example, older people are prestigious in many societies, even when they do not have the power either in their person or their political alliances to dominate others.

The existence of these fast and frugal heuristics for acquiring culture now has us deeply entangled in the adaptationist's dilemma. Easy tricks are available to improve the power of culture to evolve adaptations, seemingly simpler and less costly tricks than the individual learning and content bias that are based on ubiquitous animal capacities for learning on one's own. Darwin's intuition that imitation should be widespread seems well supported by our modeling exercises, yet we are stuck with the stubborn empirical findings that very few if any other species make anything like the use of culture that we do. Many species have simple forms of social learning that ought to be excellent foundations on which more-sophisticated forms could evolve. And, culture seems to be the very bag of tricks we've used to become the earth's dominant organism. Something quite unusual and quite remarkable must have led to our weird species. Understandably, few people think their own species is weird. Somehow being a very recently evolved species that has exploded like none other seems as right and natural to most as when we still believed that God created us in his image. A little scientific theorizing is necessary to convince us that the existence of human culture is a deep evolutionary mystery on a par with the origins of life itself. We make no pretense of having a completely satisfactory explanation for the adaptationist's dilemma of culture, but let's peck away at the strands of the problem and see if we can see a ray or two of light.

How the capacities for culture possibly evolved

We are all surprised, amused, and sometimes exhausted by the intense curiosity of young children. As Ph.D.'s who flatter ourselves as having a wide and deep fund of general knowledge, especially when we can combine our different ranges of expertise, we received some humbling lessons from Pete's firstborn child. He often was able to put his current questions to the two of us either simultaneously or sequentially (not to mention his mother, "Aunt" Joan, and other handy adults), and he frequently exhausted our collective knowledge embarrassingly quickly. Contemptuous of answers of the form "We just don't know why it happens that way," he would demand, "Then why *maybe?*"

Philosopher Robert Brandon argues that why-maybe answers play an important role in evolutionary biology (he calls them "how possibly" explanations).[55] He points out that evolutionary trajectories are so complicated that they rarely allow an exact elucidation of how and why things happen. Evolutionary processes are too complex and the paleo-environmental and fossil records are too fragmentary for us to be certain of any account of how some adaptation evolved. More than one hypothesis is usually consistent with all the data we have at hand, and several might still stand after we have all the data we are ever likely to get. Although the kinds of adaptive accounts that evolutionary biologists give to historical questions are sometimes stigmatized as "adaptive just-so stories," Brandon argues that non-adaptive accounts are equally "just so." No Darwinian account of the evolution of any lineage of organisms entirely escapes being a how-possibly explanation. Nevertheless, some how-possibly answers are better than others. They are better because they fit more of the available information, they are better grounded in theory, and they are productive of further work. While we can never be satisfied with how-possibly accounts, they can still yield appreciable progress.

The typical trajectory of the evolutionary sciences is that we begin with a simple hypothesis or two that prove to be quite wrong but in being wrong simulate a spate of further work. For a while, the number of plausible ideas grows rapidly, and the data accumulate more slowly. In this middle period of a problem, uncertainty actually appears to grow, as if the more we investigate a problem the less we are certain about any part of it. Of course, this state of affairs results from our former innocent ignorance of the magnitude of the problem. Then a pruning process begins as hard work finds fatal flaws in old, good ideas faster than new ones appear. We may never know *the* answer, but we end up *immensely* more sophisticated than when the enterprise began. Given the manifest importance of culture in human behavior, the theory of cultural evolution ought to be central to the how-possibly project. In that spirit, we offer the following why-maybe account of the origin of *Homo sapiens* in terms of the evolution of increasingly sophisticated capacities for culture.

Culture is adaptive because it provides information about variable environments

Humans, even as hunter-gatherers, adapt to a vast range of environments. The archaeological record indicates that foragers from the Pleistocene epoch occupied virtually all of Africa, Eurasia, and Australia. The data on historically known hunter-gatherers suggest that to exploit this range of habitats, humans used a dizzying diversity of subsistence practices and social systems. Consider just a few examples. The Copper Eskimos lived in the high Arctic, spending summers hunting near the mouth of the MacKenzie River and the long, dark months of the winter living on the sea ice, hunting seals. Groups were small and intensely dependent on men's hunting. The !Xo lived in the central Kalahari collecting seeds, tubers, and melons; hunting impala and gemsbok; surviving fierce heat; and living without surface water for months at time. Both the !Xo and the Copper Eskimo lived in small, nomadic bands linked together in larger patrilineal band clusters. The Chumash lived on the productive California coast around present-day Santa Barbara, gathering shellfish and seeds and fishing the Pacific from great plank boats. They lived in large permanent villages with division of labor and extensive social stratification.

This range of habitats, ecological specializations, and social systems is much greater than any other animal species. Big predators such as lions and wolves have the largest range among other animals, but lions never extended their range beyond Africa and the temperate regions of western Eurasia; wolves were limited to North America and Eurasia. The diet and social systems of such large predators are similar throughout their range. They typically capture a small range of prey species using one of two methods: they wait in ambush, or combine stealthy approach and fast pursuit. Once the prey is captured, they process it with tooth and claw. The basic simplicity of the lives of large carnivores is captured in the Gary Larson cartoon in which a *T. rex* contemplates its monthly calendar—every day has the same notation "Kill something and eat it." In contrast, human hunters use a vast number of methods to capture and process a huge range of prey species, plant resources, and minerals. For example, anthropologist Kim Hill and his coworkers have observed the Aché, a group of foragers who live in Paraguay, who take 78 different species of mammals, 21 species of reptiles, 14 species of fish, and over 150 species of birds using an impressive variety of techniques that depend on the prey, the season, the weather, and many other factors. Some animals are tracked, a difficult skill that requires a great deal of ecological and environmental knowledge. Others are called by imitating the prey's mating or distress calls. Still others are trapped with snares or traps or smoked out of burrows. Animals are captured and killed by hand, shot with arrows, clubbed, or speared.[56]

And this is just the Aché—if we included the full range of human hunting strategies, the list would be endless. The list of techniques applied to plants and minerals is similarly long and diverse. Making a living in the Arctic requires specialized knowledge: how to make weatherproof clothing, how to provide light and heat for cooking, how to build kayaks and umiaks, how to hunt seals through holes in the sea ice. Life in the central Kalahari requires equally specialized, but quite different knowledge: how to find water in the dry season, which of the many kinds of plants can be eaten, which beetles can be used to make arrow poison, and the subtle art of tracking game. Survival might have been easier on the balmy California coast, yet specialized social knowledge was needed to succeed in hierarchical Chumash villages compared to the small, egalitarian bands of the Copper Eskimo and the !Xo.

So, maybe humans are more variable than lions, but what about other primates? Don't chimpanzees have culture? Don't different populations use different tools and foraging techniques? There is no doubt that great apes do exhibit a wider range of foraging techniques, more-complex processing of food, and more tool use than other mammals.[57] However, these techniques play a much smaller role in great ape economies than they do in the economies of human foragers. Anthropologist Hillard Kaplan and his coworkers compare the foraging economies of a number of chimpanzee populations and human foraging groups. They categorize resources according to the difficulty of acquisition: *Collected foods* like ripe fruit and leaves can be simply collected from the environment and eaten. *Extracted foods* must be processed and include fruits in hard shells, tubers or termites that are buried deep underground, honey hidden in hives high in trees, or plants that contain toxins that must be extracted before they can be eaten. *Hunted foods* come from animals, usually vertebrates, that must be caught or trapped. The data show that chimpanzees are overwhelmingly dependent on collected resources, while human foragers get almost all of their calories from extracted or hunted resources.[58]

Humans can live in a wider range of environments than other primates because culture allows the relatively rapid accumulation of better strategies for exploiting local environments compared with genetic inheritance. Consider "learning" in the most general sense; every adaptive system "learns" about its environment by one mechanism or another. Learning involves a tradeoff between accuracy and generality. Learning mechanisms generate contingent behavior based on "observations" of the environment. The machinery that maps observations onto behavior is the "learning mechanism." One learning mechanism is more accurate than another in a particular environment if it generates more-adaptive behavior in that environment, and it is more general than another if it generates adaptive behavior in a wider range of environments. Typically, a tradeoff exists between accuracy and generality, because every learning mechanism requires prior knowledge about which environmental cues predict the state of the environment and what behaviors are best in each environment. The more detailed and specific such knowledge is for a particular environment, the more accurate is the learning rule. Thus for a given amount of inherited knowledge, a learning mechanism can either have detailed information about a few environments, or less-detailed information about many environments.

In most animals, this knowledge is stored in the genes, including of course the genes that control individual learning. Consider a variation on the thought experiment described in chapter 2. Pick a wide-ranging primate species, let's say baboons. Then capture a group of baboons, and move them to another part of the natural range of baboons in which the environment is as different as possible. You might, for example, transplant a group from the lush wetlands of the Okavango Delta to the harsh desert of western Namibia. Next, compare their behavior to the behavior of other baboons living in the same environment. We believe that after a little while, the experimental group of baboons would be quite similar to their neighbors. This experiment has actually been done, although not in such an extreme case. Primatologist Shirley Strum moved a group of baboons that was being threatened by humans from one site to a somewhat different one several hundred kilometers away. The baboons quickly adapted to their new home. The reason that the local and transplanted baboons would be similar, we think, is the same reason that baboons are less variable than humans: they acquire a great deal of information about how to be a baboon genetically. To be sure, they have to learn where things are, where to sleep, which foods are desirable, and which are not, but they can do this without contact with already knowledgeable baboons because they have the basic knowledge built in. But they can't learn to live in temperate forests or arctic tundra, because their learning systems don't include enough innate information to cope with those environments.

Human culture allows learning mechanisms to be both more accurate and more general, because *cumulative* cultural adaptation provides accurate and more-detailed information about the local environment. People are smart, but individual humans can't learn how to live in the Arctic, the Kalahari, or anywhere else.[59] Think about being plunked down on an Arctic beach with a pile of driftwood and seal skins and trying to make a kayak. You already know a lot—what a kayak looks like, roughly how big it is, and something about its construction. Nonetheless, you would almost certainly fail (We're not trying to dis you; we've read a lot about kayak construction, and we'd at best make a poor specimen, without doubt). Even if you could make a passable kayak, you'd still have a dozen or so similar tools to master before you could make a contribution to the Inuit economy. And then there are the social mores of the Inuit to master. The Inuit could make kayaks, and do all the other things that they needed to do to stay alive, because they could make use of a vast pool of useful information available in the behavior and teachings of other people in their population. The reason the information contained in this pool is adaptive is that a combination of learning and cultural transmission leads to relatively rapid, cumulative adaptation. Even if most individuals blindly imitate with only the occasional application of some simple heuristic, many individuals will be giving traditions a nudge in an adaptive direction, on average. Cultural transmission preserves the many small nudges, and exposes the modified traditions to another round of nudging. Very rapidly by the standards of ordinary evolutionary time, and more rapidly than evolution by natural selection alone, weak decision-making forces generate new adaptations. The complexity of cultural traditions can explode to the limits of our capacity to learn them, far past our ability to make careful, detailed decisions about them. We let the population-level process of cultural evolution do the heavy lifting of our "learning" for us.

Social learning may be an adaptation to Pleistocene climate fluctuations

The picture sketched above indicates that cumulative cultural adaptation is most advantageous when there are big differences between environments in time and space *and* when that variation arises slowly enough to make transmission and accumulation by social learning useful. If environments change too rapidly in time or space, selection will favor individual learning, but no transmission. If environments change too slowly, then ordinary organic evolution can track the fluctuations more faithfully and at less cost than a system of social learning. Humans seem to be the first species on our planet to have evolved an advanced capacity for cumulative culture, although in so doing we have proved a spectacular, though not necessarily permanent, success. Given that complex culture is adaptive, why did it evolve in the human lineage at this particular juncture of the earth's rather long biotic history?

One good how-possibly answer is that social learning is an adaptation to increased climate variation during the last half of the Pleistocene. This hypothesis provides a possible way to ease off the horns of the adaptationist's dilemma. We suspect that a sophisticated capacity for culture has only been adaptive for a short, recent bit of the earth's history and we are merely the first lineage to discover its advantages. Deteriorating climate over the last two million years favored increased behavioral flexibility, including an increased reliance on social learning, probably in many species. Already a relatively large-brained group, the primates were preadapted to evolve the cognitively taxing mechanisms of observational learning and sophisticated biasing needed to manage culture. Just storing the large cultural repertoires involved with complex, accumulated cultural adaptations may require considerable brain volume. Primates are also rather sociable as mammals go, and "learning" by cultural evolution is an intensely social phenomenon. Finally, the visual adaptation of most primates and the manipulative hands of our ancestors were likely preadaptations for imitation and the production of sophisticated tools that are the cornerstone of human economies. The evidence from the fossil and archaeological records is consistent with the hypothesis that the psychological machinery that underpins cumulative cultural change evolved over the last half million years, a period during which climates were more unstable than ever before.

Using a variety of proxy measures of past temperature, rainfall, ice volume, and the like, derived mostly from cores of ocean sediments, lake sediments, and ice caps, paleoclimatologists have recently constructed a stunning picture of climatic deterioration over the last three million years.[60] The earth's mean temperature has dropped several degrees, and the amplitudes of fluctuations in rainfall and temperature have increased (fig. 4.2).[61] For reasons that are still poorly understood, glaciers wax and wane in concert with changes in ocean circulation, carbon dioxide, methane, the dust content of the atmosphere, and changes in the average amount and distribution of precipitation. Different cyclical patterns of glacial advance and retreat involving all these variables have prevailed. A 21,700-year cycle dominated the early part of the period, a 41,000-year cycle between about 2.6 million and 1 million years ago, and a 95,800-year cycle the last million years.

Fluctuations that occur over tens of thousand of years are not likely to have driven the evolution of adaptations for social learning. Populations will adjust to such slow changes by changing their ranges and by organic evolution. However, the increased variation over such long timescales seems to be strongly associated with variation at much shorter timescales. High-resolution data for the last 80,000 years are available from ice cores taken from the deep ice sheet of Greenland. Resolution of events lasting little more than a decade is possible in ice 80,000 years old, improving to monthly resolution for events after 3,000 years ago. During the last glacial, the ice core data show that the climate was highly variable on timescales of centuries to millennia.[62] Figure 4.3 illustrates how dramatic this variability

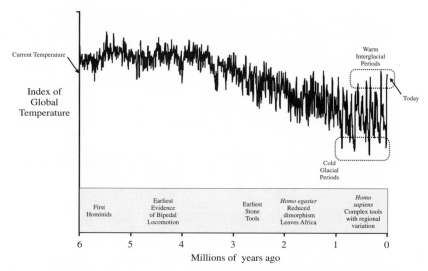

Figure 4.2. The world's climate has become colder and more variable over the last six million years. The vertical axis plots $\delta^{18}O$, the excess of ^{18}O relative to ^{16}O in samples taken from deep-sea sediments that date to different times over the last six million years. The concentration of ^{18}O in seawater increases during cold periods, because water containing the lighter isotope of oxygen, ^{16}O, evaporates more readily and is thus trapped in glacial ice. Other data indicate that during cold periods, the world was drier and the CO_2 concentration of the atmosphere was lower. (Redrawn from Opdyke et al. 1995.)

was. Even when the climate was in the grip of the ice, it briefly spiked to near interglacial warmth every thousand years or so. The intense variability of the last glacial carries right down to the limits of the decade-level resolution of the ice core data. Sharp spikes lasting a century or less are common in the Greenland record. Even more recent high-resolution data from temperate and tropical latitudes verify that the high-amplitude fluctuations seen in the ice core are global phenomena, and some of the best records suggest that most or even all of the world's climates fluctuated to the same beat recorded so beautifully in Greenland ice.[63]

Undoubtedly, oscillations such as those detected in ice cores had important impacts on evolving animal populations. The Holocene (the last relatively warm, ice-free 11,500 years) has been a period of very stable climate compared with the last glacial. Nonetheless, Holocene weather extremes have had significant effects on organisms.[64] The impact of the much greater variation that was probably characteristic of most of the Pleistocene is hard to imagine. Tropical organisms did not escape the impact of climate variation; temperature and especially rainfall were highly variable at low

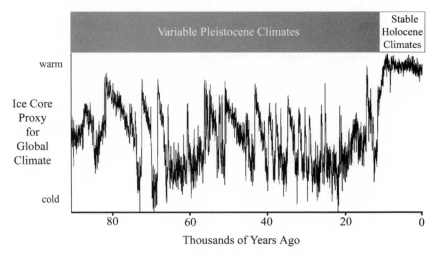

Figure 4.3. During the last glacial period, the world's climates were subject to much more variation than we have experienced in the last few thousand years. The data plotted here are from the 2-mile-long Greenland Ice Project core taken in the early 1990s. The last glacial period extended from about 110,000 years ago to 12,000 years ago, but the record in this core is only reliable back to just over 80,000 years ago. The vertical axis plots the deficiency in ^{18}O, an index of temperature, in the core. Notice that during the last glacial period, high-latitude temperature swung from glacial to nearly inter-glacial levels every thousand years or so. Other data indicate that similar fluctuations occurred at lower latitudes. Because this figure was smoothed using a 150-year averaging scheme, the actual amount of short time period fluctuations was greater than shown. (Redrawn from Ditlevsen et al. 1996.)

latitudes.[65] During most of the Pleistocene, plants and animals lived under conditions of rapid, chaotic, and ongoing reorganizations of ecological communities as species' ranges adjusted to the noisy variation in climate. Thus, for the last two and a half million years or so, organisms have seemingly had to cope with increasing variability in many environmental parameters at timescales on which strategies for phenotypic flexibility would be highly adaptive.

The Pleistocene climate deterioration is correlated with increases in brain size in many mammalian lineages besides our own. The average encephalization (brain size properly corrected for body size) of mammals has increased ever since the demise of the dinosaurs 65 million years ago.[66] However, many relatively small-brained mammals persist to the present even in orders where some species have evolved large brains. The largest increases in encephalization per unit time by far occurred over the last 2.5 million years—the increase in average encephalization during this period was larger than the increase during the previous 20 million years. Brain enlargement in the human lineage began to diverge from the trend of the other apes at the beginning of the Pleistocene, about 2 million years ago, about the same time as an abrupt increase in the amplitude of glacial fluctuations,[67] and then increased rapidly again between eight hundred thousand and five hundred thousand years ago after another increase in the amplitude of glacial fluctuations.

All other things being equal, selection should ruthlessly favor small brains, because large brains are costly.[68] Nonetheless, brain size in mammals is quite variable. Human brains account for 16% of our basal metabolism. Average mammals have to allocate about only 3% of basal metabolism to their brains, and many marsupials get by with less than 1%.[69] These differences are easily large enough to generate strong evolutionary trade-offs. In addition to metabolic requirements, there are other significant costs of big brains, such as increased difficulty at birth, greater vulnerability to head trauma, increased potential for developmental snafus, and the time and trouble necessary to fill them with usable information. In effect, all animals are under stringent selection pressure to be as stupid as they can get away with. The oft-mentioned "fact" that we actually use only a small part of our brain is a myth. Brains are a use-it-or-lose-it organ. If they have gotten bigger, they must be good for something, really good.

A recent study by comparative psychologists Simon Reader and Kevin Laland suggests that one thing they are good for is learning—both individual *and* social learning.[70] Reader and Laland surveyed the primate literature, recording the number of times that different primate species had been observed doing three different things: using tools, performing novel or innovative behavior, and engaging in social learning. They showed that all three traits are correlated with a measure of brain size. In other words, primates with bigger brains are more likely to use social learning, more likely to engage in novel behavior, and more likely to use tools. Interestingly, observations of novel behavior and social learning are correlated even after the effect of brain size is taken into account, suggesting that social learning allows more-flexible responses to novel environments.

A related study by Hillard Kaplan and economist Arthur Robson[71] supports the idea that larger brains lead to more behavioral flexibility. They showed that among primate species, larger brains (corrected for body size) are associated with a longer juvenile period and longer life span, even when other correlates of brain size, like group size, are controlled. Kaplan and Robson argue that brain size and longevity are linked in an adaptive complex. As we all know, learning takes time. You can't learn how to play chess or ski in a day—mastering mental and physical skills takes years of learning and practice. The same goes for foraging skills. This means that envi-

ronments, like the variable ones of the Pleistocene, that favor increased behavioral flexibility also favor longer juvenile periods to allow enough time
for learning. Learning and teaching culture are costly investments, and thus
increased brain size and longer juvenile periods will favor a longer life span.
Selection favors a longer life, because it allows individuals to get more benefit from what they learned during the necessary but costly extended juvenile period.[72]

According to the argument we have developed so far, humans are just
the tail of the distribution. We are the largest brained, slowest developing
member of the largest brained, slowest developing mammalian order.
However, this can't be the whole story. That increases in brain size and decreases in developmental rate are correlated with climate variation supports
the idea that fluctuating environments really do favor increased behavioral
flexibility and social learning. However, as we argued earlier, we are unique
in our ability to build up complex subsistence systems over many generations by the incremental modifications of many innovators. This capacity,
on our account, is responsible for our ability to evolve a huge range of complex cultural adaptations that in turn account for our success as a species.
But if many animal species have rudimentary to moderately sophisticated
systems for social learning and if complex culture is a highly advantageous
means of adapting to Pleistocene climatic deterioration, why is complex
culture so rare?

One interesting hypothesis is that the evolution of the cumulative cultural evolution faces a "bootstrap problem." Models show that under some
sensible cognitive-economic assumptions, a capacity for complex cumulative culture cannot be favored by selection when rare.[73] The idea is quite intuitive. Suppose that to acquire a complex tradition efficiently by imitation,
some derived cognitive machinery is required. For example, a number of
psychologists have argued that a "theory of mind" is required for observational learning.[74] The idea is that unless you can guess other people's intentions and motives, imitation is very difficult. Suppose you see our mystery device (figs. 4.1, 4.4) hanging in someone's kitchen, and later see an
identical one in a store. Are you tempted to buy? If you still don't know its
purpose, almost certainly not. If you have discovered what other people do
with it, then perhaps so. We humans automatically put ourselves inside
others' heads. If Aunt Ethel uses the mystery device in the course of making a salad in your presence, you fit its use into a scenario of Aunt Ethel
wanting to make a salad, wanting a certain ingredient in the salad, and using the mystery device to that end. Having modeled Aunt Ethel's motivations and actions, you know the function of the device and can fit it into a

Figure 4.4. It is an avocado slicer (from Progressive International Corp.). Halve the avocado, remove the pit, and then use the slicer to make long wedges in the fruit. The flat hoop makes it easy to stay near the skin, and the thin wires slice even very ripe avocado without tearing.

scenario where you personally might find the device useful, even if you never touched the mystery device or sampled the salad. The decision to buy or not is for us trivial once we've *seen* what the thing is for. As easy and automatic as this seemingly trivial mental theorizing is to us, experiments show that small children and most other animals tested either lack the capacity to see others' functional acts in this way or have only limited abilities to do so.

Suppose that the theory of mind module is necessary for rapid, accurate imitation of complex skills and that it also takes up a not-trivial amount of the resources of the brain. Suppose further that if complex, difficult-to-accumulate, culturally evolved traditions are available to imitate using the module, then the capacity to acquire them is a big fitness advantage, more than repaying the nontrivial cost. Obviously, complex traditions cannot evolve without the cognitive machinery that gives rise to cumulative cultural evolution. The rub is that complex traditions don't come out of thin air. A whole population of individuals capable of imitating has to exist and exist for some time to evolve complex traditions. This means that a rare mutant with the ability to imitate, say, because he has a better theory of mind, will observe only the behavior that can be acquired without his ability. Such a mutant will bear the costs of the module but will get no benefits.

Worse yet, as anthropologist Joe Henrich has argued, to get complex traditions, just a few individuals with the necessary cognitive complexity aren't enough; the cultural evolution of complex adaptations may require a fairly large population of imitative minds. Henrich points out that imitation is an error-prone process and that learners have a hard time getting the skills to manufacture complex artifacts down pat. In a small population, this effect will lead to the degradation of more-complex skills. However, in a large population, especially skilled or lucky toolmakers will be relatively numerous. These geniuses will improve the technology, and have the effect of preventing the degradation of the technology as their imitators spread the recovered complexity to others. Henrich's work suggests that only fairly sizable populations can sustain complex, culturally evolved artifacts and behaviors.

This result is consistent with the loss of tool complexity on Tasmania documented by the late Australian archaeologist Rhys Jones. When European explorers reached Tasmania in the nineteenth century, they collected the simplest tool kit known for any living people. When Jones got to digging on Tasmania in the 1970s, he discovered that the Tasmanians once had the full Australian tool kit, hundreds of items richer than that collected from the living Tasmanians.[75] The complexity of the tool kit began to decline when the flooding of the Bass Strait about eight thousand years ago cut the land bridge that connected Tasmania to the mainland. Yet the Tasmanian population was not tiny—at European arrival it numbered about four thousand people. Nor had the technology simplified quickly. Rather, the more-complex items, such as boats, seem to have disappeared slowly but steadily over the millennia. These data and Henrich's model suggest that surprisingly large populations are necessary to sustain a tool kit consisting of many hundreds of rather complex items against slow but inexorable decay due to small but cumulative transmission error.

If such an impediment to the evolution of complex traditions existed, evolution must have traveled a roundabout path to get the theory of mind module (or whatever) past the threshold necessary for bringing it under positive selection for the cumulative cultural adaptation. Some have suggested that primate intelligence was originally an adaptation to manage a complex social life.[76] Perhaps in our lineage the complexities of managing food sharing, the sexual division of labor, or some similar social problem favored the evolution of a sophisticated ability to take the perspective of others. Such a capacity might incidentally make imitation possible, launching the evolution of the most elementary form of complex cultural traditions. Once elementary complex cultural traditions exist, the threshold is crossed. As the evolving traditions become too complex to imitate easily they will begin to drive the evolution of still more-sophisticated imitation. This advantageous-but-can't-increase-when-rare sort of stickiness in the evolutionary processes is presumably what gives evolution its commonly contingent, historical character.[77] If such barriers exist to the evolution of a new capacity, then many species with the apparently necessary preadaptations may collect at the barrier until finally one breaks through. Other such barriers are easy to imagine. Much of the traction we get from culture comes from tools. Most apes are quadrupeds that need all four limbs for locomotion. Once our lineage became bipedal, hands could fall under selection for new functions such as making stone tools and carrying spears. Like winning the lottery, probably several such preadaptations had to come our way before natural selection could get real purchase on the capacity for complex culture.

How humans possibly evolved

With these ideas in hand, let us now turn to the evolution of the human lineage. We have two goals here. First, we want to convince you that population thinking about human culture adds quite a bit to the explanations conventionally used in paleoanthropology. Second, in chapter 6 we will argue that cultural evolutionary processes have shaped human social environments in ways that had important consequences for the genetic evolution of human psychology. Here we discuss the evidence that humans have had the capacity for cumulative cultural evolution long enough for such coevolutionary processes to be important.

The earliest hominids were bipedal, but otherwise much like contemporary apes. Genetic data indicate that the last common ancestor of humans, chimpanzees, and bonobos lived five to seven million years ago. Three different hominoid fossils date from this period, *Orrorin tugenensis, Sahelanthropus tchadensis,* and *Ardipithecus ramidus.* However, currently described specimens do not tell us whether any of these species were bipedal, or whether they are more closely related to humans or chimpanzees. Beginning roughly four million years ago, the first bipedal hominids appear in the fossil record, and when it rains it pours. For the next two million years, Africa was lousy with hominid species. The details of the taxonomy

are controversial, but most paleoanthropologists agree that there were between five and ten species belonging to three separate genera, *Australopi-
thecus, Paranthropus,* and *Kenyanthropus.* We will refer to these folks collectively as "bipedal apes," because while they were bipedal, they were still
very apelike in most other ways. Males were much larger than females, indicating that males probably invested more energy in competing for mates
than caring for offspring. Their brains were the same size as the brains of
contemporary apes (correcting for body size), and they had a relatively
short juvenile period and life span, even shorter than living chimpanzees.
They were smaller than modern humans (roughly the same size as chimpanzees), with long arms and short legs, suggesting that they still spent
quite a bit of time in the trees. Many anthropologists include the specimens
formerly included in *Homo habilis* in one of these genera because although
some of these specimens had larger brains than other early hominids, they
were otherwise apelike.[78] Paleoanthropologists have reached no consensus
about which bipedal ape species is ancestral to later hominids. Upright posture and hands did not by themselves set off a rush to complex culture as
paleoanthropologists once supposed. For a million and a half years or so of
bipedality, no evidence for artifacts exists at all.

Perhaps the bipedal apes eventually began to use chipped stone tools.
The earliest flaked stone tools have been found at Gona, a site in Ethiopia
that dates to about 2.6 million years ago. Similar crudely shaped cores and
flakes belonging to the Oldowan industry are found in many sites in Africa
that date to this period, but it is unclear which hominid species made these
tools. The bipedal apes furnish the only bones so far discovered to match
to the stones. However, *Homo ergaster* fossils have been discovered dating
to about 1.8 million years ago. Stone tools are tough objects, and a user
probably made many of them in a lifetime. Bones are much more perish-
able, and no one leaves more than one set. Thus, the stones record is denser
than the bones record. The earliest tools will typically appear in the fossil
record before the first fossil of the creature that made the tools. In any case,
since both chimpanzees and orangutans use simple tools, the bipedal apes
probably did as well, even if they didn't flake stone.

Other evidence suggests that Pleistocene bipedal apes had no more sophisticated social learning abilities than living apes. Their brain sizes and
developmental rates were similar to contemporary apes, suggesting that
their cognitive abilities and investment in learning were similar, and their
geographical ranges were limited in the same way as contemporary ape species. Thus, the tool traditions of bipedal apes were likely not transmitted by
imitation, but rather maintained by other learning mechanisms, as in con-

temporary apes. Primatologist Sue Savage Rumbaugh and archaeologist Nicholas Toth were unable to teach Kanzi, a bonobo with a considerable talent for acquiring human behaviors, to make simple stone tools. He was able to make small sharp stone flakes by flinging raw cobbles against hard concrete surfaces, and then used the flakes to open food containers. But despite much tuition, he was never able to flake cores using his hands in a controlled way.[79] Why Kanzi couldn't accomplish this task isn't clear. Perhaps his ability to imitate is deficient. Perhaps the morphology of the chimpanzee hand makes this task difficult for him.[80] Or, perhaps he is handicapped by cognitive limitations; chimpanzees seem to have a limited ability to represent causal physical relationships.[81]

Early specimens of *Homo ergaster* have been found at a number of East African sites and as far afield as Dmanisi in the foothills of the Caucasus Mountains. Anatomically similar fossils, usually called *Homo erectus,* have been found in China and Indonesia at sites that date from perhaps 1 million years ago up to less than 100,000 years ago. These creatures have larger brains than the bipedal apes, but also have larger, modern human-sized bodies, so they were only a bit brainier on average than the bipedal apes that preceded them. These hominids were fully committed terrestrial bipeds with long legs and short arms. The difference between the size of males and females was about the same as in modern people. *H. ergaster* probably developed more rapidly than modern humans. By counting growth lines in tooth enamel, biological anthropologists can accurately estimate the rate at which teeth develop, and in living primates the rate of tooth development is highly correlated with other developmental rates. Using this technique, anatomist Christopher Dean and colleagues showed that the rate of development of *H. ergaster* was similar to living apes, a little slower than the bipedal apes that preceded it, and much faster than modern humans.[82]

The earliest fossils of *H. ergaster* are associated with simple Oldowan tools, the same ones that some creature or creatures had been making for 800,000 years or so. However, beginning sometime between 1.6 million and 1.4 million years ago, a more-sophisticated tool kit, called the Acheulean industry, appears in Africa. The Acheulean is dominated by large cobbles that have been carefully reduced to a symmetrical, tear-drop-shaped hand ax. The same tool kit is found throughout Africa and western Eurasia for the next million years—not just *similar* tool kits, but statistically the *same* tool kit. Once the effects of raw materials are accounted for, the differences between the tools found at sites that are separated by a million years are, on average, no more than the differences between tools at contemporaneous sites. In East Asia, simple tools similar to the Oldowan continued to be made. Controversial evidence also suggests that hominids were able to control fire during this period.

The evidence concerning the imitative abilities of *Homo ergaster* is quite bewildering. Most scholars assume that the skills necessary to manufacture Acheulean tools were transmitted culturally in the same way that stone tool traditions are transmitted among living foragers. However, this assumption is hard to reconcile with either theory or data. Models predict that traditions among small, semi-isolated groups will rapidly diverge, so that even if functional constraints are strong, variation between groups will increase through time.[83] Both archaeological evidence from later people and ethnographic data are consistent with this prediction. How could cultural transmission *alone,* particularly if based on a relatively primitive imitative capacity, preserve such a neat, formal-looking tool as a Acheulean hand ax over half the Old World for a million years?[84] Combine this fact with *H. ergaster's* relatively small brain and rapid development, and perhaps we need to entertain the hypothesis that Acheulean bifaces were innately constrained rather than wholly cultural and that their temporal stability stemmed from some component of genetically transmitted psychology. On the other hand, the sophisticated controlled forms of the Achuelean have no parallel among the tools made by any other species of primate and demand the same sorts of manual skills that we transmit culturally.

From the point of view of cultural evolution, this already strange pattern seems even stranger. Most evolutionary scenarios connect modern humans to chimpanzees with a straight line and assume that *H. ergaster/ erectus* fell somewhere along that line. Cultural evolutionary considerations lend weight to the suspicion that the path from our quadrupedal ancestor to ourselves was more circuitous. We are getting confidently more uncertain about what was going on in the early Pleistocene, and knowing what you don't know is just as important as knowing what you do know!

Beginning roughly a half a million years ago, larger brained hominids appear in Africa and Europe. We say "roughly" because sites during this period were, until recently, extremely difficult to date accurately.[85] From the neck down these creatures were similar to *H. ergaster/erectus*—very heavily muscled and stout boned—but rather more modern from the neck up. Their brains were about the same size as ours, but their skulls were long and low, and they had large faces with prominent brow ridges. We will follow the recent practice of referring to these hominids as *Homo heidelbergensis.* The developmental rate of early *H. heidelbergensis* has not been measured directly. However, Neanderthals, which appeared in western Eurasia between 300,000 and 130,000 years ago, developed at a rate similar to modern humans. Since Neanderthals are similar to *heidelbergensis* morphologically, and used a similar stone tool kit, the slow life history that is characteristic of modern humans probably evolved during this period.

About the same time, the first uncontroversial examples of cumulative cultural adaptation begin to appear in the archaeological record, especially in Africa.[86] About 350,000 years ago in Africa, the Achuelean industry is replaced by a variety of Middle Stone Age (MSA) industries based on what archaeologists call "prepared core" technologies. To manufacture this kind of tool, the knapper first shapes a block of stone, the core, with a hammer stone, and then strikes the core so that a large flake with a predetermined shape is removed. By 250,000 years ago this technology had spread throughout western Eurasia. During this period, particularly in Africa, the amount of regional variation in tools increased dramatically. In some areas, highly refined tool industries based on long, thin stone blades appear, based on a still more-sophisticated preparation of cores. At Katanda in the eastern Congo, archaeologists recovered exquisite barbed bone spear points.[87] Untipped wooden throwing spears, weighted for accurate flight like modern javelins, have been recovered from a bog deposit in Germany.[88] Regional diversity and highly sophisticated cultural adaptations, more sophisticated than an individual could develop on their own, are the hallmarks of cumulative cultural adaptation. Signs of symbolic behavior also emerge in Africa during the latter part of this period. Red ochre, used by modern peoples for personal adornment, is found at numerous sites, even quite early ones, and ostrich-shell beads and other decorative items enter the archaeological record beginning about 100,000 years ago.[89]

A variety of genetic data suggest that modern humans evolved and spread throughout Africa during this period, and then perhaps only 50,000 years ago spread across the rest of the world.[90] The earliest modern human fossils, dating to about 160,000 years ago, have been found in Africa, and abundant evidence suggests that modern humans spread across the world about 50,000 years ago, carrying sophisticated technology with them. How much gene flow between African and Eurasian populations occurred during this period is uncertain. Mitochondrial DNA from six Neanderthals indicates that the last common ancestor of modern human and Neanderthal mtDNA lived perhaps 500,000 years ago, and good evidence shows that modern humans in Europe are not related to Neanderthals.[91] On the other hand, a sophisticated statistical analysis of all the available molecular data suggests that quite a bit of gene flow occurred between modern African and archaic Eurasian populations as the spread occurred.[92]

So far we have said nothing about language, and the reason is simple:

paleoanthropologists have no idea when human language evolved. Some anatomists think that they can identify brain structures associated with language from the skull of bipedal ape species living more than two million years ago.[93] Others, based on reconstructions of the soft anatomy of the vocal tract, argue that even very recent hominids such as the Neanderthals may have had only limited speech.[94] We cannot easily infer anything about the evolution of language from the archaeological record because whether language is necessary for cumulative cultural evolution is unclear, at least those aspects of culture that turn up in the fossil record. Archaeologist Stephen Shennan argues that stone tool technology and similar manual skills are learned by observation and that language would not be required to make them.[95] So too even with artistic productions, though many tend to assume that graphic art and language are related. One of us has a friend who is an accomplished artist, and he cannot be made to say anything about his art. He says when pressed, "You're supposed to look at it, not talk about it!"

Psychologist Merlin Donald argues that quite complex behavior can be acquired by mimicry in the absence of language.[96] Nineteenth-century accounts of the abilities of deaf-mutes to acquire many sorts of useful economic and social skills without language suggest that they could easily learn most nonlinguistic skills by observation, without any linguistic aids. Thus the increasingly sophisticated stone tools of the later Pleistocene are not beyond the abilities of mute persons with good imitative skills. Indeed, even normal speakers generally find demonstrations of such skills superior to pictures and pictures equal to a thousand words. Language is often given pride of place as the watershed between humans and other animals, and again we are much tempted to reason from modern human analogies about fossil hominids, especially big-brained ones. Some people's credulity is strained to think that rather ancient hominids didn't have at least some simple language. On this point, too, as with Acheulean hand axes, we are more impressed by the strangeness of what we do know about the lifeways of our more-distant ancestors. Reasoning from modern patterns uniformly diluted to make them "primitive" has repeatedly failed to predict the finds of the paleoanthropologists. People might have been mute until comparatively recent times.

Many scholars believe that language evolved to manage social interaction.[97] Social actors can often benefit by communicating about who did what to whom, when, and why—that is, by gossiping—and this is difficult to do without grammatically structured language. (Imagine *People's Court* with a cast consisting only of mimes!) Language is also an extraordinarily powerful device for encoding and transmitting some kinds of cultural traditions, particularly myths and stories that often carry much information about social roles and moral norms. While nineteenth-century deaf-mutes could learn simple social customs such as table etiquette, we doubt that they could manage the rules for operating a unilineal kinship system, much less a law court. The productivity of language allows humans to express a huge number of ideas and link them in patterned arrays. Some authors think that without linguistic encoding, social learning is not accurate enough to give rise to stable traditions or gradual, cumulative adaptive evolution.[98] And even if language first evolved to gossip about band politics, it could have then been elaborated, because it made more-complex cultural traditions possible by making it easy to express, memorize, and teach cultural principles verbally. Perhaps sophisticated language antedates all other forms of complex culture.

Whatever any neural reorganization, one important factor may have been that humans had larger populations than Neanderthals by about 50,000 years ago. Recall Joe Henrich's model earlier in this chapter.[99] He argues that imitation is an error-prone process because cumulative cultural evolution of complex adaptations requires sizable population. Perhaps the Neanderthals were socially unsophisticated and had relatively limited contacts with neighbors, leading to a relatively unsophisticated tool kit.

Conclusion: Why is human culture such an extraordinarily successful adaptation?

If we are right, culture is adaptive because it can do things that genes cannot do for themselves. Simple forms of social learning cut the cost of individual learning by allowing individuals to use environmental cues selectively. If you can easily figure out what to do, do it! But if not, you can fall back on copying what others do. When environments are variable and the learning is difficult or costly, such a system can be a big advantage, and most likely explains the relatively crude systems of social learning commonly found in social animals. Humans have evolved the additional capacity to acquire variant traditions by imitation and teaching, and can accurately, quickly, and selectively acquire the most common variant or the variants used by the successful. When these kinds of social learning biases are combined with occasional adaptive innovations and content biases, the result is the cumulative cultural evolution of complex, socially learned adaptations, adaptations that are far beyond the creative ability of any individual. Because cumulative cultural evolution gives rise to complex adaptations much more rapidly than natural selection can give rise to genetic adaptations, complex culture was particularly suited to the highly variable Pleistocene environments. As a consequence, humans eventually became one of the most successful species of the Pleistocene large mammal fauna.[100]

Paradoxically, humans have been even more successful in the Holocene, despite a dramatic drop in climatic variation. This is a quite surprising turn of events if we are correct that culture was originally an adaptation to Pleistocene climatic chaos. Shouldn't the quiet climate of the last eleven thousand years have led to dramatic economies of expensive nervous-system tissue, degrading the cultural system? More generally, as the influential evolutionary psychologists Leda Cosmides and John Tooby argue,

> [T]here is no *a priori* reason to suppose that any modern cultural or behavioral practice is "adaptive" . . . or that modern cultural dynamics will necessarily return cultures to adaptive trajectories if perturbed away. Adaptive tracking must, of course, have characterized the psychological mechanisms governing culture during the Pleistocene, or such mechanisms would never have evolved; however, once human cultures were propelled beyond those Pleistocene conditions to which they were adapted at high enough rates, the formerly necessary connection between adaptive tracking and cultural dynamics was broken.[101]

Tooby and Cosmides' *logic* seems sound, but, *empirically*, human populations have exploded in the last ten thousand years; we are now vastly *more* successful than we were in the Pleistocene. Another variant of the adaptationist's dilemma! One reason is that humans themselves now create rapid, large-scale environmental change comparable to the climate changes of the last glacial. For example, agriculture changes the environment for wild plants and animals and the foragers who would depend on them for subsistence. Even though weeds, pests, and diseases evolve to take advantage of the new anthropogenic environments, we readapt even faster, generating further deterioration. So long as we generally find human-modified environments more congenial than our competitors, predators, and parasites, we can thrive, if only by using cultural adaptations to stay one step ahead of onrushing pests. Humans succeed by winning arms races with species that attack our resources and us. They evolve too slowly; we outwit them by cultural counteradaptations, staying a step ahead in the race. We have done more than simply keep ahead of our own environmental deterioration; we have bounded ahead to dominate the earth to an extent perhaps not ever equaled by any single species since the origin of life. Similarly, dense human populations compete with each other, and technical and social innovations by one society tend to exert competitive pressure on their neighbors. The capacity for rapid cultural evolution thus is not just self-sustaining but has gotten progressively ever more rapid as we invent cultural devices, such as reading, writing and arithmetic, that have had the effect of speeding up cultural evolution and increasing sophistication of technology and society. At least to date! Human culture as an adaptive system evolved in response to Pleistocene environments but has subsequently upped anchor and sailed rather well on uncharted waters.

However wild cultural evolution has subsequently run, it arose by natural selection operating to build a complex adaptation in response to specific adaptive challenges. Culture is an unusual system of phenotypic flexibility only because it has population-level properties. But even in this it has numerous analogs in the history of evolution; for example, coevolving mutualisms.[102] Such coevolution sometimes precipitates spectacular evolutionary events.[103] The eukaryotic cell, derived from bacterial symbioses, is an example. We leave it for readers to decide for themselves the extent to which human gene-culture coevolution achieves a status in the history of the evolution of life akin to the rise of the eukaryotic cell. But reserve judgment until you've read chapter 6!

But this is only part of the story. Despite this extraordinary success, many of the products of cultural evolution *do* seem frankly maladaptive. Critics of Darwinian social science often lean heavily on the claim that much cultural evolution has nothing to do with adaptation. We do seem to have cut our way to our extraordinary adaptive success dragging a canoe-load of junk behind us. Some adaptationists may be discomforted by the existence of cultural maladapations, but we are not. In the next two chapters, we hope to convince you that both the baroque excesses of maladaptation and our spectacular success at organizing gigantic social systems flow directly from the processes we have outlined in this chapter and the previous one.

Chapter Five

Culture Is Maladaptive

You are engaged in maladaptive behavior right now

Many cultural anthropologists make fun of the idea that human behavior is adaptive,[1] and delight in citing examples of what seem like capricious and arbitrary differences between cultures. For example, Marshall Sahlins cites the fact that the French relish horsemeat, while Americans find it inedible as dog flesh. How could it be, he asks, that it is adaptive to eat horse in France but not in America? Moreover, such examples can be multiplied endlessly—in many societies dog meat is a delicacy. Culture, not biology, rules.

These cultural foibles may be maladaptations, or they may not.[2] But if they are, they are hardly dramatic ones. Much more hazardous to your genetic fitness is reading and writing books like the ones Sahlins writes—or more to the point, like the book you have in your hands. Most of our readers are no doubt middle-class professionals with triple-digit IQs who have (or will have) wealth beyond the imagining of most of the people who have ever lived. Most of us, however, have not used this wealth to have as many children as possible. Like other middle-class professionals, some of us have had one, two, or three children, and many us are childless. These days secular Americans average less than two children, while in Europe birthrates are even lower.[3]

Why do the modern middle classes have such low fertility? The proximate reasons are familiar to all of us. We lead busy lives. Professional work is demanding. Affluent people can afford lots of time-consuming hobbies. Travel to foreign countries, shopping for antiques, climbing mountains, excelling at dressage, and the like take lots of time and money. Since raising children also takes time and money, we limit our fertility. The ultimate reasons for this behavior are much more mysterious. Ordinarily, natural selection should favor individuals who allocate their resources so as to have as many children as they can successfully raise. Reproductive restraint in the richest populations the earth has ever seen is a striking maladaptation. From the point of view of human threats to the global ecosystem, we may applaud such restraint, but it is not the sort of behavior we expect natural selection to favor.

Most evolutionary social scientists think that such maladaptive behavior arises because the environments in which modern humans live are radically different from those in which humans evolved. Culture is shaped by the evolved information-processing properties of human brains. These were molded in Pleistocene-epoch conditions so that they would reliably give rise to adaptive behavior patterns. Pleistocene climates were very different from recent ones, and Pleistocene societies were presumably something like the hunting and gathering societies we know from the historical and ethnographic records. Natural selection, the story goes, equipped human beings with a psychology that strives for high status, and in Pleistocene foraging societies, this psychology may have often led to higher reproductive success (as seems to have been the case in simpler societies in the recent past).[4] However, in modern societies, this psychology leads to investment in professional achievement and the acquisition of expensive toys and hobbies at the expense of our reproductive success. Some versions of this hypothesis are quite sophisticated. For example, one of the most thoughtful students of this problem, anthropologist Hilliard Kaplan, argues that in past environments, investment in one's own and one's children's skills often paid big fitness dividends; consequently, human psychology is sensitive to those dividends. Modern economies have escalated these payoffs enormously in terms of material well-being, seducing parents into investing huge amounts in honing their own and their children's skills even though the payoff in numbers of children and grandchildren is negative.[5] Parents still *feel* right when they produce high-status, high-skilled offspring even if they must have only one or two children to do it, and those one or two show scant interest in converting wealth to grandchildren. Similar arguments can be marshaled to explain other important maladaptive aspects of human behavior, ranging from our propensity to overindulge in fast food to our ability to sustain cooperation in large groups of unrelated people, the topic of the next chapter. We will label this the "big-mistake" hypothesis, because it means that much of modern human behavior is a big mistake from the genes' point of view.

We think that the big-mistake hypothesis is cogent, but we doubt that it is the cause of most modern maladaptations. In this chapter, we will make the case that much human maladaptation is an unavoidable byproduct of cumulative cultural adaptation. Acquiring information from others allows people to rapidly adapt to a wide range of environments, but it also opens a portal into people's brains, through which maladaptive ideas can enter—ideas whose content makes them more likely to spread, but do not increase the genetic fitness of their bearers. Maladaptive ideas can spread because they are transmitted differently from genes. Ideas that increase the chance of becoming an educated professional can spread even if they limit reproductive success. In a modern economy, educated professionals have high status, and thus are likely to be emulated. Professionals who are childless can succeed culturally as long as they have an important influence on the beliefs and goals of their students, employees, or subordinates. The spread of such maladaptive ideas is a predictable byproduct of cultural adaptation. Selection cannot eliminate the spread of maladaptive cultural variants because adaptive information is costly to evaluate. If this *costly information hypothesis* is correct, culture capacities will evolve in ways that optimize the acquisition of adaptive information, even at the cost of an appreciable chance of acquiring evolved maladaptations.

Explaining maladaptations is important

We have sometimes been chided for paying too much attention to cultural maladaptations. The reason is understandable. Many of our evolutionary social science colleagues think that the analysis of adaptation is the most powerful tool that evolutionary methods bring to the social sciences, and they resent the ill-informed polemics of many of their critics. They struggle with social scientists who have learned their evolutionary biology from the late Stephen Jay Gould's widely known writings about adaptationist "excesses" in biology, not realizing that his alternative hypotheses have found scant empirical support.[6] As we said in the last chapter, adaptive reasoning is one of biologists' most powerful tools.

Doesn't a focus on cultural maladaptations give aid and comfort to the enemies of evolutionary analysis in the human sciences? Perhaps, but we think that the importance of understanding maladaptations outweighs any such objections. Although both the critics and friends of evolutionary theory sometimes forget the point, Darwinism's theory of maladaptation was perhaps its most important achievement. In the history of evolutionary theory, Darwin's ability to account for maladaptations was more important than his ability to account for adaptation. Natural Theology had an acceptable theory of adaptation.[7] The existence of organs of extreme perfection like eyes was the main evidence of the existence of a supernatural Power that was manifestly required to design them, or so the argument went. The crudities and approximations rife in the actual design of organisms are much harder for Natural Theology. Vertebrate eyes have their nerve net lying on top of the photosensitive rods and cones, reducing their light sensitivity and requiring a blind spot where the nerves gather and dive through the retina to form the optic nerve. Octopus eyes, otherwise quite similar in "design," are much more sensibly enervated from behind. These differences make sense in terms of the development of these independently evolved camera eyes.[8] The functionally backwards design of vertebrate eyes is only modestly maladaptive, but its transparent clunkiness betrays a history of evolution by the blind, stepwise improvement by natural selection rather that the hand of the Designer.

The same argument applies to the contemporary application of evolutionary theory to the human species. Social science has many functionalist theories that account for adaptation. However, these theories are frequently criticized for failing to account for the crude nature of social adaptations, and for their historically contingent nature.[9] If our approach is correct, adaptation and maladaptation have the same evolutionary roots. The same processes that enable us to adapt to variable environments also set up conflicts between genetic fitness and cultural success. Culture gets us lots of adaptive information, but also causes us to acquire many maladaptive traits. The big-mistake hypothesis attributes maladaptation to *individuals* misusing antique rules in novel modern environments. The costly information hypothesis attributes maladaptation to *population-level* evolutionary trade-offs that are intrinsic to cultural adaptation, and it predicts maladaptations under a much wider range of environmental circumstances. If we fail to find the predicted sorts of maladaptations that derive from the Darwinian theory of cultural evolution, the whole theory is suspect.

Why culture generates maladaptations

Biologists have traditionally said that natural selection creates well-adapted individuals, that it maximizes inclusive fitness, as the jargon goes. However, biologist Richard Dawkins points out that this is not quite right.[10] Instead, he says, think of individual genes as if they are selfish agents trying to maximize the number of copies of themselves in the next generation. Of course, genes are not really selfish agents, but selection will play upon them and favor those that behave *as if* they were. For most genes in most organisms whether you take the individual's or the gene's perspective doesn't make any difference. The process of cell division that produces eggs and sperm ensures that most genes have an equal chance of getting into any given reproductive cell. As long as this is true, all selfish genes should act in concert to help their host produce as many successful eggs and sperm as possible. In the metaphor of another distinguished evolutionary biologist, Egbert Leigh, the genome as a whole works best if genes collectively act as a "parliament" that "passes laws" to make sure that all genes have a fair chance of entering the crucial eggs and sperm, and otherwise police genomic outlaws.

The story changes when different genes reproduce by different pathways—then the selfish gene perspective is a very useful one. For example, most genes are carried on chromosomes in the cell nucleus. Individuals inherit one copy of each nuclear gene from each of their parents. A small number of genes reside in cellular organelles such as mitochondria (the energy factories of the cell) and chloroplasts (the light energy system in plant cells). Unlike nuclear genes, only females transmit organelle genes. Now, try to think like the corporate buccaneer of organelle genes—smart, selfish, conniving, and unscrupulous—what changes would you make? One appealing scheme would be to dispense with males. Since mitochondria are transmitted only by female offspring, any resources devoted to the production of males are wasted from your buccaneer's point of view. Better to trick your host into investing everything in females. Thus, the selfish gene approach predicts that selection would favor mitochondrial genes that suppress the production of male offspring. In fact, such sex-ratio-distorting genes are known to exist.[11] None known are so extreme as to produce no males, but then such extreme cases would probably become extinct before a biologist chanced upon them. The contrast becomes even starker when you think of genes carried in pathogens such as bacteria and viruses. The genes of a cold virus are expressed in your body, just like the genes on your chromosomes and your mitochondria. However, they reproduce by a completely different pathway, using the resources of your body to produce many copies of themselves. From the point of view of a selfish viral gene, it's fine to harm (or even kill) your host, as long as you leave behind enough copies of yourself.

Because such conflict can be highly destructive, the parliament of the genes favors any nuclear genes that act to reduce it. Two kinds of tactics can be effective. First, nuclear genes can restructure the inheritance system so that all genes have the same reproductive interests. The elaborate, and scrupulously fair, mechanisms of meiosis did not arise by accident. Organisms with organized nuclei, called eukaryotes, first arose as a symbiosis between different bacterial species and conflict must have been rife.[12] The bacteria that became organelles lost genes, and the bacteria that became the nucleus gained the mechanisms of meiosis. Both mechanisms probably evolved because, by reducing conflict, the remaining genes could outcompete genes in other organisms. Second, genes on chromosomes can set up mechanisms such as the immune system that prevent rogue pathogens' genes from making use of the body's resources. Of course, the selfish genes in organelles and pathogens will attempt to overcome these barriers. Nowadays, organelles have so few genes it isn't really a fair fight; the parliament's rules are only occasionally evaded. Pathogens are a completely different matter; as we all know, pathogen genes all too frequently get their way. Microbial infection is the leading cause of death in most human populations, developed countries aside.

In *The Selfish Gene,* Dawkins famously claims that the same argument applies to any replicator, particularly memes, the name he coined for the cultural analog of the gene. In spite of our reservations about the meme concept, this part of Dawkins's argument holds even if cultural variation is a poor analog of genes. If people other than parents play an important role in cultural transmission, selfish cultural variants can spread even though they reduce genetic fitness. You can often understand what kinds of cultural variants spread by thinking of them as selfish memes, even if the analogy is weak in other respects.

Suppose that people in two social roles, parents and teachers, influence the culture that children acquire.[13] Further suppose that personal characteristics affect who achieves the two roles. People who marry early have more children and therefore are more likely to be in the parental role. To become a teacher, people have to postpone reproduction in order to get an education and become teachers. Now suppose a cultural variant arises which leads people to postpone marriage. Such a variant can spread even if parents are more important than teachers in the transmission of basic values. The reason is that the amount of selection is important as well as the occupant of a role's probability of influencing the ideas of any given child. Few people attain the role of teacher. You have to do unusually well in school, earn an advanced degree, and compete for the job with other aspirants. On the other hand, most people, especially in more-traditional societies, become parents. Suppose that parents are a random sample of the population, but only people with rare views—for example, an unusual enthusiasm for intellectual endeavors that led them to postpone marriage to obtain more schooling—become teachers. In this case, learning from parents will not affect the fraction of people who postpone marriage in the next generation, but learning from professional teachers will tend to increase the frequency of late marriages. Depending on how strong the combination of relative selectivity and relative influence is, the frequency of beliefs that lead to delayed marriage will increase at a more or less rapid rate.[14]

Note that "teachers" here are just a stand-in for people occupying an influential role—substitute "superior officer," "boss," "clergy," "politician," "celebrity," or "pundit," and the logic will be the same. If holding any cultural variant makes it more likely someone will attain one of these roles, and if people in such roles play an important part in social learning, that variant will, all other things being equal, tend to spread. Army officers will cause patriotism to increase, bosses the work ethic, clergy the love of God, politicians secular ideologies, celebrities styles of popular consumption, and pundits fashions in high culture. Note also that as beliefs leading to delayed marriage enter the population, parents will begin to teach them as well as teachers. In *Huckleberry Finn,* Huck's unlettered Pap threatens to beat him for going to school and taking on airs, but the ex-schoolgirls Aunt Polly and Miss Watson try their level best to get Huck to pay attention to book learning.

The selfish meme effect is quite robust. Nothing in the argument depends on cultural variants being discrete, genelike particles. It works exactly the same if "memes" were continuously varying and children adopted a weighted average of their parents' and teachers' beliefs. Anytime selective forces shape which cultural variants spread, the same basic logic will arise.

Why genes don't win the coevolutionary contest

Why, you ask, doesn't natural selection favor the evolution of genes that protect their own interests by limiting the influence of people other than parents? Or, alternatively, why doesn't natural selection structure the psychology of social learning so that we pay attention to the behavior of non-parents, but only learn what is good for our genetic fitness? The answers to these questions are at the heart of our debate with much of the rest of the Darwinian social science community.

Many evolutionary social scientists believe that the possibility of selfish cultural variants can be safely ignored. At each step in the evolution of the hominid lineage, they argue, selection would have edited the emerging psychological machinery that governed the acquisition of culture to ensure that maladaptive cultural variants were of minimal importance. As a result, selection would not likely favor a psychological system that led to the frequent spread of selfish cultural variants.[15] In ancestral conditions, our evolved psychology would protect us from selfish cultural variants. Modern environments are a different matter, but as we've said, evolutionists generally favor noncultural explanations for maladaptive behavior in complex societies. In the last chapter, we ourselves were enthusiastic users of this sort of adaptationist reasoning. If it was OK then, what is wrong with it now?

There is nothing wrong with adaptationist reasoning in general. The problem lies in applying it correctly to the evolution of culture. We agree with our colleagues that culture is shaped by psychological predispositions which are products of natural selection, and that these predispositions will frequently lead to the spread of adaptive cultural variants. However, the conclusion that evolved biases alone will determine the outcome of cultural evolution does not follow from these two premises. The reason evolved biases will not prevent the evolution of selfish cultural variants is that the structural features which allow such beliefs to proliferate *are the same features* that give rise to the adaptive benefits of cultural transmission. The nub of the matter is that selection can't get rid of cultural maladaptation without giving up the ability to rapidly track varying environments.

Adaptations always involve tradeoffs.[16] No herbivore can be as fleet as a gazelle, as tall as a giraffe, and as powerful as an elephant. Inescapable biophysical tradeoffs ground magical organisms such as gigantic, flying, fire-breathing dragons. Pigs can't fly; even if they had optimally designed wings, they'd be too heavy.[17] Imitation is an adaptive information-gathering system, but it involves tradeoffs. Culture gets humans fast cumulative evolution on the cheap, but *only* if it also makes us vulnerable to selfish cultural variants. Four interrelated tradeoffs conspire to weaken the grip of genetically determined biases on cultural evolution. First, people other than parents are a crucial source of adaptive information. Second, content biases cannot be made too restrictive without becoming too costly or sacrificing the adaptive flexibility that social learning provides. Next, fast and frugal adaptive heuristics such as conformist and prestige biases have specific, unavoidable, maladaptive side effects. Finally, rogue cultural variants evolve devious strategies to evade the effects of content biases. Because the rate of cultural adaptation is rapid compared with genetic evolution, rogue variants will often win arms races with genes.

Learning from people other than your parents is adaptive

Most Americans (at least most American parents) mistakenly think that parents are the main source of their children's beliefs and values. True, children normally form close bonds to parents, and in some cultures, parents make strenuous efforts to shape their children's beliefs. True also that beliefs and attitudes of children and parents are often quite similar. However, much evidence indicates that parents play at best a minor role in many domains in determining the final cultural variants their children adopt.[18] Behavior genetic studies indicate that most of the similarity between the personality traits of parents and children is due to genetic inheritance, not vertical cultural transmission.[19] At the same time, these studies also detect a large amount of "environmental" variation that is *not* shared within families. Children learn a lot from one another, and from adults other than their parents. In some domains—language, for instance—peers are much more important than parents. Immigrant children in the United States usually learn English from their peers, and come to prefer it over their native tongue. When people move from one region to another, their children usually use the local dialect rather than their parents'.[20] In other domains, transmission from nonparental adults to children is also influential, particularly when formal education is important.

Since even moderate amounts of nonparental influence can allow genetically maladaptive cultural variants to spread, why hasn't selection shaped the psychology of social learning so that children preferentially attend to their parents (instead of the reverse, if our experience as parents of teenagers is any indication)?

The reason is simple. Social learning is about collecting adaptive information from the surrounding social environment. Increasing the size of the sample increases your chance of acquiring useful information, because a larger sample makes all kinds of biased transmission more effective. These forces, like selection, depend on variation, and the more models surveyed, the more variation the bias has to work with. This is easiest to see for what we call content bias—an ability to judge the utility of a cultural variant directly on its merits. Mom may be an inefficient or poorly informed gatherer, and an aunt, grandmother, in-law, or friend may be much better. But if you can only learn from Mom, you are stuck with her way of doing things. By searching more widely you increase the chance that you will observe something worth learning. Anthropologist Barry Hewlett has documented how young boys learn to hunt among the Aka "Pigmies" of central Africa.[21] Boys learn most of their hunting techniques from their fathers, but as boys get older and more independent they become willing to depart from Dad's ways, though really everyone is hunting just like Dad. However, crossbows were a recent innovation at the time of Hewlett's study, and most fathers did not know how to make and use them. Crossbows were useful, so boys learned to use them from those who knew how to use them, regardless of relatedness. The "perceived advantage" (content bias, in our terms) is one of the strongest correlates of the successful spread of an innovation.[22] The same basic logic holds for conformist- and prestige-biased transmission. In each case, alternative variants are compared by some rule, and the preferred variant is selected at better than chance levels. Increasing the sample size of variants observed increases the chance that you acquire the best variant available in the population.

Biases are costly, and therefore imperfect

As far as many evolutionary social scientists are concerned, Richard Dawkins is way up in the pantheon of contemporary evolutionary thinkers. (For sure, he makes most Top Five lists.) Nonetheless, most place little stock in Dawkins's argument about rogue memes, regarding it as an imaginative device for explaining the nature of replicators, rather than a serious proposal about human cultural evolution. Instead, they tend to think that all forms of learning are processes whereby the organism exploits statistical regularities in the environment so as to develop behaviors that are well suited to the present environment. Over time, selection shapes psychology (and other processes as well) so that it uses predictive cues to generate adaptive behavior. Social learning is just another learning mechanism that exploits cues available in the social environment. As a result, to oversimplify just a bit, most evolutionary social scientists expect people to learn things that were good for them in the Pleistocene and perhaps also in the smaller-scale human societies that resemble those of the Pleistocene. Adaptation arises from the information-processing capacities built into the human brain by natural selection acting on genes. These mechanisms may give rise to maladaptive behaviors nowadays, but it's got nothing to do with culture and everything to do with the fact that "environments" are far outside the parameters to which our innate decision-making talents are calibrated.

This argument neglects an important tradeoff. Selection cannot create a psychology that gets you only the adaptations and always rejects maladaptive variants, because selection cannot generate accurate general-purpose learning mechanisms at a feasible cost. Why not? Think of using the taste of a substance as a guide to whether it is edible. Many toxic plants have a bitter taste, and accordingly we tend to reject foods that taste bitter. On the other hand, many toxins do not taste bitter, so bitterness is no infallible guide to edibility. Further, many bitter plants, such as acorns, can be rendered edible by cooking or leaching. Further still, some bitter-tasting plant compounds have medicinal value. People can actually grow fond of some bitter-tasting food and drink. Think gin and tonic. A bitter taste is only a rough and ready guide to what is edible and what is not. In principle, you could do much better if you had a modern food chemist's laboratory on the tip of you tongue, one that could separately sense every possible harmful and helpful plant compound, rather than having just four very general taste senses. Some animals are much better at these things than humans—we have a very poor sense of smell, for example. But the number of natural organic compounds is immense, and selection favors compromises that *usually* result in adaptive behavior and don't cost too much. A fancy sense of smell requires a long muzzle to contain the sensory epithelium wherein all those fancy sensory neurons are deployed, and plenty of blood flow to feed them. Bitter taste is a reasonably accurate and reasonably general screening device, but to get the good, you have to risk adopting the bad because the evaluative machinery the brain deploys to exercise the various biases is necessarily limited. Let's see why.

John Tooby and Leda Cosmides define an adaptation as "a reliably developing structure in the organism, which, because it meshes with the recurrent structure of the world, causes the solution to an adaptive problem."[23] They give behavioral examples such as inbreeding avoidance, the avoidance of plant toxins during pregnancy, and the negotiation of social exchange. Evolutionary psychologists are prone to wax eloquent over marvelous cognitive adaptations created by natural selection. And they are right to marvel; everyone should. Natural selection has created brains and sensory systems that easily solve problems that stump the finest engineers. Making robots that can do anything sensible in a natural environment is exceedingly difficult, yet a tiny ant with a few thousand neurons can meander over rough ground hundreds of meters from its nest, find food, and return in a beeline to feed its sisters. Humans are able to solve many astoundingly difficult problems as they go through daily life because natural selection has created numerous adaptive information-processing modules in their brains. Notably, the best examples involve tasks that have confronted every member of our lineage in every environment over tens of millions of years of evolution, things such as visual processing. The list of well-documented examples that apply to humans alone is short, and once again these psychological adaptations provide solutions to problems that every human if not every advanced social vertebrate faces—things such as learning language, choosing a good mate, and avoiding cheaters in social exchange.

Cultural evolution also gives rise to marvelous adaptations. However, they are typically solutions to problems posed by *particular* environments. Consider, once again, the kayaks built and used by the Inuit, Yupik, and Aleut foragers of the North American Arctic. By Tooby and Cosmides' definition, kayaks are clearly adaptations. These peoples' subsistence was based on hunting seals (and sometimes caribou) in Arctic waters. A fast boat was required to get close enough to these large animals to reliably hit and kill them with an atlatl dart.[24] Kayaks are a superb solution to this adaptive problem. Their slim, efficient hull design allowed sustained paddling at up to seven knots. They were extremely light (sometimes less than fifteen kilograms), yet strong and seaworthy enough to safely navigate rough, frigid northern seas.[25] They were also "reliably developing"—every successful hunter built or acquired one—until firearms allowed hunting from slower, but more stable and more widely useful umiaks. For at least eighty generations, people born into these societies acquired the skills and knowledge necessary to construct these boats from available materials—bone, driftwood, animal skin, and sinew.

Certainly, no evolved "kayak module" lurks in the recesses of the human brain. People have to acquire the knowledge necessary to construct a kayak using the same evolved psychology that people use in other environments to master other crucial technologies. No doubt, learning any craft *requires* an evolved "guidance system." People must be able to evaluate alternatives, to know that boats that don't sink and are easy to paddle are better than leaky, awkward designs. They have to be able to judge, to some significant degree, whose boats are best, and when and how to combine information from different sources. The elaborate psychological machinery that allows children to bootstrap general knowledge of the world is also clearly crucial. People can't learn to make kayaks unless they already understand something about the properties of materials, how to categorize plants and animals, and so on and on. This guidance system is not "domain general" in the sense that it allows people to learn *anything*. It is highly specific to life on earth, in a regime of middle-sized objects, relatively moderate temperatures, living creatures, manual skills, and small social groups. However, it is domain general in the sense that nothing in our evolved psychology contains the specific details that make a difference in the case of kayaks—knowledge of the dimensions, materials, and construction methods that result in constructing a fifteen-kilogram craft that safely skims across the Arctic seas, making a living for its occupant, instead of an inferior vessel that leads to death by drowning or hypothermia. These crucial details were stored in the brains of each generation of Inuit, Yupik, and Aleut peoples. They were preserved and improved by the action of a population of evolved psychologies, but employing mechanisms that are equally useful for preserving a vast array of other kinds of knowledge.

Such widely applicable learning mechanisms are necessarily more imperfect and error prone than highly constrained, domain-specific ones. As Tooby and Cosmides have emphasized, broad general problems are much more difficult to solve than simple constrained ones.[26] A kayak is a highly complex object, with many different attributes or "dimensions." What frame geometry is best? Should there be a keel? How should the components of the frame be joined? What kind of animal provides the best skin? Which sex? Harvested at what time of year? Designing a good kayak means finding one of the very few combinations of attributes that produces a highly specialized boat. The combinations of attributes grow geometrically as the number of dimensions increases, rapidly exploding into an immense number. The problem would be much easier if we had a kayak module that constrained the design so that we would have fewer choices to evaluate. However, evolution cannot adopt this solution because environments are changing far too quickly and are far too spatially variable for selection to shape the psychologies of Arctic populations in this way. The same learning psychology has to do for kayaks, oil lamps, waterproof clothing, snow houses, and all the other tools and crafts necessary to survive in the Arctic. It also has to do for birch bark canoes, reed rafts, dugout canoes, planked rowboats, rabbit drives, blowguns, hxaro gifts, and the myriad marvelous, specialized, environment-specific technologies that human hunter-gatherers have culturally evolved.

For the same reason that evolution cannot "design" a learning device that is both general purpose and powerful, selection cannot shape social learning mechanisms so that they reliably reject maladaptive beliefs over the whole range of human experience. A young Aleut cannot readily evaluate whether the kayaks he sees his father and cousins using are better than alternative designs. He can try one or two modifications and see how they work, and he can compare the performance of the different designs he sees. But small samples, many dimensions of variability, and noisy data will severely limit his ability to choose the best design. What a bias gains in generality, it has to give up in accuracy. The repeated action of weak domain-general mechanisms by a *population* of individuals connected by cultural inheritance over many generations can generate complex adaptations like kayaks, but individuals must adopt what they observe with only marginal modifications. As a result, we may often adopt maladaptive behaviors if population-level processes like selection on nonparentally transmitted variation have somehow favored them.

In the last chapter, we showed that when determining the best cultural variant is difficult, selection favors heavy reliance on imitating others. The natural world is complex and variable from place to place and time to time. Is witchcraft effective? What causes malaria? What are the best crops to grow in a particular location? Does prayer affect natural events? The relationship between cause and effect in the social world is often equally hard to discern. What sort of person should one marry? How many husbands are best? Tibetan women often have two or three. What mixture of devotion to work and family will result in the most happiness or the highest fitness? Students of the diffusion of innovations note that "trialability" and "observability" are some of the most important regulators of the spread of ideas from one culture to another.[27] Many important cultural traits, including things such as family organization, have low trialability and observability and are generally conservative. We act as if we know that sensible choices about such behaviors are hard to make and that we are liable to err if we try to depart far from custom.

As the effects of biases weaken, social learning becomes more and more like a system of inheritance. Much of an individual's behavior is thus a product of beliefs, skills, ethical norms, and social attitudes that are acquired from others with little if any modification. To predict how individuals will behave, one must know something about their cultural milieu. This does not mean that the evolved predispositions that underlie individual learning become unimportant. Without them, cultural evolution would be uncoupled from genetic evolution. It would provide none of the fitness-enhancing advantages that normally shape cultural evolution and produce adaptations. However, once cultural variation is heritable, it can respond to selection for behaviors that conflict with genetic fitness. Selection on genes that regulate the cultural system may still favor the ability and inclination to rely on imitation, because it is beneficial on average. Selection will balance the advantages of imitation against the risk of catching pathological superstitions. Our propensity to adopt dangerous beliefs is part of the price we pay for the marvelous power of cumulative cultural adaptation. A saying might go, "if you evolve the adaptation, you have to pay its costs."

Adaptive biases have specific, unavoidable, maladaptive side effects

You might think that weak biases would just be a recipe for accepting a variety of more or less random beliefs, and while this may be true of some simple heuristics, other biases lead to systematic, predictable pathologies, a fact that allows us to check for their existence and importance.

Conformist bias can lead to the evolution of maladaptive self-sacrifice

Recall from the last chapter that conformist rules such as "imitate the most common variant" are adaptive in any environment that also favors social learning. If a social learner has difficulty determining the best way to behave, doing what everybody else is doing is probably safe. Conformity has an important side effect: it tends to reduce the amount of variation within groups and increase and preserve variation between groups. This can, in turn, increase the importance of group selection, and if cultural rules arise that cause individuals to sacrifice their own interests for the good of the group, group selection can cause the frequency of individually costly but group-beneficial traits to increase.[28]

Suppose that two groups differ in religious belief. In one group, most people believe in a god who punishes the wicked; in the other group, most people are worldly atheists. Further, suppose that believers engage in individually costly but group-beneficial behavior—they are more honest in business transactions, less prone to hedonistic excess, and more generous and charitable. (Their religious beliefs don't have to make them angels—just a little more group oriented than their competitors.) Finally, suppose that other parts of their evolved psychology cause people to prefer deception, self-indulgence, and selfishness, and as a result, a content bias causes atheism to spread. If the content bias were the only force acting, the group benefits associated with religious belief could not spread, because atheists would quickly come to dominate. However, if people are also predisposed to imitate the majority, believers may remain common in the first group, simply because they are already common. People act as if they looked around and thought to themselves, "Everybody believes, so there must be gods who punish the wicked." As a result, the two groups will remain different, and over the long run, the group of believers that is wealthier, healthier, and more stable will tend to replace the group of atheists.[29]

We have to be very careful with our definitions of *fitness* to keep this argument clear. If cultural group selection operates successfully, the benefits of group-adapted beliefs may raise everyone's reproductive success. Nevertheless, selection acting on genes will continue to favor atheists who take the benefits of living in a better society but evade paying the costs. Group-selected institutions may even arrange payoffs to discriminate against selfish atheists and other deviants from community orthodoxy; for example, by establishing punishment systems like the Inquisitions.[30] Even when such systems are powerful, selection acting on genes will favor any new variant that can evade the prevailing system of punishment. Thus, selection on genes still favors the evolution of individually advantageous traits, even if the collapse of religious belief would harm the reproductive success of atheists themselves in the long run, and even if none of the variants currently in the population can escape the punishment system.

Group selection on cultural variation has been an important force in human evolution. Conformist bias and rapid cultural adaptation conspire to generate oodles of behavioral variation between groups. The conformist effect overcomes the critical problem with group selection. In the case of a genetic system of inheritance, variation between groups tends to evaporate quickly in the face of modest amounts of migration. In the case of altruistic traits, selection within groups against altruists also reduces between-group variation for altruism. The existence of large-scale cooperation in human societies invites a group-functional interpretation, and perhaps the peculiarities of the cultural system of inheritance are responsible. We develop this argument in more detail in the next chapter.

The prestige-biased force can lead to "runaway" cultural evolution

Darwin believed that sexual selection was responsible for the maladaptive elaboration of secondary sexual characters such as the spectacular tails of peacocks.[31] Males with conspicuous tails have more offspring even though they are more subject to predation, because peahens prefer males with spectacular tails. In essence, Darwin thought that evolutionary fads in sexual attractiveness often led to the evolution of maladaptive fads in feathers, fur, and bugs' ears. However, he did not explain why females should have such faddish preferences. The pioneering evolutionary theorist R. A. Fisher showed that there need not be any *adaptive* explanation.[32] Fisher's insight was to see that the male offspring of females who preferred showy males would tend to have both the genes for showy tails *and* the genes that caused females to prefer such males. Thus, if female choice increases the frequency of genes leading to showy tails, it may also increase the genes that cause females to prefer such tails. This will lead to progressively stronger selection for showy males that will further increase the preference for such males. The process feeds back on itself in an explosive spiral that can cause a trait originally correlated with fitness to become wildly exaggerated. This subject remains controversial in evolutionary biology, but in theory this mechanism can operate; moreover, it seems to account for otherwise mysterious characters such as the peacock's tail, the bower-bird's bowers, and the elaborate penises of many insects.[33]

Prestige-biased transmission can work in a similar way. Remember that prestige bias occurs when individuals choose models based on indicators of prestige. Suppose that people have beliefs (not necessarily conscious ones) that cause them to imitate the actions of pious people—people who devote time and resources to religious rites, are conspicuously abstemious, and are charitable. This process will cause more people to act piously, and will also increase the propensity to imitate the pious, because people who do will acquire from them beliefs about who should be imitated, and the most pious people will prefer more piety than the population as a whole. The resulting dynamic is closely analogous to runaway sexual selection.[34] We have argued that many phenomena, ranging from maladaptive fads and fashions to group-functional religious beliefs to symbolically marked boundaries between groups, might result from the properties of prestige bias.[35]

The exaggeration of traits signaling status in human societies is virtually a truism. For example, on the island of Ponpae in the Pacific, a man's prestige is partly determined by his contribution of very large yams to periodic feasts.[36] Prize yams require up to a dozen men to carry, and their cultivation is inefficient from the point of view of food production. We imagine an evolutionary scenario in which, at the beginning, people just brought their best produce to the feast, and the size and number of yams were straightforward indicators of farming ability. Then, as the idea that the best people would contribute the biggest yams took hold, families began to devote special effort to grow big yams, and the custom of growing giant yams took off. In California, where we live, the twelve-man yam comes to mind when we see a Hummer II rolling down a Los Angeles boulevard.[37]

Cultural systems can defend against adaptive biases

Finally, cultural systems often evolve clever defenses against the action of our evolved psychology.[38] The nonparentally transmitted parts of culture are analogous to microbes. Our immune system evolved to kill microbial pathogens, but it also allows us to acquire helpful symbionts. As we know all too well, microbial pathogens are common, despite the sophistication of the immune system. One reason is that we are not the only players in this game. Natural selection helps parasites trick our immune system. Since microbial populations have short generation times and large populations, parasite adaptation can be very rapid. The psychology of social learning is like an immune system in that it is adapted to absorb beneficial ideas but resist maladaptive ones. And like the immune system it is not always able to keep up with rapidly evolving cultural "pathogens."

Consider, for example, Christian theology. It paints a picture of eternal rewards and punishments that is convincing to the faithful. If biases are viewed as a rough-and-ready method of weighing fitness benefits and costs, a system that adds imaginary costs and benefits puts a thumb on the scale. Believers may behave in ways that cause them to perpetuate the faith at a cost to their fitness. Blaise Pascal, the pioneering sixteenth-century mathematician and scientist, wrote a famous defense of faith based on the laws of probability that he codiscovered. In his famous wager, he invites us to weigh the finite pleasures and pains of life on earth against the infinite rewards of heaven and the infinite punishments of hell: "[T]here is an infinity of infinitely happy life to be won, one chance of winning against a finite number of chances of loss, and what you are staking is finite," concluding, "Wager then without hesitation that He is."[39] This sophisticated argument is frequently used to persuade nonbelievers and to reassure believers tempted by doubts. Pascal himself abruptly retired from secular pursuits in 1654 and spent the rest of his life defending Jansenism, an austere, Calvinist-tainted brand of Catholicism, which was eventually suppressed by the Church.[40] We ourselves are not concerned with any fitness Pascal lost in the service of his beliefs, but we regret that he was lost to science.

Pascal is in good company. Christian believers over the centuries include many awe-inspiring intellects.[41] Greek philosophy inspired early Christian theologians, most notably St. Augustine. Isaac Newton was at least as proud of his theology as his science. Proofs of the existence of God were a staple of Pascal's philosophical contemporaries, such as Leibniz and Descartes. Modern science has the advantage of being a large, prestigious, well-funded community of highly trained rational skeptics. Even then, scientists work hard to keep "disciplines" like paranormal psychology and creation science in check. Individual skeptics can hardly be expected to make much headway against belief systems that have been buttressed by the best efforts of a succession of able thinkers.

Summing up: If information costs are high, maladaptive beliefs will spread

We submit that any feasible, adaptive social learning psychology will leave plenty of scope for rogue variation. Paying attention to only Mom and Dad throws away too much valuable information, so adaptive evolution will favor learning from lots of people. But, like opening your nostrils to draw breath in a microbe-laden world, nonparental cultural information will tend inevitably to be laden with maladaptive ideas. From the gene's "point of view," a bias that picks *the* fitness-optimizing trait from a large pool of potential "teachers" in every Pleistocene environment would be great. But the tradeoffs inherent in learning and cognition make such biases unattainable, just as biomechanical tradeoffs prevent the evolution of fire-breathing dragons and flying pigs. The adapted mind is constrained by the prohibitive cost of vetting every cultural variant for its contribution to fitness. Our main conclusion in the last chapter was that culture is adaptive because *populations* can quickly evolve adaptations to environments for which *individuals* have no special-purpose, domain-specific, evolved psychological machinery to guide them. Rigid control of cultural evolution would make the cultural evolutionary system slow and clunky. In the wildly varying environments of the Pleistocene, individuals were better off relying upon fast and frugal social learning heuristics to acquire pretty good behaviors RIGHT NOW rather than await the perfect innate or cultural adaptation to an environment that that would be gone before perfection could evolve. Such heuristics leave space for selfish cultural variants to seep into the population—just the price of doing business in a highly variable environment where information is costly.

This way of thinking is human evolutionary psychology done right. The comparative psychology of social learning demonstrates that humans are able to learn complex tasks by observing others. This capacity is, apparently, distinctively human; no other species is known to depend on such a large repertoire of complex, highly evolved traditions. The evidence of chapter 4 shows how culture, because of its population-level properties, can act as a potent problem-solving device. Human cultural diversity is ample testimony to the power of culture to solve the problem of living nearly anywhere in the world. As cognitive psychologists have argued so persuasively, general-purpose, problem-solving devices at the individual level are ineffective by comparison.

Much human psychology relies on clever but simple heuristics for managing cultural transmission. Culture, then, is a sophisticated cognitive *and* social system evolved to finesse the problem that information costs preclude a general-purpose, problem-solving system inside every individual's head. The scientific enterprise itself is the ultimate example of culture's capacity to solve extraordinarily difficult problems. Given the right social institutions, quite fallible individual intellects can gradually reveal the deepest secrets of the universe.[42] The price we pay for our promiscuous lust for adaptive information is playing host to sometimes spectacularly pathological cultural variants.

Witchcraft is a simple example of maladaptive cultural variants

Pascal Boyer provides a good example of how a widely useful, general-purpose learning heuristic can sometimes lead us astray. Boyer argues that people apply "abductive reasoning" to the acceptance of supernatural ideas (and probably to much else).[43] Abductive reasoning is a form of induction in which a premise is deemed to be true if the implication of that premise is observed.[44] Arctic Americans used kayaks to hunt sea mammals. They were very successful doing so with primitive weapons. Thus, kayaks are the optimal boat for Arctic sea mammal hunting with atlatls. *Plausible.* But: People pray to gods for health and prosperity. Many sick people get well and many economic ventures succeed. People who do not pray often get sick. Prayers are answered. Thus, gods do intervene on behalf of the faithful! *Not so plausible.* Abductive reasoning ignores the cases in which praying did not result in a cure and cures occurred without prayer. Alternative hypotheses are not considered; many times prayers for good health must precede bad health. We live in a very complex world. False disconfirmations of hypotheses are common due to the operation of countervailing causes. A really good understanding of the natural world requires time-consuming observations, elaborate calculations, and controlled experiments, and these rigorous inductive methods are too costly for everyday use. Even though abduction is far from logically or empirically guaranteed to succeed, it often discovers real causal and correlative patterns, and it is easy to apply. However, if people are armed with the wrong hypotheses, abduction can easily lead them to adopt false and often deleterious beliefs.

Many religious ideas seem to be good for people's mental health and for creating strong communities.[45] However, the adaptive virtue of ritually handling rattlesnakes is hard to fathom. Some of the southern Pentacostalists who engage in this practice are bitten, and some die.[46]

Other supernatural beliefs seem to be deleterious. For example, witchcraft beliefs are very common in societies at all levels of organization. Anthropologist Bruce Knauft studied a simple horticultural society in New Guinea, the Gebusi, who had an elaborate system of highly formalized witchcraft inquests. Despite their elaboration, the inquests depend on abductive inference, and "evidence" to support accusations was very easy to "discover." For example, witches supposedly worked their magic by making bundles of twigs and leaves. Witchcraft investigators easily found "evidence" of such bundles in the litter of decaying twigs and leaves on the forest floor. Before contact with Europeans, the Gebusi executed many people for practicing witchcraft, and these executions ranked alongside malaria as one of the leading causes of death. Despite other institutions designed to increase "good company," witchcraft suspicions handicapped the Gebusi's ability to resist the depredations of a neighboring tribe, the Bedamini; Gebusi society was paralyzed by witchcraft accusations and the fear of them.[47]

The sociologist and historian of religion Rodney Stark recounts a similar story for the wave of witch executions that took place in Europe during the Reformation. Both Protestants and Catholics found compelling theological justifications for the possibility of black magic. If God is benevolent, then some powerful evil force must exist that can be blamed for the rough nature of life on earth. If humans could gain access to the benevolent powers of God through prayer, then magic or devil worship ought likewise to be effective at calling up the evil forces. This argument was widely held by the most sophisticated thinkers of the day. These beliefs led to a steady trickle of witch trials in which most defendants confessed and promised to abstain, but a few were executed. Destructive outbreaks of witch killings sometimes occurred in small communities where unsophisticated local authorities accepted the unsupported testimony of children and confessions under torture. The initial victims would readily implicate others to save themselves from further torture. Killings often went unchecked until the authorities had executed some 5% to 10% of the community. By that time accusations began to be made against solid citizens, and the episode became self-limiting. Most of the destructive outbreaks occurred in the politically fragmented Rhineland, where sophisticated higher authorities had a difficult time intervening.[48]

Superstitious beliefs and elaborate, potentially costly rituals exist in many societies. Nineteenth-century scholars felt very free to attribute maladaptive superstitions to "primitives." Later, anthropologists of various schools became enamored with functionalist explanations of many kinds. In the late twentieth century, scholars became sensitive to the possibility that superstitious beliefs are common in advanced societies. For example, journalist Dorothy Rabinowitz details how eerily the ritual child abuse cases of the 1980s and 1990s in the United States resemble the witchcraft persecutions of an earlier era. Seemingly sophisticated prosecutors, such as former U.S. Attorney General Janet Reno, believed what in hindsight were ludicrous accusations made by suggestible children.[49] Of course, the functions of beliefs are sometimes not easy to discern, and much work needs to be done before any sweeping generalizations are warranted.

The modern demographic transition may result from the evolution of selfish cultural variants

The contemporary drop in birthrates, which started in the developed countries but is now occurring in most of the world, attracts considerable attention from demographers. For the most part, they portray the phenomenon in positive terms. It is a concomitant of the economic changes that make people in the industrial world prosperous and prevent an undesirable overpopulation of the world. The global environment aside, this decline in birthrate represents a failure to maximize individual genetic fitness and requires an explanation. The Catholic Church's distaste for birth control is much closer to the prediction of ordinary evolutionary theory. From the perspectives of the Pope and natural selection, the wealth of modern societies is wasted on consumerist lifestyles dedicated to crass materialism. Imagine the alarm that a virulent fertility-reducing pathogen would cause at the Centers for Disease Control and Prevention, especially if the newer strains were beginning to cause population decline over wide areas of the globe. So the Vatican must feel.

The demographic transition is at least partly caused by the increased nonparental cultural transmission associated with modernization. Modern economies require educated managers, politicians, and other kinds of professionals who typically earn high wages and achieve high status. Accordingly, competition for such roles is fierce. People who delay marriage and child rearing in order to invest time and energy in education and career advancement have an advantage in this competition. High-status people have a disproportionate influence in cultural transmission, so beliefs and values that lead to success in the professional sector will tend to spread. Because these beliefs will typically lead to lower fertility, family size will drop.

Consider the situation for the mass of people in premodern agrarian societies. In pretransition populations, most people are illiterate or poorly educated and live in relatively isolated villages. The elites to whom the average person is exposed—landowners, priests, military officers, government officials—gain their status by right of birth, not merit. That is, hereditary aristocracy, to which ordinary people cannot aspire, dominates the prestige system. The family is the most significant social institution for the majority of the population, the primary unit of production, consumption, and socialization. When cultural transmission is vertical, selection on cultural variation will tend to favor the same behavior that selection on genes would favor—large, economically successful families. Very often a strong familial ethic encourages reproduction in order to increase the power of one's lineage or clan. Childless couples are pitied. A large and prosperous family is the greatest achievement to which ordinary men and women can aspire.

Large families under the supervision of able men and women mobilize family labor to prosper. The less able find it hard to assemble the resources necessary to marry or more difficult to support the children they have, and are more prone to become victims of the many risks to survival in traditional circumstances. Death rates are always high and spike upward during famines, plagues, wars, and natural disasters. The well-managed family is the key to survival and reproductive success in the scramble to recover from catastrophes or to become established on frontiers, and in the tight competition in a dense population near carrying capacity.[50] *Relatively* little conflict will arise between the fitness of culture and genes in such circumstances.[51] Genetic biases and cultural norms conspire to adapt reproductive behavior to changing situations. Frontiers—eighteenth- and nineteenth-century America was economist Thomas Malthus's own example—favored sheer maximization of offspring number as the critical resource, land, was not in short supply. In densely populated lands—as in Ireland in the nineteenth and early twentieth centuries—delayed marriage, and other expedients produced very low birthrates and prevented families from becoming paupers.

Premodern demographic systems were no doubt complex even in the stationary and slowly growing populations of the Old World. Economist-demographer Ansley Coale notes that many combinations of mortality and fertility yield approximately zero population growth, the norm in most premodern circumstances. For example, the birthrate in China was higher and life expectancy lower than in the countries of northwestern Europe, although all had essentially zero population growth.[52] Anthropological demographer William Skinner argues that Eurasian "family systems," the normative patterns of marriage, postmarital residence, numbers and sexes of children, and inheritance of family resources are highly variable and have large impacts on all demographic variables.[53] He provides many examples of premodern societies using fertility control, infanticide, fosterage, and adoption to obtain offspring sets of desired size and composition. Some of this variation includes behavior detrimental to fitness that we will discuss a little later in this chapter, but nearly all traditional family systems were capable of rapid growth, resources permitting, and maintaining high populations in resource-constrained times and places.

The evolution of modern industrial societies embodies two linked but imperfectly correlated revolutions. One is a revolution in production due to industrialization that boosts the material standard of living. This phase of modernity lowers death rates by raising the material standard of living and by related innovations in public health and medicine. It also provides the technical means to more easily control conception. The second is a revolution in the structure of the transmission of ideas of all sorts. Literacy rates rise as schooling becomes nearly universal. Production activity is transferred from family-dominated farms to factories and offices controlled by entrepreneurs and managers rather than a hereditary elite. The role of government in people's everyday lives increases, and bureaucratic reforms make government offices competitive posts open to aspiring educated men (and eventually women). High literacy and the industrialization of printing led first to the emergence of print mass media and later to ongoing innovations in the broadcast media and the film industry. In the contemporary world, cheap electronics brings entertainment produced in Hollywood, Mexico City, Sao Paulo, and Mumbai to the remotest villages. The rise of mass media and universal education suddenly exposed people to much more nonparental cultural influence than had been experienced in more-traditional societies. Proportionally, the scope for the spread of cultural variation in conflict with genetic fitness increased.

Demographers have noted the association of the demographic transition with the rise of modern industrial societies since the pioneering work of A. M. Carr-Saunders and others before World War II.[54] Most discussions cite a long list of correlates between economic and social modernization. Economics provides the most ambitious theoretical framework for dissecting the causal pathways from modernization to fertility decline.

Economists have considered the costs and benefits of having children under different circumstances, and then attempted to test various hypotheses by examining correlations between economic variables and observed changes in fertility. For example, the shift from farm to factory work plausibly reduces the value of child labor, especially if factory work requires an educated workforce. With less need for family labor in production and the necessity to pay school fees, the benefits of children will decline and their costs will increase; ergo, fertility will fall. Most of these models assume preferences to be fixed, and the shift in fertility is assumed to stem from changes in opportunities and constraints arising from the Industrial Revolution in production. The model is cogent, but the empirical data suggest a rather more-complex causal process.

The most ambitious test of the economic model was the Princeton European Fertility Project led by Ansley Coale.[55] This study investigated the fertility decline in over six hundred administrative units in Europe over the last two centuries. For most districts, Coale and his coworkers could estimate the time paths of fertility, proportion of women married, and marital fertility. The results show a striking disjunction between economic development and the onset of fertility decline. For example, the provinces that show the earliest sustained declines in fertility are in France, where the onset of the transition dates to about 1830. The onset of the decline in Britain and Germany occurred fifty years later, and some German districts maintained high fertility until 1910–20. These trends challenge the economists' simple model in which fertility declines follow increased industrialization. France experienced an early and extreme *social* modernization, but the pace of *economic* modernization was much slower than in Britain and Germany.

Fertility patterns show a striking effect of culture—all across Europe, culturally distinctive areas began their fertility decline around the same time. For example, French-speaking areas in Belgium experienced the onset of the transition in the 1870s, while the transition was delayed in Flemish-speaking areas by as much as forty years. Hungary's transition was much earlier than the rest of Austro-Hungary, Catalonia's much earlier than the rest of Spain, and Brittany and Normandy's nearly a century later than most of the rest of France. This result should not be altogether surprising, even to economists. Modernity's emphasis on individualism and rationality has created new demands for political rights as well as demands for efficient economic organization. These pressures are filtered through the preexisting variation in values, beliefs, skills, and environments of particular regions. The common systemic features of modernity maintain a loose correlation across domains such as industrial production, literacy, and demography, but if historical differences in culture are important, each cultural region will experience any transitions at its own pace. Unfortunately, the Princeton European Fertility Project was not designed to collect the kind of data needed to understand the role cultural processes play in the fertility transition. Indeed, the demographers traditionally focus on correlations between fertility and macrosocial variables that preclude a fine-grained analysis of the causal processes that underlie them, especially if the causal processes are evolutionary.

Modern low fertility does not maximize fitness

Before proceeding under the assumption that the demographic transition is fitness reducing, we need to be sure that it actually is. Evolutionary biologists have long known that there is an evolutionary tradeoff between the quantity and quality of offspring. In his classic study, ornithologist David Lack demonstrated that the optimal clutch size in European starlings is smaller than the maximum clutch size, because parents who lay a large number of eggs fledge fewer than those who lay an intermediate number. Similarly, if parents have just enough property to endow one child with a farm that will support a family, they should do that, rather than dividing their property among their children and giving none of them enough to make a good living. Perhaps the modern focus on producing a few healthy, well educated but expensive children just reflects a fitness-optimizing tradeoff of quality for quantity.[56] The basic idea is that when offspring quality is as important as quantity, fitness needs to be counted at the level of grandchildren, not children. Those who produce lots of runty, ill-educated starvelings may have more children, but those who produce a smaller number of healthy offspring will have more grandkids.

Anthropologists Jane Lancaster and Hilliard Kaplan tested this explanation of modern low fertility in a large study of the reproductive histories of men in Albuquerque, New Mexico.[57] Anglos were found to be typically more affluent than Hispanics and have fewer children. However, Lancaster and Kaplan could find no evidence that these findings reflected an adaptive tradeoff between quantity and quality. Anglos invest more in fewer children but have fewer grandchildren than Hispanics, not more. Hispanic men have larger numbers of offspring and grand-offspring than Anglos even when economic factors were controlled statistically. These ethnic differences are like the results of the European Fertility Project. The inverse relationship between resources and fertility among the modern middle class is almost certainly also an inverse relationship between wealth and fitness. The continuing decline of fertility to below replacement levels in many parts of Europe (both richer and poorer parts) is unlikely ever to find a fitness-enhancing explanation.

The nonparental transmission hypothesis predicts a diverse array of rogue cultural variants

What does the nonparental transmission hypothesis predict about patterns of fertility decline? As we have seen, all the forces of cultural evolution can support the spread of rogue cultural variants under the right circumstances. The change in the relative importance of nonparental transmission in the modern period is progressive and became massive with the development of cheap mass media. And the more nonparental transmission, the greater the opportunity for maladaptive variants to spread. Innate and cultural-bias heuristics adapted to a lower rate of nonparental transmission would be ill equipped to manage a flood of newly evolved beliefs and attitudes. Selfish cultural variants should exploit a diversity of strategies in these suddenly vulnerable populations. At the same time, variation in values among groups exposed to the same variants will translate into different rates of "infection." Natural selection on vertically transmitted elements of culture will favor pronatalist values directly, and pronatalist values will tend to confer a measure of resistance to the "infection."

In what follows, we present evidence that the successful strategies of selfish cultural variants affecting fertility are indeed diverse, and that both preexisting and newly evolved values with pronatalist effects provide some resistance to fertility-reducing effects.

Beliefs leading to the demographic transition exploit content biases

The clearest examples of cultural ideas that exploit content biases in our psychology are the basic products of the industrial and information revolutions. Modernity has made us consumers who spend a lot of time and money buying and using modern products. The desire for material possessions and creature comforts is fitness enhancing in traditional societies, and this desire is strong in almost all societies; the basic acquisitive impulses are likely innate. After the first few innovators adopt the new item, they serve as demonstrators for the rest of us, and before long another "necessity," such as the telephone or television, is born. We can remember a time when personal computers and mobile phones were undreamed-of devices. In other cases, industrial goods spread only to those rich enough (gourmet food products) or interested enough (mountaineering gear) to use them.

Economist Gary Becker has built a rational-choice explanation for the low fertility of the affluent along these lines.[58] The elite can earn fine salaries, given professional effort. Wealth permits us to consume many luxuries, but this takes time. Our work and our patterns of consumption crowd out our ability to raise children. In contrast, the poor, whose wages are low and who are unable to afford time-consuming hobbies, find raising children an enjoyable way to spend their time. Just as the rich consume less beans and beer than the poor because they can afford steak and champagne, so too do the affluent spend more time earning money and indulging costly hobbies at the expense of having children. Our preferences for children, costly luxuries, and time-consuming hobbies need not differ from person to person or time to time. As the economy undergoes a major structural change, budgets expand and universal preferences merely lead to the substitution of more-preferred for less-preferred items in our consumption set. As the decision-making force of direct bias becomes very strong, the adoption of "traits" such as using an electric toaster becomes cultural only in a quite trivial sense. As we have said before, the rational-choice model is a limiting case of cultural evolution. Note that Becker's model is a covertly cultural evolutionary one in which all the evolutionary action is occurring offstage in the innovations that cause economic growth.

That said, modern economies certainly do produce a plethora of goods and services that appeal mightily to our preferences—universal, culture specific, idiosyncratic, and deviant. Modern business management is aimed at making as direct a connection between our preferences and the industrial production system as highly trained minds can devise. They are successful, and the results no doubt impact our fitness. Here as anywhere, those who wish to box off culture and focus on environment-contingent decisions can find plenty of phenomena that approximate the rational-choice model. Done with eyes peeled, there is nothing wrong with such efforts. The difficulty comes when rational-choice theorists lose track of which shell hides the cultural pea—particularly if the cultural pea is evolving. If we are correct, we cannot depend on humans to have common preferences across societies or stable preferences over time.

However, the pressures and distractions of modern life cannot be the only cause of reduced fertility, because Americans and other citizens of industrialized countries still have plenty of time for childrearing.[59] Sociologists John Robinson and Geoffrey Godbey have collected data on nationwide samples of Americans from 1965 to the present using rather detailed time diaries, the data from which are quite different from data based on people's recollections. Similar data exist for some European countries and Japan. Americans report that they work more than they actually do and underestimate their leisure time, leading to many media stories about overworked Americans.[60] The educated and affluent do work more than those less educated and less affluent, but they also exaggerate their work hours to a greater extent. The truth is that work hours have fallen since 1965. Hours worked by women have increased about three hours per week because of their increased employment outside the home, but the average amount of time worked by all Americans has fallen by more than three hours per week. Hours spent on housework have also fallen substantially for both men and women, mainly because there are more single people and smaller numbers of children. As a result, Americans have about five hours more free time per week than they had in 1965. This increase, however, has been entirely offset by a five-hour-per-week increase in TV viewing, which now amounts to about fifteen hours per week for the average adult.[61]

Our incomes are also ample to support larger families. Baby boom cohorts are earning substantially more than their parents—when income is adjusted to reflect the number of children in the household and the cost of living, Boomers are 50% better off than their parents. This advantage reflects sharply lower rates of childbearing and higher numbers of women in the workforce. Boomers have the financial resources to match or exceed the fertility of their parents, but choose to work more and have fewer children.[62]

Aside from sheer consumerism, there are many obvious reasons to decide to participate in modern economic institutions. Advanced medicine, better hygiene, inexpensive food, and improved shelter, contribute positively to basic components of fitness. Other things, such as reduced dependence on often whimsical or despotic family leaders, must seem like a great boon even as they remove incentives and aids to child rearing. In principle, people could carefully evaluate modern beliefs and attitudes and selectively adopt fitness-enhancing ones. In fact, as is argued below, a few cultures have created this kind of rigorously selective system. The means by which they do so are of great theoretical import.

Beliefs leading to the demographic transition exploit prestige biases

Much of the wealth of the industrial revolution flows to those who vie strenuously for competitive positions in education, business, the arts, medicine, the mass media, and government bureaucracies. Little trickles down to people occupying traditional roles, especially to those in the traditional rural trades. As we argued above, natural selection has shaped the psychology of social learning so that we are predisposed to imitate people with prestige and material well-being. Imitators using prestige as an indicator character will tend to cause people to acquire the whole modernist corpus of values and attitudes. Modern people not only respect wealth itself but the career achievements that give rise to prosperous lifestyles. Free and inexpensive education reduces the barriers to competing for such careers. Not everyone can realistically aspire to great wealth, but a great mass of people can aspire to become respected by their professional colleagues. One of our mothers was in the habit of bragging, without irony, of her son's rather abstract and obscure achievements, "He's well known in his field."

Such new strivings reduce the desire to have children. This change is most dramatic for women. In traditional societies, women derive the bulk of their self-respect and social status from raising children and performing other domestic tasks. In most traditional cultures, a strict sexual division of labor substantially limits women's ability to compete for the most prestigious roles; those are almost entirely monopolized by men. Formal schooling radically alters this pattern. One of the strongest correlates of the beginning of the demographic transition is women's access to education.[63] In school, girls are exposed directly to teachers (frequently women) and indirectly to others occupying prestigious modern roles. The ability of moderns to display wealth and sophistication gives such roles considerable attraction. Further, girls learn that they can succeed in doing schoolwork; what's more, if they are in coeducational schools, they discover that they are actually a little better at schoolwork than boys. Naturally enough, many school-taught girls come to aspire to paid work, to earn the money for participating in the modern economy.

Natural selection on cultural variation influences the demographic transition

The power of the modern prestige system to spread the demographic transition by prestige bias depends on a reliable correlation between achievement in modern roles and small-family norms. If people from small families have an advantage in school achievement and in subsequent competition for prestige in modern social roles, then people who occupy the roles carrying the most weight in nonparental transmission will tend to come disproportionately from small families. The late demographer Judith

Blake presented strong evidence of a tradeoff between family size and intellectual and educational achievement.[64] To test the hypothesis that larger sibships dilute parental resources, she surveyed a wide range of data collected mainly from large-scale U.S. surveys from the 1950s to the 1980s. The effect of family size is consistent across a variety of dependent measures. Large families have a consistent negative effect on intelligence and educational achievement. Children in large sibships (seven-plus children) receive two or three years less education than children in sibships of one or two children. Only children and those in sibships of two generally have the same years of education, but in larger sibships there is a linear decline in number of years of education. The difference between sibships of one and seven is greater than the difference between black and white averages or between successive generations.[65] The effect of sibship size on intelligence, especially in terms of verbal ability, is fairly large, even when father's education (as a partial control for innate aspects of intelligence) was controlled for statistically. Youths' educational aspirations are directly and indirectly affected by sibship size, which in turn negatively affects a wide variety of extracurricular pursuits, such as amount of time spent in cultural activities and reading.

Direct observations of child-rearing practices also indicate that mothers may devote less time per child to children in larger families, supporting the quantity/quality tradeoff.[66] The kinds of supportive, nonpunitive, engaged middle-class child rearing styles that produce children who perform well in school are doubtless more time consuming than punitive or negligent styles that produce less scholarly children.[67] If you want to improve your kids' genetic fitness, for goodness sake don't help them with their homework!

Data from modernizing situations suggest that fertility norms and other correlates of modern culture are transmitted in schools and workplaces.[68] In the United States, sociologists Melvin Kohn and Carmi Schooler investigated the psychological impact of work environment. Men in professional jobs with considerable self-direction promoted this attitude among coworkers.[69] The same sorts of influences presumably operated in nineteenth-century Europe, where the demographic transition started.

Considerable evidence suggests that people who get advanced education tend to be from small families. Occupants of education-intensive roles will tend to have small families, articulate a preference for small families, and correctly attribute their professional success and the expected success of their children to limiting family size.

Enhanced channels of communication currently cause demographic transitions to begin at lower socioeconomic levels

Demographers John Bongaarts and Susan Watkins show that the demographic transitions now occurring in most nations in Latin America and Asia are quite different from the earlier transitions in Europe.[70] Contemporary transitions occur more rapidly and are starting at ever lower levels of socioeconomic development as measured by the United Nations Population Division Human Development Index, a weighted average of life expectancy, literacy, and Gross Domestic Product/capita. The most likely explanations for these changes are innovations that link local communities to national and international influence earlier in the development process. As Bongaarts and Watkins put it, development multiplies the channels of communication between traditional local communications networks and modernizing institutions. For example, friends and relatives discussing issues of interest in informal settings are the retail market for new ideas about contraception and fertility. As long as these markets remain closed, transitions do not occur. The process of development brings new ideas into the market via education, migration, and other forms of contact with the modernizing sector, the wholesalers of new ideas.

In recent decades, three forms of wholesale exposure to new ideas have become much more important at the local level. First, inexpensive electronic media now expose quite remote villagers to entertainment programming produced both nationally and in the developed countries. Second, most national governments have adapted neo-Malthusian policies. Local health workers and other government change agents promote contraception and extol the advantages of small families. Third, international nongovernmental organizations such as Planned Parenthood supplement national neo-Malthusian policies with their own propaganda campaigns. Bongaarts and Watkins regard the legitimization of local discussion concerning the possibility of deliberate fertility reduction as the first important step on the road to widespread adoption of family size reduction. Consider the effect of nationally produced soap operas. They portray prosperous, attractive people leading modern urban lives. Extended, overt discussion of birth control may be rare in such entertainment, but the steamy romances portrayed and the scant presence of children imply it. One of us frequently travels in rural Mexico. Very often the staff of roadside and small-town restaurants are glued to the tube watching telenovelas. The explicit neo-Malthusian propaganda of governments and nongovernmental organizations fills in any blanks left by the entertainment sector.

The diffusion of innovations is by no means a simple or automatic process.[71] However, the exposure to modern ideas through a diversity of channels will eventually begin to strike cords unless local informal communications networks have powerful biases against modernism. The multiplication of these links in the last few decades is having the effect the nonparental transmission hypothesis predicts—earlier and more-rapid declines in fertility.

Rare subcultures are successfully resisting the demographic transition

In modern societies, some subcultures have persistently higher birthrates than others. Groups such as conservative Protestants, Catholics, and Orthodox Jews with strong pronatalist ideologies and significant social and material support for large families have delayed and to some extent mitigated the impact of modern attitudes toward family. As late as the 1960s, Catholic women with parochial high school and college educations desired a child more than Catholics with nonsectarian educations, and nonsectarian Catholics desired more children than Protestants.[72] Sociologists Wade Roof and William McKinney's data show that Catholics and conservative Protestants still hold a reproductive edge on other religious denominations.[73] On the other hand, formerly high birthrates in Catholic Italy have fallen well below replacement in recent years. The Muslim countries of the Middle East and North Africa have higher birthrates than most of the developing world, but most have now begun their transition. No modern transition has reversed itself once begun.[74] As we saw above, ethnicity, not income, provides the best explanation for differences in fertility among Albuquerque men.

Here we focus on two Anabaptist groups that retain very high birthrates, the Amish and the Hutterites. We think that these subcultures are the exceptions that prove the rule. Despite substantial wealth, people in these societies have not gone through the demographic transition, because Anabaptist customs block those same features of cultural evolution that make almost all modern societies susceptible to it.

The birthrates of Anabaptist groups rival those of the highest birthrate pretransition nations, while their death rates are at the levels characteristic of industrial societies.[75] Consequently, their population growth rates are exceedingly high. The Amish population increased from about 5,000 in 1900 to about 140,000 in 1992. In recent years, the population has been doubling every twenty years. The Hutterite rate of increase was a little above 4% per year, giving a doubling time of seventeen years. Hutterite and Amish losses to apostasy are not known with any certainty. Conversion to conventional conservative Protestant churches seems to be a growing problem, though these losses pose no immediate threat to the viability of Anabaptist communities. These societies are prosperous, but they have greatly restricted luxury consumption in order to support very high population growth.

Anabaptists are not relentless procreators; they are perfectly capable of reducing fertility in response to economic constraints. In recent years high land prices have greatly affected both Hutterite and Amish societies. Hutterite total fertility rate has fallen from over nine children in the first fifteen years after World War II to only a little over six in the early 1980s as the creation of new colonies has become more difficult.[76] The Amish have responded to land price increases by taking up other occupations, including factory work and nonfarming family businesses, especially handicraft manufacture for sale to tourists, rather than reducing fertility.

Anabaptists are descendants of sixteenth-century German Protestants that rejected the institutional linkage of religion to the state. Advocates of adult baptism and pacifism, their radical espousal of a church free from state inference resulted in vigorous persecution by state authorities in Europe, but small groups persevered and a few eventually emigrated to the United States (Amish and Mennonites, eighteenth century) and Canada (Hutterites, nineteenth century). Although Anabaptists are no longer proselytizing, they continue to stress farming as a way of life. In many respects, they still resemble the sixteenth-century central European peasant societies from which they are derived. Hutterites have a communal economic system, whereas the Amish are independent family farmers.

While the archaic features of Amish life—buggies and horse-drawn farm equipment—are well known, it is a mistake to think of these groups as isolated from the modern economy. Hutterites use modern equipment but are conservative about incorporating modern conveniences into their home lives. Telephones are generally forbidden, for example. However, both groups are actually quite tightly integrated into the modern economy. They purchase many supplies from the larger economy and sell much in exchange. Moreover, their high birthrates require the accumulation of substantial amounts of capital to expand their land base to accommodate children. Their enterprises must be as efficient conventional operations, if not more so, to support rapid population growth. Thus, the cultural separation of Anabaptists is maintained despite their high degree of economic connection with the larger world. Likewise, Anabaptist culture is conservative and to some degree insulated from popular culture, but it is neither fossilized nor completely isolated from the influences of their host cultures.

Successful Anabaptist sects have cultural beliefs and practices that strongly bias their acquisition of culture from their host societies. For every route of exposure to fertility-reducing beliefs, there is a corresponding defense.

Anabaptist patterns of cultural transmission are nonmodern

The Amish originally sent their children to rural public schools. This is still the norm in Hutterite communities. In both cases they often had sympathetic teachers, sometimes Anabaptists themselves, in part because compact settlements often meant that children attended schools where they were a large if not dominant group. Amish and Hutterites believe that an eighth-grade education is sufficient for the Anabaptist style of life, and feel that older children should attend to practical chores and participate in community and spiritual life. They also perceive that exposure to offensive modern ideas is much greater and more dangerous in high school than in grammar school. In the 1960s and '70s, the "enrichment" of U.S. public school curricula with innovations such as movies became common, and compulsory attendance laws came to conflict with Amish desires to end education early. A U.S. Supreme Court decision in 1972 endorsed the Amish right to end schooling at age fourteen, and the Amish began a parochial school system that today educates many of their children. This system, and the lack of exposure to television and movies, means that Anabaptist youngsters (and adults for that matter) have a much smaller exposure to modern ideas than other children.

Anabaptist families are very traditional. The sexual division of labor is strong, and fathers are important authority figures. Boys learn "manly" skills and attitudes from their fathers and other adult males, very often relatives. Girls learn "womanly" skills from their mothers and other community women. Women are encouraged to find their main satisfactions in raising children and managing the household economy. Men also take much pride in their families and their abilities to provide their sustenance. If, as the demographers' data strongly suggest, the attraction of girls to modern occupations via schooling is a potent force in the demographic transition, the curtailed, conservative education and highly traditional family structure limit the exposure of Anabaptist girls to modernizing influences.

The patterns of education and family life followed by Anabaptists provide a measure of protection from the cultural forces that drive the demographic transition. But only a measure: other rural and conservative groups have been drastically if belatedly affected, while Anabaptists still have extraordinary growth rates. More-active mechanisms must play a role.[77]

Anabaptists retained the asceticism of the early Calvinist churches

In Hutterite theology, great emphasis is placed on the concept of *Gelassenheit,* a mental state of oneness with God to the exclusion of worldly concerns. Anabaptist theology holds that the corrupt world of the flesh is doomed to death and that only believers can expect the reward of eternal life. The world of the spirit is emphasized as much to the exclusion of the world of the flesh as possible. Note that these ideas go back to the sixteenth century. They were not invented to avoid the demographic transition, nor is that an articulated reason for their maintenance. To the extent that such values are operative, the gadgets, comforts, and recreations that the rest of us take for granted have little appeal. Some modern items of consumption do filter into Anabaptist societies, but they are relatively few. Television, that great thief of time, is shunned. Modern technology is thoroughly scrutinized and is adopted if it reasonably fits into the objectives of Anabaptist communities as defined by their religious values. For example, Amish rejection of automobiles is not unthinking traditionalism. Rather, it derives from a careful analysis. Cars are avoided because even the most basic are luxuriously appointed by Amish standards. They have radios that would tempt drivers with secular ideas, and would allow people to live far from their fellow community members. The most ascetic branches of the Amish suffer the lowest losses due to apostasy. Maintaining a high standard of asceticism is an important tool in the defense against the flow of ideas from the world of the flesh. Anabaptist values immunize them against the spread of time-consuming hobbies and a taste for expensive gadgets. For the hearts and minds of the Anabaptists, the industrial designer and the advertising executive appeal in vain.

Anabaptists are socially separate from their host societies

The original separation of the Anabaptists was based on doctrinal differences with fellow Protestants. Believers wanted to protect themselves from the influence of a sinful world. Persecution by states in Europe required a high level of commitment on the part of people who stood by their faith. Symbolic markers of separateness evolved. Anabaptists wear distinctive dress, speak archaic German dialects, and accord status within communities according to criteria derived from their theology. The prestige system of Anabaptists is distinct and different from that of the host society. This prestige system defines as sinful the status gained by success in the host society and discourages anything beyond necessary contact with worldly individuals. Within the Anabaptist community, several institutions minimize competitive status seeking that might lead to sacrificing reproductive success. An all-male executive committee consisting of preachers, top economic managers, and the settlement schoolteacher head Hutterite communities. A bishop, two preachers, and a deacon lead Amish church districts (25–35 families). Men who most exemplify Amish mores are nominated for these prestigious roles, but among those nominated, the choice of role is by lot. The emphasis is on preventing men from competing for office and preventing successful candidates from feeling too proud or mighty, a state dangerous to their souls. Since communities are small, a fairly high proportion of men will occupy prestigious positions by late middle age. Norms of modesty prevent these leaders from claiming too much authority. Since many men will achieve positions of respect and authority, selection on any selfish cultural variants would be weak. Organization above the level of local communities is weak, and no supracommunity roles exist to tempt the ambitious to sacrifice family for the pursuit of high office.

Adherents to Anabaptist ways come to have a high degree of self-confidence in their beliefs. When exposed to the wonders of modern science and technology, most have no regrets or doubts. The power of science is great, no doubt, as most Anabaptists would admit; and they gratefully avail themselves of modern medical advances. But the power of God is greater, they say. Thus, the prestige-bias mechanism affects Anabaptists only weakly and is counterbalanced by a very salient system that favors Anabaptist norms.

Anabaptists demand conformity to community norms

Anabaptist child-rearing styles are rather archaic and stress respect for parental and teacher authority. Behavior that does not conform to community standards is curbed by authority figures, starting with parents who demand old-fashioned obedience from children. In these small communities deviant behavior is conspicuous. The tradition of adult baptism makes full membership in the community conditional upon a solemn act of personal commitment to the community's values. Among Hutterites, the applicant must demonstrate an excellent knowledge of Anabaptist theology and undergo a rigorous questioning by elders concerning past behavior and future intentions. Of course, the attractions of the sinful world of the flesh do make an impression on Anabaptist youngsters, especially young adults. Life is austere and tedious in their communities. In addition, the communities do not always function smoothly; conflicts and dissention weaken people's resolve.

As is typical of deviance in any society, young men make up the bulk of delinquents. Among the Amish, a period between age sixteen and the early twenties intervenes between the time of strict parental control and the baptismal commitment to the church. The Pennsylvania "Dutch" term for this stage of life is *rumspringa*.[78] During rumspringa, many Amish young adults sample the pleasures of the world with little interference from parents or the church. The Anabaptist doctrine of adult baptism emphasizes the free commitment of adults to the church, and rumspringa serves to emphasize that the renunciation of worldly life is *voluntary*. After baptismal vows are taken, the community actively and formally shuns serious deviants. Their own families are expected to refuse contact with them, while contact with even seriously deviant young adults during rumspringa is their own affair. Defectors can return, and many do, with a full confession and rededication to community practices. The high degree of conformity expected in Anabaptist communities prevents the seepage of host-society values by piecemeal adoption of innovations. In effect, social change is restricted to changes that are approved by the community collectively.

Can Anabaptists resist modernization in the end?

Both Hutterites and Amish are subject to strong modernizing forces. As we mentioned earlier, the economic viability of the Anabaptists' traditional, expansionist farming system is threatened by the acceleration of industrialized farming and rising land prices. Farm industrialization forces Anabaptist farmers to accept many innovations to remain economically viable, and these innovations threaten the separation of Anabaptist communities. Telephones that become necessary for business are tempting to use for social calls. More-sophisticated machinery requires more education. High land prices force many Amish to turn to nontraditional occupations. Serving tourists and working in non-Amish factories generate daily contact with outsiders. In the case of the Hutterites, proselytizing conservative Christian ministers welcome apostates with a theologically friendly alternative lifestyle that is much less austere. Perhaps all Anabaptist communities will eventually follow the path of the New Order Amish, whose generally less-strict rules invite more-rapid penetration of many modern techniques and who suffer high defection rates.

In one scenario, Anabaptist separatism could vanish, and these sects would merge into mainstream conservative Protestantism. However, this scenario is by no means certain. For example, the extensive tourist industry on Martha's Vineyard increased rather than decreased the Vineyarders' sense of social distance from mainland New Englanders.[79] Anabaptists have maintained separateness in the face of persecution and temptation from host societies for four and a half centuries. Perhaps Anabaptists will curtail their rate of reproduction to fit the limited power of their farming economies to expand while retaining other archaic customs. Thus, even if their birthrates fall somewhat, they may remain fitness optimizing given more-severe economic constraints. Or perhaps the new economic niches that the Amish are pioneering will keep demographic expansion rapid and permit the retention of conservative lifeways. To date, a substantial shift from farming to wage labor and the tourist trade does not seem to be causing problems for the Amish. However, the Anabaptist adaptation is predicated on a fine balance between cultural separation and economic engagement with modern society.

The Anabaptist case illustrates the manifold power of modern fertility-reducing beliefs and values to spread by highlighting how comprehensive an adaptation—or, in this case, a preadaptation—must be to resist them. Innovations in communication and transport have had the unintended consequence of unleashing the evolution of maladaptive cultural variants that seep into cultures by a number of routes. So far, only the Anabaptists and a few similar groups, like ultra-Orthodox Jews, seem to have much resistance to modernity's infections. Anabaptism is like a tightly made kayak navigating the turbulent modernist sea. It looks so fragile but survives because it doesn't leak despite the enormous stresses it faces. One serious cultural leak anywhere, and it's gone. Anabaptism's evolutionary future or futures are impossible to predict. In the meantime, you can't help but admire the beauty of the design!

Cultural evolution explains the cultural complexity of the demographic transition

Given the examples of the highly resistant Anabaptists, partially resistant Catholics, conservative Protestants and Muslims, and the precocious transitions today in many developing countries, that the demographic transition in Europe varied greatly by culture area is not so surprising. The modernizing of the economy and of social roles are complex processes, no doubt influenced by preindustrial cultural variation. The modernization phenomenon, including the demographic transition, is driven by the fact that economic and social modernization are coupled, albeit sloppily. Social modernization can race ahead of industrial production, as in France, or industrialization can lead more slowly to social modernization, as in Britain. Social modernization creates educated individualists who easily adapt to running factories even if their first aspiration is commerce or public service. Industrialization creates a demand for laborers and managers with education and individualistic motivations. However, the process can proceed some way on the stock of such individuals produced by traditional education systems. Aristocratic elites can shift from government service to business; middle-class clergy, doctors, and lawyers can provide managerial talent; and traditional craftsmen, with a little help from a mathematically literate middle-class manager, make acceptable engineers. In the long run, the synergy between social and economic modernization creates a strong correlation between the two, but with enough slop to preserve considerable variation between different modernized societies.

As industrial production and social modernization began to spread from their heartlands in Britain and France, respectively, they met very different patterns of resistance and acceptance. The strength and effectiveness of resistance depended on how beliefs, values, and economic activities structured patterns of nonparental transmission of culture and generated forces that favored or resisted modern ideas. Anabaptists represent one extreme in terms of receptivity to modernism. Catholics and conservative Protestants in the United States illustrate a much more moderate, but still significant, resistance to modernism generally and the demographic transition specifically. The modern Third World includes cases in which the mass media and primary education for women are sufficient to induce the onset of a rapid fertility transition but also includes conservative Muslim societies where relatively high fertility rates persist, perhaps because these societies tend toward traditional, highly gendered roles for women.

Conclusion: Culture is built for speed, not comfort

All adaptations involve compromises and tradeoffs. Flight allows birds easy escape from many kinds of predators, and it makes long-distance migration practical. However, birds operate under many design constraints necessary to make flight possible in the low-density, low-viscosity medium of air. For example, their bones must be light but rigid—constraints are met by the fact that their bones are hollow tubes that, while light and rigid, are very delicate, failing catastrophically when bent, like aluminum lawn furniture.

In this chapter, we have argued that cultural maladaptations arise from a design tradeoff. Culture allows rapid adaptation to a wide range of environments, but leads to systematic maladaptation as a result. To turn the Willie Dixon blues classic on it head, culture is built for speed, not comfort.[80] Learning mechanisms depend critically on preexisting knowledge. If you already know a lot about a problem, learning can be easy and efficient. If you don't know much, learning can be impossible. This fact creates a severe problem for learned adaptation to environments that undergo big changes in short periods of time. Because natural selection cannot keep up with rapid environmental change, it cannot endow individuals with an evolved psychology tailored to their current environment; it can only endow them with a knowledge of the common statistical features of a whole range of environments. We think culture (both its psychological basis *and* its pool of transmitted ideas) is an adaptation that evolved to solve this problem. Accurate teaching and imitation combined with relatively weak general-purpose learning mechanisms allow populations to accumulate adaptive information much more rapidly than selection could change gene frequencies. This capacity has great benefits, allowing human foragers to adapt to a far wider range of environments than any other animal species. However, just as flight requires fragile, hollow bones, cultural adaptation entails design compromises. In creating a simulation of a Darwinian system using imitation instead of genes, natural selection created conditions that allow selfish cultural variants to spread. If our argument from the empirical cases is correct, we do see just the sort of selfish variants this hypothesis predicts.

Our culture is a lot like our lungs. They both work great for their evolved functions, but they also make us susceptible to infection by pathogens. You would be a lot less likely to catch either a serious respiratory disease or a selfish cultural variant if you kept away from other people as much as possible. We have evolved to take much greater risks with both sorts of diseases, because contact with others has many benefits. Culture gives us the ability to imitate things essential to human life, but it also makes us take up bits that cripple and kill—not unlike like the air we breathe.

The big-mistake hypothesis represents the most serious alternative attempt to account for human maladaptation. It holds that most of the information necessary to construct what we call culture is latent in genes shaped by Pleistocene environments. Its proponents argue that this information is

organized into decision-making systems evolved to produce adaptive behavior during the Pleistocene epoch. In the post-Pleistocene, they argue, a
sudden acceleration of cultural change transformed "environments" so that
they are now far outside the ranges of evolved decision-making systems.
Different evolutionary social scientists have different ideas about just where
and how often big mistakes will occur. For example, John Tooby and Leda
Cosmides seem to believe that little post-Pleistocene behavior can be reliably predicted by adaptive considerations.[81] Human behavioral ecologists,
by contrast, cite considerable evidence that traditional Holocene societies
often seem to behave quite adaptively compared to modern societies.[82] In
either case, explanation rests on a direct interaction between individual
minds and the "environment," not on the evolutionary dynamics of culture.

Distinguishing between the big-mistake and explicitly cultural evolutionary explanations for maladaptive behavior is important for two reasons.
First, the cultural hypothesis makes systematic predictions about the details of how cultural maladaptations arise. The generic "big-mistake" hypothesis makes no such predictions, and concrete variants of it, like Kaplan's explanation of the demographic transition, have an ad-hoc quality.
Indeed, since the ways that a complex, highly evolved adaptation can go
wrong are huge, the big-mistake hypothesis is inherently ad-hoc. Ad-hoc
explanations are not necessarily wrong; environments outside the range
in which a species has evolved are quite likely to result in a miscellany of
breakdowns of adaptations. Humans are not the best candidates to exemplify such breakdowns, because we are a species that is superbly adaptable
to variable environments, as our explosive success during the Holocene
testifies. In the test case here, we think that the details of the demographic
transitions fit better with our account than with explanations that rely only
on preferences for wealth and prestige that have turned maladaptive in
modern environments. It provides a general theory of maladaptations that
gets details right.

Second, the two hypotheses make very different predictions about
Pleistocene hunter-gatherer environments. The big-mistake hypothesis predicts that the behavior of Pleistocene foragers should have been adaptive
most of the time. By contrast, our hypothesis predicts that as soon as cultural transmission became significant, selection on culture capacities would
have begun to favor nonparental transmission, and, inevitably, rogue cultural variants would appear. We are willing to entertain the hypothesis that
modern societies have a higher frequency of maladaptive cultural variation
given that the ratio of nonparental to parental cultural influence has increased so dramatically. The use of mass media for advertising fitness-

reducing distractions has evolved into a fine art, but on the other hand, literacy and science have scotched many harmful superstitions by making the adaptive component of content biases more powerful.

The sharpest test of these two hypotheses would come from the existence, or not, of Pleistocene maladaptations of the sort predicted from considerations of cultural evolution. Of course, this is difficult. Behavior in contemporary foraging societies is useful but imperfect, since Holocene environments are so different from those of the late Pleistocene. The low resolution of the paleoanthropological record makes direct tests difficult. One mechanism that might permit truly large-scale and durable deviations from fitness optimization is gene-culture coevolution. Once cultural traditions create novel environments, environments that can affect the fitness of alternative *genetically* transmitted variants, genes and culture are joined in a coevolutionary dance. In the extreme case, culturally determined social traditions can select for genotypes favorable for the perpetuation of the cultural tradition.[83] Since a population of human beings is necessary to make culture work, such coevolutionary maladaptations will tend to be self-limiting and hence hard to observe based on the skimpy Pleistocene evidence. The most detectable maladaptations would be those strange ones that actually *increase* the average fitness of populations even though *selection on genes* will act against them. Human cooperation is a potential example. Humans are quite adept at cooperating in large groups with strangers and near strangers, while the theory of selection on genes suggests that cooperation should be restricted to relatives and well-known nonrelatives. As we remarked earlier, the conformity bias offers a possible mechanism to generate stable variation at the group level on which selection might act to favor in-group cooperation. Could the human aptitude for cooperation be an example of one of these seemingly paradoxical adaptive maladaptations? Can we have any confidence that human patterns of cooperation reach back into the Pleistocene? We turn to these topics in the next chapter.

Chapter Six

Culture and Genes Coevolve

Milk was once marketed in the United States with the slogan, "Every Body Needs Milk." Catchy, but it's not true. Most people not only don't need milk, they can't tolerate it. The majority of the world's adults lack the enzyme necessary to digest lactose, the sugar in milk, and if they drink milk, the lactose is fermented by bacteria rather than absorbed by the gut, leading to uncomfortable attacks of flatulence and diarrhea. That we didn't know this until the 1960s is testimony to how scientists are blinkered by their cultural background—most nutritionists came from countries where adult lactose malabsorption is rare. It is also testimony to how small a role evolution plays in biomedical science, because even a little adaptationist thinking would have suggested that it is the ability to digest milk that is abnormal, not the reverse. Milk has always been baby food for mammals, and lactose only occurs in mother's milk. Thus, adult mammals had no need for the enzyme that cleaves lactose. Unsurprisingly, ever frugal natural selection shut down the production of this enzyme after weaning in almost all mammal species. The majority of people exhibit the standard mammalian developmental pattern; they can digest milk as infants but not as adults. The real evolutionary puzzle is why in some human populations most adults can digest lactose.

In the early 1970s, geographer Fredrick Simoons suggested that the ability to digest lactose evolved in response to a history of dairying.[1] The people of northwest Europe have long kept cows and consumed fresh milk. Dairying was carried to India by "Aryan" invaders, and has been practiced by pastoralists in western Asia and Africa for millennia. In each of these regions, most adults can drink fresh milk. Mediterranean dairying people traditionally consume milk in the form of yogurt, cheese, and other products from which the lactose has been removed. Some adults in these populations can digest lactose while others cannot. Dairying is rare or absent in the rest of the world, and few Native Americans, Pacific Islanders, Far Easterners, and Africans are lactose absorbers. Simoons's hypothesis was controversial at the time, but subsequent genetic data confirm that adult lactose digestion is controlled by a single dominant gene, and careful statistical work indicates that a history of dairying is the best predictor of a high frequency of this gene. Moreover, calculations indicate that there has been plenty of time for this gene to spread since the origin of dairying.[2]

The evolution of adult lactose digestion is an example of "gene-culture coevolution." Biologists developed the term *coevolution* to refer to systems in which two species are important parts of each other's environments so that evolutionary changes in one species induce evolutionary modifications in the other.[3] This can lead to an intricately choreographed coevolutionary dance, often with surprising results. For example, normally predatory ants often tend aphids, protecting them from predators. The aphids reward their ants by exuding sugar-rich honeydew, which the ants collect.

The evolving pools of cultural and genetic information carried by human populations are partners in a similar swirling waltz. Genetic evolution created a psychology that allows the cumulative cultural evolution of complex cultural adaptations. In some environments, this process led to the evolution of the dairying traditions. This new culturally evolved environment then increased the relative fitness of the gene that allows whole-milk consumption by adults. As that gene spread, it in turn may have changed the environment-shaping cultural practices, perhaps favoring more whole-milk consumption, or more serendipitously, giving rise to the evolution of ice cream.

We think that gene-culture coevolution has also played an important role in the *genetic* evolution of human psychology. If genetically maladaptive cultural variants are an inevitable consequence of cumulative cultural adaptation, then the pools of cultural and genetic information carried by human populations each respond to their own evolutionary dynamic. Natural selection, mutation, and drift shape gene frequencies, while natural selection, guided variation, and a variety of transmission biases mold the distribution of cultural variants. However, these two processes are not independent. Each partner in the coevolutionary dance influences the evolutionary dynamics of the other. Genetically evolved psychological biases steer cultural evolution in genetic fitness-enhancing directions.[4] Culturally evolved traits affect the relative fitness of different genotypes in many ways. Consider just a few examples:

- Culturally evolved technology can affect the evolution of morphology. For example, modern humans are much less robust than earlier hominid species. Paleoanthropologists have argued that this change was due to the cultural evolution of effective projectile hunting weapons.[5] Before projectile weapons, robust genotypes were favored because people killed large animals at close range, but once they could be killed at distance, selection favored a less robust (and less expensive) physique.
- The availability of valuable culturally evolved information may lead to selection for enhanced capacities for acquiring and using that information. Language provides the canonical example. There is no doubt that the human vocal tract and auditory systems have been modified to enhance our ability to produce and decode spoken language, and we seem to have special-purpose psychological machinery for learning the meaning of words and grammatical rules. Selection could not have produced these derived features in an environment without spoken language. The most plausible explanation is that simple culturally transmitted language arose first, and then selection favored a special-purpose throat morphology to generate speech sounds and a special-purpose psychology for learning, decoding, and producing speech, which in turn gave rise to a richer, more-complex language, and led to yet more modifications of the traits that allow language acquisition and production.
- Culturally evolved moral norms can affect fitness if norm violators are punished by others. Men who cannot control their antisocial impulses are exiled to the wilderness in small-scale societies and sentenced to prison in contemporary ones. Women who behave inappropriately in social circumstances are unlikely to find or keep husbands.[6] In this chapter, we will argue that coevolutionary forces have radically reshaped innate features of human social psychology.

Gene-cultural coevolution can generate such significant genetic changes because it has been going on for a long time. Dairying has been a force in populations with high frequencies of adult lactose digestion for some three hundred generations. In chapter 4, we presented evidence that the capacity for the cumulative evolution of complex cultural adaptations is roughly half a million years old. This means that complex cultural traditions have been exerting coevolutionary selective pressures on human gene pools for about twenty thousand generations. In this amount of time, culturally evolved environments could have had dramatic coevolutionary effects on the evolution of human genes.

We hope that the idea of gene-culture coevolution seems intuitive and plausible to most of our readers. Be warned, however, that you are being invited to start down what many evolutionary social scientists believe is a garden path. Researchers in this tradition emphasize that cultural evolution is molded by our evolved psychology, but not the reverse. As psychologist Charles Lumsden and evolutionary biologist E. O. Wilson put it, genes have culture on a leash.[7] Culture can wander a bit, but if it threatens to get out of hand, its genetic master can bring it to heel. We think that this is only half the story. As we argued at length in the last chapter, heritable cultural variation responds to its own evolutionary dynamic, often leading to the evolution of cultural variants that would not be favored by selection acting on genes. The resulting cultural environments then can affect the evolutionary dynamics of alternative genes. Culture is on a leash, all right, but the dog on the end is big, smart, and independent. On any given walk, it is hard to tell who is leading who.

Better to think of genes and culture as obligate mutualists, like two species that synergistically combine their specialized capacities to do things that neither one can do alone.[8] Humans by themselves cannot convert grass into usable food. Cows by themselves cannot drive away lions and wolves. The cow-human mutualism works to the advantage of both. However, such mutualisms are never perfect. Humans will always be tempted to take more milk at the expense of calves, and cows will always be subject to natural selection favoring shorting the humans to feed their offspring. Each caters to the whimsical biology of the other so long as there is a net payoff to the cooperation. Humans chauvinistically see themselves as controlling domestication. A cow might as well flatter herself on how clever she is to elicit so much work on her behalf from her humans. The relationship between genes and culture is similar. Genes, by themselves, can't readily adapt to rapidly changing environments. Cultural variants, by themselves, can't do anything without brains and bodies. Genes and culture are tightly coupled but subject to evolutionary forces that tug behavior in different directions.

Biologists John Maynard Smith and Eörs Szathmáry point out that mutualisms have played an important role in the evolution of major transitions in levels of biological organization.[9] The origin of eukaryotic cells provides a good example.[10] Until about two billion years ago, the world's biota was dominated by prokaryotes, organisms without nuclei or chromosomes, like modern-day bacteria. Then, eukaryotes arose as a result of a close symbiosis between prokaryote species; one of these species eventually evolved to become the nucleus and others became cellular organelles such as mitochondria and chloroplasts. The larger and functionally more-complex eukaryotic cells that resulted from the coevolution of these mutualists were able to outcompete prokaryotes in some existing adaptive niches and enter many new ones.

In the remainder of this chapter, we will argue that the symbiosis between genes and culture in the human species has led to an analogous major transition in the history of life—the evolution of complex cooperative human societies that radically transformed almost all the world's habitats over the last ten thousand years.

Gene-culture coevolution and human ultrasociality

Human societies are a spectacular anomaly in the animal world. They are based on the cooperation of large, symbolically marked in-groups. Such groups have economies based on substantial division of labor and compete with similarly marked out-groups. This is obviously true of modern societies, in which enormous bureaucracies like the military, political parties, churches, and corporations manage complex tasks, and in which people depend on a vast array of resources produced in every corner of the globe. But it is also true of hunter-gatherers, who have extensive exchange networks and regularly share food and other important goods outside the family and the residential group.

In most animal species, cooperation is either absent or limited to very small groups, and there is little division of labor.[11] Among the few animals that cooperate in large groups are social insects such as bees, ants, and termites, and the naked mole rat, a subterranean African rodent. Multicellular plants and many forms of multicellular invertebrates can also be thought of as complex societies made up of individual cells. In each of these cases, however, the cooperating individuals are genetically related. Typically, the cells in a multicellular organism are members of a genetically identical clone, and the individuals in insect and naked mole rat colonies are siblings.

Thus we have another evolutionary puzzle. Our ancestors six million years ago in the Miocene presumably cooperated in small groups mainly made up of relatives, as contemporary nonhuman primates do. There was no trade, little division of labor, and coalitions were limited to a small number of individuals. As we will argue below, these patterns are consistent with our understanding of how natural selection shapes behavior. Sometime between then and now, something happened that caused humans to cooperate in large, complex, symbolically marked groups. What caused this radical divergence from the behavior of other social mammals?

We think that gene-culture coevolution provides the most likely solution to this puzzle. There are two parts to this argument. First, cultural adaptation potentiates cultural evolution of cooperation and symbolic marking. Human culture allows rapid, cumulative evolution of complex adaptations and is particularly adaptive in variable environments. Such rapid adaptation has radically increased the amount of heritable cultural variation between human groups, which means that intergroup competition (always present) gives rise to the cumulative evolution of cultural traits that enhanced the success of groups. Since larger, more-cooperative, and more-coherent groups should outcompete smaller, less cooperative groups, group selection could give rise to culturally transmitted cooperative, group-oriented norms, and systems of rewards and punishments to ensure that such norms are obeyed. Stable variation between groups can also lead to the evolution of symbolic markers that allow individuals to choose whom to imitate or whom to interact with.

Second, culturally evolved social environments favor an innate psychology that is suited to such environments. In culturally evolved social environments in which prosocial norms are enforced by systems of sanction and reward, individual selection will favor psychological predispositions that make individuals more likely to gain social rewards and avoid social sanctions. Similarly, in a world made up of coherent, culturally distinct, symbolically marked groups which demand loyalty from their members, individual selection will favor psychological adaptations that allow people to parse the groups that make up their social world, and identify with the appropriate ones.

As a result, people are endowed with two sets of innate predispositions, or "social instincts."[12] The first is a set of ancient instincts that we share with our primate ancestors. The ancient social instincts were shaped by the familiar evolutionary processes of kin selection and reciprocity, enabling humans to have a complex family life and frequently form strong bonds of friendship with others. The second is a set of "tribal"[13] instincts that allow us to interact cooperatively with a larger, symbolically marked set of people, or tribe. The tribal social instincts result from the gene-culture co-evolution of tribal-scale societies by the process described above. Consequently, humans are able to make common cause with a sizable, culturally defined set of distantly related individuals, a form of social organization that is absent in other primates.[14]

In the remainder of this chapter, we will describe and defend this hypothesis. First, we provide a brief primer on the theory of the evolution of cooperation. Our goal is to convince you that human sociality is indeed a puzzle, and provide necessary background for understanding our coevolutionary account and a competing hypothesis from evolutionary psychology. We then describe in more detail how gene-culture coevolution has given rise to tribal social instincts. Next, we summarize data from psychological studies that suggest that such instincts actually exist. Then, we present ethnographic and historical evidence that suggests that the recent hunter-gatherer societies exhibit tribal-scale social organization. Finally, we use the evolution of complex societies as a natural experiment to test the hypothesis.

Cooperation is usually limited to kin and small groups of reciprocators

When we were graduate students during the late 1960s and early 1970s, biology texts quite commonly explained animal behaviors in terms of their benefit to the species. Alarm cries helped defend the social group against predators, and sexual reproduction maintained the genetic variation necessary for the species to adapt. A key advance in biology forty years ago was to show that such explanations are mostly wrong. Natural selection does not normally lead to the evolution of traits that are for the good of the species, or even the social group. Selection usually favors traits that increase the reproductive success of individuals, or sometimes individual genes; and when a *conflict* occurs between what is good for the individual and what is good for the group, selection usually leads to the evolution of the trait that benefits the individual.

Selection favors cooperation among kin

The big exception to this rule occurs when groups are made up of genetic kin—then selection can favor behavior that reduces fitness of the individual performing the behavior as long as it causes a sufficient increase in the fitness of the group. Consider a very Prussian species in which individuals all live in groups of exactly 9 drawn from the global population. Further suppose that there are two types: helpers and egoists. The helpers perform a prosocial behavior that increases the fitness of each of the other 8 individuals in their group by $\frac{1}{4}$ unit, but decreases the fitness of helpers by $\frac{1}{2}$ unit. This behavior is clearly group beneficial—it increases the average fitness of each of the 8 other group members by $\frac{1}{4}$, so the net increase in group fitness due to the behavior is $8 \times \frac{1}{4} - \frac{1}{2} = 1\frac{1}{2}$ fitness units.

People untrained in evolutionary biology often think that behaviors that produce group benefits will be favored by natural selection. But group benefits are not enough. Suppose groups are formed at random. Then each prosocial act has the same average effect on the fitness of helpers and egoists. This means that prosocial behavior has no effect on the *relative* fitness of helpers and selfish types, because helpers behave as saints, helping good guys and bad guys indiscriminately. In which case, no change in the frequency of these two types in the population will occur due the *receipt* of altruism. At the same time, the *costs* of performing prosocial behavior fall solely on helpers, and thus decrease their fitness relative to egoists.

Now suppose that groups are made up of full siblings. Full siblings share 50% of their genes, so helpers will find themselves in groups in which, on average, 4 of the other 8 members carry the helping gene. The other 4 carry a random sample of genes from the population. Now, the prosocial act increases the relative fitness of 4 individuals with the prosocial gene $4 \times \frac{1}{4} = 1$ fitness unit, at a cost of only $\frac{1}{2}$ fitness units. Selection can favor this behavior, because the benefits of prosocial acts are nonrandomly directed toward others who carry the same gene.

This simple example illustrates a fundamental evolutionary principle: costly group-beneficial behavior cannot evolve unless the benefits of group-beneficial behavior flow nonrandomly to individuals who carry the genes that give rise to the behavior. Altruism toward kin can be favored by selection because kin are similar genetically. The late, great evolutionary biologist W. D. Hamilton worked out the basic calculus of kin selection in 1964[15] and deduced many of its most important effects on social evolution. As you have seen, full siblings can count on sharing half their genes through common descent, and can therefore afford to help a sibling reproduce so long as the fitness payoffs are twice the costs. More-distant relatives require a higher benefit-cost ratio.[16] This principle, often called Hamilton's rule, successfully explains a vast range of behavior (and morphology) in a very wide range of organisms.[17]

Selection can favor cooperation among small groups of reciprocators

When animals interact repeatedly, past behavior also provides a cue that allows nonrandom social interaction. Suppose that animals live in social groups and the same pair of individuals interacts over an extended period of time. Often, one member of the pair has the opportunity to help the other, at some cost to itself. Suppose that there are two types: defectors who do not help, and reciprocators who use the strategy "Help on the first opportunity. After that, help your partner as long as she keeps helping you, but if she doesn't help, don't help her anymore." Initially, partners are chosen at random, so that at the first opportunity, reciprocators are no more likely to be helped than defectors. However, after the first interaction, only reciprocators receive any help, and if interactions continue long enough, the high fitness of reciprocators in such pairings will be enough to cause the average fitness of reciprocators to exceed that of defectors.

Beyond this basic story, there is little agreement among scientists about how reciprocity works. The contrast with kin selection theory is instructive. The simple principle embodied by Hamilton's rule allows biologists to explain a wide range of phenomena. Despite much work, evolutionary theorists (including yours truly) have not managed to derive any widely applicable general principles describing the evolution of reciprocity. Worse, evidence that reciprocity is important in nature is scanty;[18] only a handful of studies provide evidence for reciprocity, and none of them are definitive.[19]

Despite its many problems, theoretical work does make one fairly clear prediction that is relevant here: reciprocity can support cooperation in small groups, but not in larger ones.[20] Instead of assuming that individuals interact in pairs, suppose that individuals live in groups, and each helping act benefits all group members. For example, the helping behavior could be an alarm cry that warns group members of an approaching predator, but makes the callers conspicuous and thereby increases their risk of being eaten. Suppose there is a defector in the group who never calls. If reciprocators use the rule, only cooperate if all others cooperate, this defector induces other reciprocators to stop cooperating. These defections induce still more defections. Innocent cooperators suffer as much as guilty defectors when the only recourse to defection is to stop cooperating. On the other hand, if reciprocators tolerate defectors, then defectors can benefit in the long run.

Theoretical work suggests that this phenomenon will limit reciprocity to quite small groups, and while no good empirical data exist, it does fit with everyday experience. We know that reciprocity plays an important role in friendship, marriage, and other dyadic relationships. We eventually stop inviting friends over to dinner if they never return our invitations; we become annoyed at our spouse if she does not take her turn watching the children; and we change auto mechanics if they repeatedly overcharge us for repairs. But cooperation in larger groups cannot be based on the same principle. Each one of a thousand union members does not keep walking the picket line because she is afraid that her one defection will break the strike. Nor does each Enga warrior maintain his position in the line of battle because he fears that his desertion will precipitate wholesale retreat. Nor do we recycle our bottles and newspapers because we fear our littering will doom the planet.

Some authors have emphasized that punishment takes other forms such as reduced status, fewer friends, and fewer mating opportunities[21]—what evolutionary biologist Robert Trivers calls "moralistic punishment."[22] While moralistic punishment and reciprocity are often lumped together, they have very different evolutionary properties. Moralistic punishment is more effective in supporting large-scale cooperation than reciprocity for two reasons. First, punishment can be targeted, meaning that defectors can be penalized without generating the cascade of defection that follows when reciprocators refuse to cooperate with defectors. Second, with reciprocity, the severity of the sanction is limited by the effect of a single individual's cooperation on each other group member, an effect that decreases as group size increases. Moralistic sanctions can be much more costly to defectors, so that cooperators can induce others to cooperate in large groups even when they are rare. Cowards, deserters, and cheaters may be attacked by their erstwhile compatriots, shunned by their society, made the targets of gossip, or denied access to territories or mates. Thus, moralistic punishment provides a much more plausible mechanism for the maintenance of large-scale cooperation than reciprocity.

However, two problems remain.[23] First, why should individuals punish? If punishing is costly and the benefits of cooperation flow to the group as a whole, administering punishment is a costly group-beneficial act, and therefore, selfish individuals will cooperate but not punish. The Enga man who punishes a coward suffers a cost to himself and provides a benefit to other members of his clan. The Enga woman who shuns a deserter may forgo an otherwise desirable marriage partner while helping to ensure that cowards do not become common among the Enga. Thus, as long as the effect of the punishment administered by a single individual will have little effect on the outcome of the battle, selfish individuals will not punish. Second, moralistic punishment can stabilize *any* arbitrary behavior—wearing a tie, being kind to animals, or eating the brains of dead relatives. Whether the behavior produces group benefits is of no significance. All that matters is that when moralistic punishers are common, being punished is more costly than performing the correct behavior, whatever it might be. When any behavior can persist at a stable equilibrium, then the fact that cooperation is a stable equilibrium does not tell us whether it is a likely outcome.

While much of the debate about moralistic punishment has focused on the first problem, we think the second presents a bigger obstacle to the evolution of cooperation in large groups. If moralistic punishment is common, and punishments sufficiently severe, then cooperation will pay. Most people may go through life without having to punish very much, which in turn means that a predisposition to punish may be cheap compared with a disposition to cooperate (in the absence of punishment). Thus, relatively weak evolutionary forces can maintain a moralistic predisposition, and then punishment can maintain group-beneficial behavior. However, if evolutionary change is driven only by individual costs and benefits, then moralistic punishment can stabilize cooperation, but it can also stabilize anything else. Societies do often seem to use moralistic punishment or its threat to enforce social conventions of no apparent utility of any kind, such as wearing ties to work. Since cooperative behaviors are a tiny subset of all possible behaviors, punishment does not explain why large-scale cooperation is so widely observed. In other words, moralistic punishment may be necessary to sustain large-scale cooperation, but it is not sufficient to explain why large-scale cooperation occurs.

Selection among large, partially isolated groups is not effective

Group selection may be the number one hot-button topic among evolutionary biologists. The controversy ignited in the early 1960s, when ornithologist V. C. Wynne-Edwards published a book that explained a number of puzzling avian behaviors in terms of the benefit to the group.[24] For example, he thought that the great, whirling evening displays of thousands of roosting starlings allowed the birds to census population size and control their birthrates to avoid overexploiting their food supply. While this kind of explanation was not unusual in those days, Wynne-Edwards was much clearer than his contemporaries about the process that gave rise to such group-level adaptations: groups that had the display survived and prospered, while those that didn't overexploited their food supply and perished.

The book generated a storm of controversy, with biological luminaries such as David Lack, George Williams, and John Maynard Smith penning critiques explaining why this mechanism, then called group selection, could not work.[25] At the same time, Hamilton's newly minted theory of kin selection provided an alternative explanation for cooperation. The result was the beginning of an ongoing and highly successful revolution in our understanding of the evolution of animal behavior, a revolution that is rooted in carefully thinking about the individual and nepotistic function of behaviors.

In the early 1970s, an eccentric retired engineer named George Price published two papers that presented a genuinely new way to think about evolution.[26] Up until that time, most evolutionary theory was based on an accounting system that kept track of the fitness of different genes. To understand the evolution of a particular trait, one needed to know how the behavior of others affects each individual carrying a particular gene and average this over all situations in which individuals find themselves (just as we did above in explaining kin selection and reciprocity). Price argued that it was *also* fruitful to think about selection occurring in a series of nested levels: among genes within an individual, among individuals within groups, and among groups; and he invented a very powerful mathematical formalism, now called the Price covariance equation, for describing these processes. Using Price's method, kin selection is conceptualized as occurring at two levels: selection *within* family groups favors defectors, because defectors always do better than other individuals within their own group, but selection *among* family groups favors groups with more helpers, because each helper increases the average fitness of the group. The outcome depends on the relative amount of variation within and between groups. If group members are closely related, most of the variation will occur between groups. This is easiest to see if groups are composed of clones (as in colonial invertebrates such as corals). Then there is almost no genetic variation within groups; all the variation is between groups, and selection acts to maximize group benefit.

Price's multilevel selection approach and the older gene-centered approaches are mathematically equivalent. One approach may be more heuristic or mathematically tractable for particular evolutionary problems than the other, but if you do your sums properly, you will come up with the same answer either way.[27] Adopting the multilevel formalism does not imply that animals are more or less likely to do things for the good of the group, because these two approaches are equivalent.

The multilevel selection approach has led to a renaissance in group selection in recent years which has generated new wrangling between those who thought that they had killed group selection and those who, thinking in multilevel terms, see nothing wrong with it.[28] This argument is mainly about what kinds of evolutionary processes should be *called* group selection. Some people use *group selection* to mean the process that Wynne-Edwards envisioned—selection between large groups made up of mostly genetically unrelated individuals—while others use *group selection* to refer to selection involving any kind of group in a multilevel selection analysis, including groups made up of close kin.

The real scientific question is, what kinds of population structure can produce enough variation between groups so that selection at that level can have an important effect? The answer is fairly straightforward: selection between large groups of unrelated individuals is not normally an important force in organic evolution. Even very small amounts of migration are sufficient to reduce the genetic variation between groups to such a low level that group selection is not important.[29] However, as we will explain below, the same conclusion does not hold for cultural variation.

Among primates, cooperation is limited to small groups

The punch line is that evolutionary theory predicts that cooperation in nonhuman primates and other species that have small families will be limited to small groups. Kin selection results in large-scale social systems only when there are large numbers of closely related individuals; social insects in which a few females produce a mass of sterile workers, and colonial invertebrates are examples of such exceptions. Primate societies are nepotistic, but cooperation is mainly restricted to relatively small kin groups. Theory suggests that reciprocity can be effective in such small groups but not in larger ones. Reciprocity may play some role in nature (though many experts are unconvinced), but there is no evidence that reciprocity has played a role in the evolution of large-scale sociality. All would be well if humans did not exist, because human societies, even those of hunter-gatherers, are based on groups of people linked together into much larger, highly cooperative social systems.

Rapid cultural adaptation potentiates group selection

So why aren't human societies very small in scale, like those of other primates? We believe that the most likely explanation is that rapid cultural adaptation led to a huge increase in the amount of behavioral variation among groups. In other primate species, there is little heritable variation among groups, because natural selection is weak compared with migration. This is why group selection at the level of whole primate groups is not an important evolutionary force. In contrast, there is a great deal of behavioral variation among human groups. Such variation is the *reason* why we have culture—to allow different groups to accumulate different adaptations to a wide range of environments. By itself, such variation is not enough to give rise to group selection. For group selection to be an important force, some process that can maintain variation among groups must also operate. We think that there are at least two such mechanisms: moralistic punishment and conformist bias. Let's see how they work.

Variation is maintained by moralistic punishment

As we explained earlier, moralistic punishment can stabilize a very wide range of behaviors. Imagine a population subdivided into a number of groups. Cultural practices spread between groups either because people migrate, or because they sometimes adopt ideas from neighboring groups. Two alternative, culturally transmitted moral norms exist in the population, norms that are to be enforced by moralistic punishment. Let's call them norm X and norm Y. These could be "must wear a business suit at work" and "must wear a dashiki to work," or "A person owes primary loyalty to kin" and "A person owes primary loyalty to the group." In groups where one of the two norms is common, people who violate the norm are punished. Suppose that people's innate psychology causes them to be biased in favor of norm Y, and therefore Y will tend to spread. Nonetheless, if norm X somehow becomes sufficiently common, the effects of punishment overcome this bias, and people tend to adopt norm X. In such groups, new immigrants whose beliefs differ from the majority (or people who have adopted "foreign" ideas) rapidly learn that their beliefs get them into trouble and adopt the prevailing norm. When more believers in norm Y arrive, they find themselves to be in the minority, rapidly learn the local norms, and maintain norm X despite the fact that it does not fit best with their evolved psychology.

This kind of mechanism only works when the adaptation occurs rapidly, and is not likely to be an important force in genetic evolution. Evolutionary biologists normally think of selection as weak, and although there are many exceptions to this rule, it is a useful generalization. So, for example, if one genotype had a 5% selection advantage over the alternative genotype, this would be thought to be strong selection. Suppose that a novel, group-beneficial genotype has arisen, and that it has, through a chance event, become common in one local group where it has a 5% advantage over the genotype that predominates in the population as a whole. For group selection to be important, the novel type must remain common long enough to spread by group selection, and this is only possible if the migration rate per generation is substantially less than 5%.[30] Otherwise, the effects of migration will swamp the effects of natural selection. But this is not very much migration. The migration rate between neighboring primate groups is on the order of 25% per generation. While migration rates are notoriously difficult to measure, most likely they are typically high among small local groups that suffer frequent extinction. Migration rates between larger groups are much lower, but so, too, will be the extinction rate.

Variation is maintained by conformist social learning

A conformist bias can also maintain variation among groups. We argued in chapter 4 that natural selection can favor a psychological propensity to imitate the common type. This propensity is an evolutionary force that causes common cultural variants to become more common and rare variants to become rarer. If this effect is strong compared with migration, then variation among groups can be maintained.

As before, think of a number of groups linked by migration. Now, however, assume that the two variants affect religious beliefs: "believers" are convinced that moral people are rewarded after death and the wicked suffer horrible punishment for eternity, while "nonbelievers" do not believe in any afterlife. Because they fear the consequences, believers behave better than nonbelievers—more honestly, charitably, and selflessly. As a result, groups in which believers are common are more successful than groups in which nonbelievers are common. People's decision to adopt one cultural variant or the other is only weakly affected by content bias. People do seek comfort, pleasure, and leisure, and this tends to cause them to behave wickedly. However, a desire for comfort also causes thoughtful people to worry about spending an eternity buried in a burning tomb. Since people are uncertain about the existence of an afterlife, they are not strongly biased in favor of one cultural variant or the other. As a result, they are strongly influenced by the cultural variant that is common in their society. People who grow up surrounded by believers choose to believe, while those who grow up among worldly atheists do not.

The difference between moralistic punishment and conformist learning is illustrated by the different answers to the question, given that people have grown up in a devout Christian society, why do they believe in the tenets of the Christian faith? If cultural variation is maintained mainly by moralistic punishment, those who do not adopt Christian beliefs in a devout Christian society are punished by believers, and people who do not punish such heretics (say, by continuing to associate with them) are themselves punished. People adopt the prevalent belief because it yields the highest payoff in readily measurable currencies, inclusive of the cost of being punished. If cultural variation is maintained mainly by conformist transmission and similar cultural mechanisms, young people adopt the tenets of Christianity because such beliefs are widely held, fit with certain content-based biases, and are difficult for individuals to prove or disprove. (Of course, any mixture of conformity and punishment is also possible; the answer is quantitative, not qualitative.)

Conformist transmission can potentiate group selection only if it is strong compared with opposing content biases, and this can occur only if individuals have difficulty evaluating the costs and benefits of alternative cultural variants. In some cases this is not very difficult—should you cheat on your taxes or fake illness to avoid military service? The threat of punitive action may be sufficient to keep taxpayers and conscripts honest. However, many beliefs have effects that are hard to judge. Will children turn out better if they are sternly disciplined or lovingly indulged? Is smoking marijuana harmful to one's health? Is academia a promising career option? These are difficult questions to answer, even with all of the information available to us today. For most people at most times and most places, even more basic questions may be very difficult to answer. Does drinking dirty water cause disease? Can people affect the weather by appeals to the supernatural? The consequences of such difficult choices often have profound effect on people's behavior and their welfare.[31]

Heritable variation between groups + intergroup conflict = group selection

In *On the Origin of Species*, Darwin famously argued that three conditions are necessary for adaptation by natural selection: there must be a "struggle for existence" so that not all individuals survive and reproduce; there must be variation so that some types are more likely to survive and reproduce than others; and the variation must be heritable so that the offspring of survivors resemble their parents.

Darwin usually focused on individuals, but the multilevel selection approach tells us that same three postulates apply to *any* reproducing entity —molecules, genes, and cultural groups. Only the first two conditions are satisfied by most other kinds of animal groups. For example, vervet monkey groups compete with one another, and groups vary in their ability to survive and grow, but—and this is the big but—the causes of group-level variation in competitive ability aren't heritable, so there is no cumulative adaptation.

Once rapid cultural adaptation in human societies gave rise to stable, between-group differences, the stage was set for a variety of selective processes to generate adaptations at the group level. As Darwin said,

> It must not be forgotten that although a high standard of morality gives but a slight or no advantage to each individual man and his children over other men of the same tribe, yet that an increase in the number of well-endowed men and an advancement in the standard of morality will certainly give an immense advantage to one tribe over another. A tribe including many members who, from possessing in a high degree the spirit of patriotism, fidelity, obedience, courage, and sympathy, were always ready to aid one another, and to sacrifice themselves for the common good, would be victorious over most other tribes; and this would be natural selection.[32]

Darwin's is the simplest mechanism: intergroup competition. The spread of the Nuer at the expense of the Dinka discussed in chapter 2 provides a good example. Recall that the Nuer and Dinka are two large ethnic groups living in the southern Sudan. During the nineteenth century, each consisted of a number of politically independent groups. Cultural differences in norms between the two groups meant that the Nuer were able to cooperate in larger groups than the Dinka. The Nuer, who were driven by the desire for more grazing land, attacked and defeated their Dinka neighbors, occupied their territories, and assimilated tens of thousands of Dinka into their communities.

This example illustrates the requirements for cultural group selection by intergroup competition. Contrary to some recent critics,[33] there is no need for groups to be sharply bounded, individual-like entities. The only requirement is that there are persistent cultural differences between groups, and these differences must affect the groups' competitive ability. Winning groups must replace losing groups, but losers need not be killed. The members of losing groups just have to disperse or be assimilated into the victorious group. If losers are resocialized by conformity or punishment, even very high rates of physical migration need not result in the erosion of cultural differences.

This kind of group selection can be a potent force even if groups are usually large. For a group-beneficial cultural variant to spread, it must become common in an initial subpopulation. The rate at which this occurs through random driftlike processes will be slow in sizable groups.[34] However, it only needs to occur once. Several processes might supply the initial variants. Even if groups are usually large, occasional bottlenecks that reduce group size could allow a group-favoring variant to arise by chance. Environmental variation in even a few subpopulations could provide the initial impetus for group selection. Small, deviant groups, if successful, can grow into large ones, as often happens with religious sects. Whatever their source, differences between societies in contact, like those of the Nuer and Dinka, are often quite substantial; we have noted many other examples.

Group competition is common in small-scale societies. Contrary to some romanticized accounts, ethnographic and archaeological data indicate that raiding and warfare are frequent in foraging societies.[35] For example, data collected by pioneering anthropologist A. L. Kroeber and his students during the first half of the last century indicate that warfare was very common among hunter-gatherers in western North America during the nineteenth century, often exceeding four armed conflicts per year. However, the data from hunter-gatherers are far too poor and too influenced by contact with colonial powers to estimate how often such conflicts resulted in group extinction. Better data come from highland New Guinea, which provides the only large sample of simple societies studied by professional anthropologists before these societies experienced major changes due to contact with Europeans. Although they were horticulturalists rather than hunter-gatherers, New Guinea peoples lived in simple tribal societies much as many hunter-gatherers did, and intergroup competition was still ongoing, or at least quite fresh in informants' minds, when ethnographers arrived.

Anthropologist Joseph Soltis assembled data from the reports of early ethnographers from highland New Guinea. Many studies report appreciable intergroup conflict and about half mention cases of social extinction of local groups. Five studies contained enough information to estimate the rates of extinction of neighboring groups (table 6.1). The typical pattern is for groups to be weakened over a period of time by conflict with neighbors and finally to suffer a sharp defeat. When enough members become convinced of the group's vulnerability to further attack, members take shelter with friends and relatives in other groups. The group thus becomes socially extinct, even if mortality rates are well below 100%. At the same time, suc-

Table 6.1 Extinction rates for cultural groups from five regions in New Guinea

Region	Number of groups	Number of social extinctions	Num-ber of years	% groups extinct every 25 years	Source
Mae Enga	14	5	50	17.9%	Meggitt 1977
Maring	13	1	25	7.7%	Vayda 1971
Mendi	9	3	50	16.6%	Ryan 1959
Fore/Usurufa	8–24	1	10	31.2%–10.4%	Berndt 1962
Tor	26	4	40	9.6%	Oosterwal 1961

From Soltis et al. 1995.

cessful groups grow and eventually fission. The social extinction of groups was common (table 6.1). At the these rates of group extinction, it would take between 20 and 40 generations, or 500 to 1,000 years, for an innovation to spread from one group to most of the other local groups.

These results imply that cultural group selection is a relatively slow process. But then, so are the actual rates of increase in political and social sophistication we observe in the historical and archaeological records. New Guinea societies were no doubt actively evolving systems,[36] yet the net increase in their social complexity over those of their Pleistocene ancestors was modest. Change in the cultural traditions that eventually led to large-scale social systems like the ones that we live in proceeded at a modest rate. These estimates can explain the five-thousand-year lag between the beginnings of agriculture and the first primitive city-states, and the five millennia that transpired between the origins of simple states and modern complex societies.

Group-beneficial cultural variants can spread because people imitate successful neighbors

Intergroup competition is not the only mechanism that can lead to the spread of group-beneficial cultural variants—a propensity to imitate successful neighbors can play a role. Up to this point, we have mainly focused on what people know about the behavior of members of their own local group. But people also often know something about the norms that regulate behavior in neighboring groups. They know that we can marry our cousins here, but over there they cannot; or anyone is free to pick fruit here, while individuals own fruit trees there. Now suppose that one set of norms causes people to be more successful than alternative norms. Both theory and empirical evidence suggest that people have a strong tendency to imitate the successful. Consequently, the better norm will spread because people imitate their more-successful neighbors.

You might wonder if this mechanism can really work. It requires enough diffusion between groups so that group-beneficial ideas can spread; and at the same time, there can't be too much diffusion, or the necessary variation between groups won't be maintained. Is this combination possible? We wondered the same thing, so we built a mathematical model of this process. Our results suggest that group-beneficial beliefs spread in a wide range of conditions.[37] The model also suggests that such spread can be rapid. Roughly speaking, it takes about twice as long for a group-beneficial trait to spread from one group to another as it does for an individually beneficial trait to spread within a group. This process is much faster than simple intergroup competition because it depends on the rate at which individuals imitate new strategies, rather than the rate at which groups become extinct.

The rapid spread of Christianity in the Roman Empire may provide an example of this process. Between the death of Christ and the rule of the emperor Constantine, a period of about 260 years, the number of Christians increased from only a handful to somewhere between six million and thirty million people (depending on whose estimate you accept). This sounds like a huge increase, but it turns out that it is equivalent to a 3%–4% annual rise, about the growth rate of the Mormon Church over the last century. According to sociologist Rodney Stark,[38] many Romans converted to Christianity because they were attracted to what they saw as a better quality of life. In pagan society the poor and sick often went without any help at all. In contrast, in the Christian community charity and mutual aid created "a miniature welfare state in an empire which for the most part lacked social services."[39]

Such mutual aid was particularly important during the epidemics that struck the Roman Empire during the late imperial period. Unafflicted pagan Romans refused to help the sick or bury the dead, sometimes leading to anarchy. In Christian communities, strong norms of mutual aid produced solicitous care of the sick, thereby reducing mortality. Both Christian and pagan commentators attribute many conversions to the appeal of such aid. For example, the emperor Julian (who detested Christians) wrote in a letter to one of his priests that pagans need to emulate the virtuous example of the Christians if they wanted to compete for their souls, citing "their moral character even if pretended" and "their benevolence toward strangers."[40] Middle-class women were particularly likely to convert to Christianity, probably because they had higher status and greater marital security within the Christian community. Roman norms allowed concubinage, and married men freely engaged in extramarital affairs. In contrast, Christian norms required faithful monogamy. Pagan widows were required to remarry, and when they did they lost control of all their property. Christian widows could retain property, or, if poor, would be sustained by the church community. Demographic factors were also important in the growth of Christianity. Mutual aid led to substantially lower mortality rates during epidemics, and a norm against infanticide led to substantially higher population growth among Christians.

In order to spread by this mechanism, practices have to be relatively easy to observe and to try out.[41] Evangelizing religions such as Christianity and Islam are at pains to help potential converts learn the new system and to welcome awkward neophytes. Even so, most modern conversions, and presumably ancient ones, are of fellow family members, close friends, and other intimate associates.

Rapid cultural adaptation generates symbolically marked groups

One of the most striking features of human sociality is the symbolic marking of group boundaries.[42] Some symbolic markers are seemingly arbitrary traits, such as distinctive styles of dress or speech, while others are complex ritual systems accompanied by elaborately rationalized ideologies. It is a commonplace that social relations are regulated by norms embedded in a group's sanctified belief system.[43] Even in simple hunting and gathering societies, symbolically marked groups are large. Ethnicity, the canonical example of symbolic marking, is diverse and difficult to define. Ethnicity grades into class, region, religion, gender, profession, and all the myriad systems of symbolic marking humans use to regulate (among other things) the scope of altruistic norms.

Considerable evidence indicates that symbolic marking is not simply a byproduct of a similar cultural heritage. Kids acquire lots of traits from the same adults, and if cultural boundaries were impermeable, akin to species boundaries, this would explain the association between symbolic markers and other traits. For example, if Mexican immigrant kids in California never imitated anyone except ethnic Mexicans and if Anglo Californians were similarly conservative, the persistence of an ethnic boundary would be easy to explain. However, there is a great deal of evidence that ethnic identities are flexible and ethnic boundaries are porous.[44] Chicano kids in California learn good English and adopt many other Anglo customs. Anglo Californians in turn learn at least a few words of Spanish, prefer salsa to ketchup, bash piñatas at birthday parties, and acquire a smattering of other Mexican customs. The movement of people and ideas between groups exists everywhere and will tend to attenuate group differences. Thus, the persistence of existing boundaries and the birth of new ones suggest that other social processes resist the homogenizing effects of migration and the strategic adoption of ethnic identities.

The persistence of marked boundaries may be a consequence of rapid cultural adaptation. First, notice that symbolic marking allows people to identify in-group members. In-group marking serves two purposes. First, the ability to identify in-group members allows selective imitation. When cultural adaptation is rapid, the local population becomes a valuable source of information about what is adaptive in the local environment. It's important to imitate locals and avoid learning from immigrants who bring ideas from elsewhere. Second, the ability to identify in-group members allows selective social interaction. As we have discussed, rapid cultural adaptation can preserve differences in moral norms between groups. Best to interact with people who share the same beliefs about what is right and wrong, what is fair, and what is valuable so as to avoid punishment and reap the rewards of social life. Thus, once reliable symbolic markers exist, selection will favor the psychological propensity to imitate and interact selectively with individuals who share the same symbolic markers.

The second and less obvious step is to see that these same propensities will also create and maintain variation in symbolic marker traits.[45] Suppose that there are two groups; call them red and blue. In each group a different social norm is common, the red norm and the blue norm. Interactions among people who share the same norm are more successful than interactions among people with different norms. For example, suppose that the norm concerns disputes involving property, and people with shared norms resolve property disputes more easily than people whose norms differ. These groups also have two neutral but easily observable marker traits. Perhaps they are dialect variants. Call them red-speak and blue-speak. Suppose red-speak is relatively more common in the red group, and blue-speak in the blue group. Further suppose that people tend to interact with others who share their dialect. Individuals who have the more-common combination of traits, red-norm and red-speak in the red group and blue-norm–

blue-speak in the blue group, are most likely to interact with individuals like themselves. Since they share the same norms, these interactions will be relatively successful. Conversely, individuals with the rare combinations will do worse. As long as cultural adaptation leads to the increase of successful strategies, the red-marked individuals will become more common in the red group and the blue-marked individuals will become more common in the blue group. The real world is obviously much more complicated, but, nonetheless, the same logic should hold. As long as people are predisposed to interact with others who look or sound like themselves, and if that predisposition leads to more-successful social interaction, then markers will tend to become correlated with social groups.

The same basic logic works for markers that allow people to imitate selectively.[46] People who imitate others with the locally more-common marker have a higher probability of acquiring locally advantageous variants. If people imitate both the marker and the behavior of the marked individuals, then individuals with the locally common marker will, on average, be more successful than people with other markers. This will increase the frequency of locally common markers, which in turn means that they become even *better* predictors of whom to imitate. If a sharp environmental gradient or a sharp difference in local norms exists, differences in marker traits will continue to get more extreme until the degree of cultural isolation is sufficient to allow the population to optimize the mean behavior.[47]

Many people think that ethnic markers arise because they allow altruists to recognize other altruists.[48] The problem with this idea is that symbols are easy to fake. Talk is cheap and so is hair dye. Advertising that you are an altruist is a dangerous proposition, because it's so easy for bad guys to signal that they are good guys. If you wear a big *A* on your chest, you are liable to attract false friends who take the benefits of your good heart, returning nothing. Indeed, sociopaths seem to be quite good at simulating good-guy behavior in the pursuit of their predatory schemes.[49] What *can* evolve are markers signaling that you are a member of a group that shares cooperative norms that are enforced by moralistic punishment. Then, behaving altruistically is in your own self-interest, and advertising that you are a member of a moral community does not expose you to merciless exploitation by sociopaths, because the moralists in your community will punish those who victimize you. Wearing the badge of a community whose altruism is protected by moral rules and moralistic punishment supplements cheap talk with a big stick.[50]

Tribal social instincts evolved in social environments
shaped by cultural processes

This new social world, a result of rapid cultural adaptation, drove the evolution of novel social instincts in our lineage. Cultural evolution created cooperative, symbolically marked groups. Such environments favored the evolution of a suite of new social instincts suited to life in such groups, including a psychology which "expects" life to be structured by moral norms and is designed to learn and internalize such norms; new emotions, such as shame and guilt, which increase the chance the norms are followed; and a psychology which "expects" the social world to be divided into symbolically marked groups.[51] Individuals lacking the new social instincts more often violated prevailing norms and experienced adverse selection. They might have suffered ostracism, been denied the benefits of public goods, or lost points in the mating game. Cooperation and group identification in intergroup conflict set up an arms race that drove social evolution to ever greater extremes of in-group cooperation. Eventually, human societies diverged from those of other apes and came to resemble the hunting-gathering societies of the ethnographic record. We think that the evidence suggests that about one hundred thousand years ago, most people lived in tribal-scale societies.[52] These societies were based on in-group cooperation where in-groups of a few hundred to a few thousand people were symbolically marked by language, ritual practices, dress, and the like. Social relations were egalitarian, political power was diffuse, and people were ready to punish transgressions of social norms, even when personal interests were not directly at stake.

But why should selection favor new prosocial motives? People are smart, so shouldn't they just calculate the best mix of cooperation and defection given the risk of punishment? We think the answer is that people aren't smart enough for evolution to trust them with the necessary calculations. For example, there is ample evidence that many creatures, including humans, overweight the present in decision making. For example, most people offered the choice between $1,000 right now and $1,050 tomorrow grab the $1,000. On the other hand, if offered the choice of $1,000 in 30 days or $1,050 in 31 days, most people choose to wait. But this means that when 30 days have passed, people regret their decision. This bias can cause individuals to make decisions that they later regret, because they weigh future costs less in the present than they will weigh the same costs in the future.[53] Now suppose that, as we have hypothesized, cultural evolution leads to a social environment in which noncooperators are subject to punishment by others. In many circumstances the reward for noncooperation can be enjoyed right away, while the cost of punishment will be suffered later; and thus people who overvalue immediate payoffs may fail to cooperate, even though it is in their own interest to do so. If generally cooperative behavior is favored in most social environments, selection may favor genetically transmitted social instincts that predispose people to cooperate and identify within larger social groupings. For example, selection might favor feelings such as guilt that make defection intrinsically costly, because this would bring the costs of defection into the present, where they would be properly compared with the cost of cooperation.

These new tribal social instincts were superimposed onto human psychology without eliminating those that favor friends and kin. Thus, there is an inherent conflict built into human social life. The tribal instincts that support identification and cooperation in large groups are often at odds with selfishness, nepotism, and face-to-face reciprocity. Some people cheat on their taxes, and not everyone pays back the money he borrows. Not everyone who listens to public radio pays her dues. People feel deep loyalty to their kin and friends, but they are also moved by larger loyalties to clan, tribe, class, caste, and nation. Inevitably, conflicts arise. Families are torn apart by civil war. Parents send their children to war (or not) with painfully mixed emotions. Highly cooperative criminal cabals arise to prey upon the production of public goods by larger scale institutions. Elites take advantage of key locations in the fabric of society to extract disproportionate private rewards for their work. The list is endless. The point is that humans suffer these pangs of conflict; most other animals are spared such distress, because they are motivated only by selfishness and nepotism.

Some of our evolutionist friends have complained to us that this story is too complicated. Wouldn't it be simpler to assume that culture is shaped by a psychology adapted to small groups of relatives? Well, maybe. But the same friends almost universally believe an equally complex coevolutionary story about the evolution of the language instinct. The Chomskian principles-and-parameters model of grammar[54] holds that children have special-purpose psychological mechanisms that allow them to rapidly and accurately learn the grammar of the language they hear spoken around them. These mechanisms contain grammatical principles that constrain the range of possible interpretations that children can make of the sentences they hear. However, sufficient free parameters exist to allow children to acquire the whole range of human languages.

These language instincts must have coevolved with culturally transmitted languages in much the same way that we hypothesize that the social instincts coevolved with culturally transmitted social norms. Most likely, the language instincts and the tribal social instincts evolved in concert. Initially, languages must have been acquired using mechanisms not specifically adapted for language learning. This combination created a new and useful form of communication. Those individuals innately prepared to learn a little more protolanguage, or learn it a little faster, would have a richer and more-useful communication system. Then selection could favor still more-specialized language instincts, allowing still richer and more-useful communication, and so on. We think that human social instincts very similarly constrain and bias the kind of societies that we construct, with important details left to be filled in by the local cultural input.[55] When cultural parameters are set, the combination of instincts and culture produces operational social institutions. Human societies everywhere have the same basic flavor, if the comparison is with other apes, say. At the same time, the diversity of human social systems is quite spectacular. Like the language instincts, the social instincts coevolved with such institutions over the last several hundred thousand years.

So much for theory. What is the evidence that such instincts actually exist?

Altruism and empathy

Lots of circumstantial evidence suggest that people are moved by altruistic feelings, which motivate them to help unrelated people even in the absence of rewards and punishments.[56] People give to charity, often anonymously. People risk their own lives to save others in peril. Suicide bombers give their lives to further their cause. People give blood.

The list of examples is long. Long, but not long enough to convince many who are skeptical about human motives. For these people all examples of altruism are really self-interest in disguise. Charity is never anonymous; the right people know who gave what. Heroes get on Letterman. Resources are lavished on the families of suicide bombers. You get a sticker to wear when you give blood. Or, in the words of the bioeconomist Michael Ghiselin, "Scratch an altruist and watch a hypocrite bleed."[57] The possibility of covert selfish motives can never be excluded in these kinds of real-world examples.

In recent years, however, experimental work by psychologists and economists has made it a lot tougher to hang on to dark suspicions about the motives behind good deeds. In these experiments, the possibility of selfish reward is carefully excluded. Nevertheless, people still behave altruistically. Psychologist Daniel Batson thinks that empathy is the key to altruism.[58] Once it is engaged, helping behavior is motivated by a genuinely unselfish desire to relieve the victim's suffering. He doesn't doubt (nor do we!) that egoistic motives are quite important. The question is whether empathy-driven altruism is also important. Batson executed a series of experiments designed to explore the role of empathy in altruistic behavior. Participants were divided into experimental and control groups. Experimenters encouraged an empathetic response in the experimental group by asking them to write an account of the experiment from the point of view of its victim. Controls were asked to view the situation objectively. Then the experimental conditions were manipulated to test whether participants in the empathy condition were more likely to provide aid. In one experiment, for example, "Elaine," a sham participant-victim, was purportedly to suffer a series of ten moderately painful shocks—not a pleasant thing to experience or to witness someone else experience. Some real participants were told they would escape watching Elaine's suffering after two shocks; other participants would purportedly have to observe all ten. Then all real participants were told, just before the shocks to Elaine were to start, that she is unusually sensitive to shocks due to a traumatic childhood experience, and finds them exceedingly uncomfortable. The experimenter expresses concern about this, and offers the real participants the chance to continue the "experiment" in place of Elaine. The shocks will be uncomfortable for them, but not nearly as painful as for Elaine.

Batson reasoned that if helping is motivated by the selfish desire to avoid viewing someone else suffering, the ability to leave after only two shocks should reduce the tendency to offer to take Elaine's place. On the other hand, if participants had a genuine desire to help the victim, even subjects allowed to leave after two shocks should offer to help. In the control, low-empathy condition, difficulty of escape had a dramatic effect on helping, raising the proportion helping from about one in five to about three in five participants offering to take Elaine's place. This suggests that people expected to feel quite unpleasant while watching Elaine's suffering, and they offered to help when this was the most effective way to avoid their own discomfort. In the empathy condition, the difficulty of escape made no significant difference in helping; nearly everyone offered to help. In this case, people's empathy for the victim seemed to be the overriding factor in their response.

Batson also produced evidence that people are motivated by a sincere desire to help, not just a desire to earn self-administered psychological brownie points. In experiments in which the desire to help was aroused and then frustrated because someone else provided the help, participants who saw help provided but didn't have to provide it themselves had the greatest mood increase, and those prevented from helping when no one else provided the help had the lowest mood. Once empathy is engaged, people apparently have a genuinely unselfish desire to help. The attitude seems to be "It's a dirty job, but someone's got to do it." Attitudes like this crop up in the reminiscences of combat soldiers, to take an extreme case. Few veterans are eager for the next fight; they expect the whole experience to be hateful. But they do their duty.

These kinds of experiments did not convince most economists, game theorists, and others in the rational-choice camp. First, psychologists routinely lie to their subjects—Elaine was not really going to be shocked. Since subjects are often drawn from psychology classes and have presumably done the assigned reading, they may not believe what experimenters tell them. Maybe most of Batson's subjects suspected that "Elaine" was the experimenter's confederate. Second, the costs and rewards are vague and hard to measure. Subjects said their mood was elevated, but how do we really know they were telling the truth? Finally, the effects of reciprocity and reputation are not usually carefully controlled. Subjects may expect to meet Elaine again on campus and get some reward for their help. A psychology of altruism may just be a proximal mechanism for forming reciprocal bonds.

Such skepticism led economists to design their own experiments in which these kinds of effects were controlled for. The Dictator Game provides a good example. Participants are recruited to the laboratory, and all are paid a "show-up" fee. Then some participants are given a sum of money, the endowment. Usually this is a modest sum, say, ten dollars, but in some experiments the endowment is much larger. Each participant who receives an endowment is offered the opportunity to give some (or all) of it to a second participant. Participants make their choice and then walk out of the lab with whatever money they have decided to keep. The interaction is totally anonymous. Neither participant ever sees the other or is told anything about the other, and in some experiments even the experimenter does not know what the individual participants do. As to the game's outcome, economic theory makes an unambiguous prediction: selfish, money-maximizing players should keep all the money.

The Dictator Game has been played hundreds of times in many different settings. University students in the United States, Europe, and Japan typically keep about 80% of their endowment and give away 20%. Older nonstudents (aka grown-ups) give much more, sometimes averaging an even split. The Dictator Game has also been played in a number of small-scale, non-Western societies; offers in these societies vary more than offers in Western societies, but even then most participants give some money away.[59] The news couldn't be much worse for the view that people have purely selfish motives.

Moralistic punishment and reward

A great deal of circumstantial evidence also suggests that we are inclined to punish fellow group members who violate social norms, even when such punishment is costly. Road rage is a classic example. Think about how you feel if somebody cuts you off, or makes an illegal left turn in front of you. If you are like most people you get annoyed, perhaps very annoyed, and want to punish the rule breaker, even though you know you'll never see the person again. Or, think about how you feel when someone cuts in line while you wait for a movie. Most people get quite angry, even if they are near the front of the line and are sure to get a good seat. Such emotions can give rise to voluntary, informal punishment of people who break social rules. But in complex societies, it's hard to know whether such punishment plays a significant role in maintaining social norms because police and courts also act to punish rule breakers. Many simple societies lack formal legal institutions, so the only kind of punishment is informal and voluntary. In small-scale societies, considerable ethnographic evidence suggests that moral norms are enforced by punishment.[60]

A series of experiments by economist Ernst Fehr and his coworkers at the University of Zurich provide strong evidence that many people are willing to punish rule breakers, even when it doesn't profit them in any way.[61] One certain-to-be-classic experiment is based on the public-goods game often used by experimental economists. As usual in experimental economics, participants are anonymous and are paid real money. For each round of the game, participants are randomly divided into groups of four, and each participant is given a sum of money that he can keep or contribute to a common pool. The experimenters increase all contributions to the common pool by 40% and divide it equally among all players in the group. If one player contributes $10, for example, the experimenter increases it to $14, and gives $3.50 to each player. Then, new groups of four are formed at random, and they repeat the same procedure. This procedure continues over a series of trials.

In this game players do best, on average, if everyone contributes all his resources to the common pool; but an individual is best off if she contributes nothing while everyone else contributes everything. This selfish individual gets to keep her stake and reaps a share of the rewards from the suckers who do contribute. The participants in Fehr's experiment behaved much like the participants in many previous public-goods games: initially, many participants contributed to the common pool, but over time contributions declined until by the tenth round, participants contributed almost nothing.

But Fehr did not stop there. In another treatment, each round consisted of two stages. The first stage was a public-goods game like the one just described. In the second stage, the contributions of each player in the group were posted (without revealing the player's identity). Then participants could reduce any player's payoff at some cost to themselves. Since groups were randomly re-formed each time period, there was no possibility that punishment could induce a player to behave differently toward the person who behaved punitively. Nonetheless, many participants punished low contributors to the common pool, and as a result contributions rose over time so that by the tenth round, most participants contributed their entire endowment. Postgame interviews indicate that participants were motivated by moral emotions described above, and Fehr reports that some participants were quite angry about the bad behavior of others.

One of the frequent criticisms of these kinds of experiments is that people don't really believe they are playing a one-shot game with strangers; our psychology is simply not set up to deal with this possibility, so we always behave as if our neighbors were watching. Perhaps so, but Fehr's experiment suggests that some of the neighbors watching us take sadistic pleasure in punishing our transgressions, or at least feel obligated to exert considerable effort to punish. Worrying about what unselfishly moralistic neighbors will do is an entirely reasonable precaution for humans! Even if these impulses are really designed for the repeated game in small groups, they nevertheless seem to misfire readily in the anonymous, nonrepeated case. We submit that cultural rules capitalize on this tendency and routinize the misfiring, if misfiring it is.

Unless these experiments are highly misleading, even strangers with whom your will never interact again are liable to be nice to you unless you are not nice to them. Many ordinary things we do depend on this being so. Take travel. Solitary individuals can travel through strange cities and usually come to no harm as long as they behave themselves. We've traveled through Third World cities where our pocket money and personal valuables were worth a small fortune in local terms, and where the police were inefficient and corrupt. We usually had a good time. We remember whispered advice from storekeepers, hotel clerks, and officious matrons when we were inadvertently doing something risky, such as choosing the wrong bar to go into. To take a more-extreme case, recall the video from the August 1998 embassy bombing in Kenya or the 9/11 attack in New York City in which streams of wounded were helped away from the bomb site, often by others nearly as bloody as themselves. Disasters of all kinds yield similar footage: people other than highly trained, paid emergency services personnel will come to your rescue if need be.

Evidence for social instincts relevant to symbolically marked groups

Finally, there is much evidence that symbolic markers of group boundaries motivate important behavior. Tribal instincts cause people to use symbolic markers to define the boundaries of in-groups and establish, for example, who is eligible for empathy, who should excite suspicion, and, in some horrible cases, who should be killed.[62]

Evidence suggests that ethnolinguistic boundaries among foragers are symbolically marked, and that stylistic marks of group membership are highly salient. Anthropologist Polly Wiessner collected arrow points from a number of Kalahari San Bushmen groups, including groups unknown to the !Kung San, the people she studied. Wiessner asked !Kung San men for comments on the distinctive styles.[63] Confronted by unfamiliar arrow points, !Kung men guessed that their makers were very different people from themselves. They reported that they would be alarmed to find these points in their territory, because they certainly would have been lost by people unknown to the !Kung and therefore potentially dangerous. On the other hand, exchange of stylistically familiar beadwork and other valuables within groups is used to build up a notion of the !Kung social universe and to build a web of relationships that link people within the ethnolinguistic unit. In simple band-scale societies like the !Kung, the institutions that link members of a tribe are informal but very important. In a harsh and unpredictable world, succor in times of disaster may often mean the difference between life and death. Using gift exchanges, ceremonial activities, and rules of exogamy to create a large group of trusted friends and affines is an effective form of insurance. These data, together with the appearance of stylistic artifacts at least one hundred thousand years ago, indicate that expressive symbolic displays have been part of human strategies for managing social life for a respectable period.[64]

At the proximal psychological level, the "minimal group" experimental system developed by social psychologist Henri Tajfel provides interesting insights into the cognitive mechanisms involved in the use of symbols to demarcate groups, and the actions people take based on group membership.[65] In social psychological experiments, as in real life, members of groups favor one another and discriminate against out-groups. The social psychologists in Tajfel's tradition were interested in separating the effects of group membership per se from the personal attachments that form in-groups. Social psychologist John Turner, for instance, contrasts two sorts of hypotheses to explain group-oriented behavior.[66] Functional social groups might be composed entirely of networks of individuals that are linked by personal relationships, objective shared fate, or other individual-centered ties. Groups could be a collection of individuals bound together by mutual interpersonal attraction reflecting some degree of functional interdependence and mutual aid. The alternative hypothesis is that identity symbols alone are sufficient to induce humans to accept membership in a group, acting positively toward in-group members and negatively toward out-groups.

In his prototypical experiments, Tajfel told participants that they were participating in a test of aesthetic judgment. They were shown pictures of paintings by Paul Klee and Wassily Kandinsky, and asked to indicate which they preferred. Then the participants were divided into two groups, supposedly on the basis of their art preference, but in fact at random. The participant's task was then to divide a sum of money among members of her own group or the other group. Participants discriminated in favor of the in-group members: people gave more money to people who (supposedly) shared their own preference for Klee or Kandinsky. The most plausible evolutionary interpretation of these results is that people react to symbolic badges of group membership because in the evolutionary past they marked important social units. When experimenters take away any information about the nature of groups, they may expose the "default settings" of in-group psychology. Looked at this way, minimal group experiments suggest that people are well primed to make quick and intuitive judgments about behavior appropriate to life in symbolically marked groups. In the politically complex world outside the lab, where many groupings are potentially salient, people attempt to make sensible decisions about what cues to take seriously in any given circumstance, and socially learned determinants play a role alongside whatever genetic dispositions exist.

Recent field experiments by psychological anthropologist Francisco Gil-White in Mongolia suggest that humans use the same cognitive strategy for classifying ethnic groups as they use for classifying species of plants and animals. Much evidence suggests that people believe that individual members of a given species have important hidden properties in common—essences—and that these essences are transmitted from parents to offspring. These essences are immutable, so for example, if a zebra is transformed so that it looks and behaves exactly like a horse, even small children will insist that it is still a zebra. Because people intuitively believe the essences are important, they readily generalize what they observe about one individual of a species to all members.

Gil-White's experiments suggest that our folk theory of ethnicity is also essentialist. He interviewed Mongols and Kazakhs, the most numerous ethnic groups in the area where he worked, asking them questions designed to see if they thought that Kazakhs possessed inalterable features in common that distinguished them from Mongols. When asked if a Kazakh child adopted at birth and raised by a Mongol mother and father was of Mongol or Kazakh ethnicity, most respondents replied "Kazakh." Neither biologists nor anthropologists regard essentialism as the proper basis for a taxonomy of either species or cultures, but for everyday purposes, it may be sufficient. Gil-White thinks that Kazakhs and Mongols are distinguished mainly by differences in customs that would make everyday intimate interactions unpleasant. Customs of family life, food, hygiene, hospitality, and formality of everyday intercourse differ between the two groups in ways that would make social interactions awkward. For example, polite reserve is the centerpiece of Mongol hospitality, while the Kazakhs take delight in rough teasing, which they fully expect to be reciprocated by their guests. Gil-White, whose first hosts were Mongol, reports that he took several days to adapt to Kazakh teasing even though his own personal style is more in accord with theirs than that of the Mongols. These are the sorts of differences that are likely to arise by rapid cultural evolution and motivate the evolution of a regard for ethnic markers.[67]

We think there can be little doubt that humans give great emotional salience to large, impersonal groups (Protestant Irish, Serb, Jew, German, Hutu, Tutsi, etc.), and under the right circumstances, they undertake desperate deeds on the behalf of such groups. When such group identities become highly salient, individuals in one group will turn their hearts against former friends and neighbors in the other group with appalling frequency. So few Germans went out of their way to protect Jewish friends in Nazi Germany that they are counted as heroes.[68] So few Euro-Americans turned out to aid Japanese-American internees during World War II that the few who did are well remembered by those who benefited. If groups are always built on the foundation of dyadic ties, we would find it hard to explain how loyalties to large and necessarily abstract groups could override the ties of per-

sonal friendship to create the atrocities that too commonly result from ethnocentrism. Even after long periods of relative dormancy, group identity
can make strong claims on our emotions. And there is always the awful possibility that an aggressive out-group may suddenly, for reasons of its own,
target one as belonging to a previously weakly relevant group, as has happened recently to Bosnian Muslims and, in the mid-twentieth century, to
German Jews. Not unlikely, a long history of conflicts between symbolically
marked groups led to the evolution of in-group sentiments that are all too
easily turned to the service of conflict with out-groups. Nonetheless, relatively relaxed relations between different ethnic groups are more common
than genocidal hostility.[69]

The scale of Pleistocene societies is consistent with the social instincts hypothesis

Many in the evolutionary social science community are skeptical that culture has much to do with social emotions such as empathy and ethnocentrism. Instead, they think that the human social instincts evolved in small
foraging groups in which kinship and reciprocity favored the evolution of
cooperative behavior.[70] While variants of this argument are many, we think
the most convincing one goes something like this: Until the spread of agriculture over the last ten thousand years, humans probably lived in relatively small groups. In such a world, ordinary natural selection could favor
psychological mechanisms such as empathy and moralistic anger because
groups were small, and many of the potential recipients of altruism were
kin or members of small, reciprocal social networks. Motives that generated
unconditional altruism toward strangers in large, anonymous modern societies (or in the experimental economics laboratory) were favored during
the period when our social psychology evolved because no interaction in
a small hunter-gatherer group would actually be anonymous. Ties of kinship and reciprocity within groups are stronger than kinship ties among
groups, and as a result, neighboring groups competed for territory or other
resources. If neighboring groups of interrelated families had differences in
dialects, customs, or artifacts on a quite fine scale, selection might favor a
rule: "Be nice to people who talk like you, dress like you, and act like you.
Be suspicious of everyone else." When agriculture made much larger, culturally homogeneous social groups possible, these social emotions gave rise
to tribal-scale social organization. The cultural similarity once characteris-

tic of the small bands came to apply to a much larger group, and the emotions appropriate to the kin group scaled up accordingly. This is another variant of the "big-mistake hypothesis" we discussed in the last chapter. If it is correct, almost everything in modern life—trade, religion, government, and science—is a mistake from the viewpoint of the selfish gene.[71]

The relative plausibility of the tribal social instincts hypothesis and this big-mistake hypothesis depends on the scale of Pleistocene foraging societies. The tribal social instincts hypothesis requires that these societies already had fairly complex social organization in which sizable groups of people shared moral norms and symbolic group makers. The tribal social instincts are an adaptation to tribal social life. In contrast, the big-mistake hypothesis is more plausible if forager societies were considerably smaller. Theory strongly suggests that reciprocity, especially in the production of public goods such as cooperation in warfare and enforcement of moral rules, can only evolve in very small groups,[72] and kin groups are necessarily small given human reproductive biology.

So the question is, what were Pleistocene foraging societies like? Unfortunately, this is a hard question to answer. Ethnographic work gives us a detailed, sometimes quantitative picture of the economy and social organization of contemporary foragers. However, the ethnographic sample of foraging societies is biased toward groups living in unproductive environments like the Kalahari, central Australian deserts, and the Amazonian rain forest. We know from historical accounts, particularly from western North America, that foragers in more-provident environments had more complex social organization than those studied ethnographically.[73] That the spectacular cave art of late Pleistocene Europe is reminiscent of elaborate rituals associated with complex societies[74] provides circumstantial evidence that at least some Pleistocene societies were similarly complex. However, any claims about the nature of the social life of bygone hunter-gatherers should be taken with a grain of salt. Historical accounts are of uncertain quality, and the elderly men and women interviewed by ethnographers in the early twentieth century lived their entire lives in communities that had already been influenced by modern societies. Another problem is that we don't know how to project the ethnographic and historical samples back into the Pleistocene. The climate of the last 11,500 years is warmer, wetter, and much less variable than the climates that prevailed during most of the Middle and Upper Pleistocene.

With these problems in mind, let's try to estimate the range of social organization of late Pleistocene foragers as best we can using descriptions of the foraging societies that persisted into the modern period. The band-level societies of the Great Basin in North America, the Kalahari Desert in South Africa, and the desert of central Australia are among the simplest in the ethnographic and historical record.[75] The Great Basin societies were composed of autonomous family bands with minimal and informal tribal institutions, yet there was generalized propensity to be more cooperative with speakers of one's own and closely related languages. Bands often came together for socializing or for communal enterprises such as rabbit and antelope drives. Thus, even in the simplest foraging societies known, there is significant tribal-scale cooperation. Kinship and friendship may have been sufficient to account for social organization at the band level, but at the tribal level, principles of social organization unique to humans were widespread, consistent with the presence of tribal instincts.

Other band-level societies have marked tribal institutions. For example, the !Kung San of southern Africa have a system of gift exchange (involving artistic productions like those known from the late Pleistocene) that welds the small residential bands into a tribe composed of a much larger number of people.[76] Like a modern nation in miniature, the whole tribe never gathers in one place, but there is normally a clear sense of who belongs to the tribe and who does not. People maintain contacts with members of other bands, because in times of subsistence emergencies, they can call on other members of their tribe living in other bands for permission to forage on their territories or receive emergency aid. Anthropologist Aram Yengoyan suggests that peoples of the desert in central Australia, living in the poorest environments on the continent, have more-elaborate institutions to maintain solidarity with other bands than those living in more-provident environment. Precarious subsistence in the desert means that one often has to appeal to poorly known acquaintances and distant relations for aid.[77]

Tribal institutions in such simple band-level societies are modest. There is no discernable superstructure of government, not even an informal council of influential people. Surrounded by powerful neighbors, the !Kung are not warlike, but within-group rates of violence are quite high, because self-help coercion is the only mechanism for punishing transgressors.[78] The most egalitarian and least politically sophisticated foragers have problems maintaining internal peace and rallying responses to external threats, despite vigorous efforts to maintain friendly ties with as many people as possible.[79] More broadly, however, the great majority of ethnographically known foraging societies make war, and military cooperation was likely an important function of tribal institutions in Pleistocene societies.[80]

At the other end of the spectrum, some ethnographically known foragers lived in complex, hierarchical societies. For example, societies on the
Northwest Coast of North America, such as the famous Kwaikiutl, had
large, permanent settlements, substantial division of labor, hierarchal social
systems, hereditary political ranks, and extensive large-scale warfare—all
characteristics usually associated with agricultural subsistence. Their elaborate art rivals that of the Pleistocene caves, suggesting that Upper Pleistocene hunter-gathers may have had similar sociopolitical sophistication.
While some of this complexity may have arisen in response to trade stimu-
lated by the arrival of Europeans, there is much historical and archaeological evidence for the existence of complex foraging societies in many other
areas.[81] It is quite plausible that the societies of Upper Paleolithic Europe
might have achieved similar complexity. Much as the rich marine resources
of the Northwest Coast supported locally dense populations that created
the population base for complexity, the harvest of migratory big game at favorable sites might also have supported large populations.[82]

In between these extremes, a variety of ethnographically or historically
known foraging societies might be proposed as approximating the central
tendency of the late Pleistocene. Good candidates might be the North
American Plains groups that specialized in big-game hunting. Their environments resemble the cold, semi-arid environments that were more common in the last glacial period, and the focus of the economy on large mammals was probably more like Pleistocene foraging economies than the
plant-focused subsistence strategies of groups like the !Kung. Some historical information is available for Plains societies before the introduction of
the horse in the eighteenth century. Much more is available from the succeeding two or three generations as fur traders established regular contact
with the groups.[83] The Blackfeet came from a purely foraging ancestry, unlike many other Plains tribes of the horse era who were formerly farmer-
hunters. The core of their subsistence was hunting bison. Several families
cooperated to construct traps for the herds and to drive the animals into
them. Successful drives yielded lots of meat, but failures were common.
Likely, unsuccessful groups often had to depend on the generosity of successful ones, motivating bands to maintain tribal-scale affinities for insurance purposes, as do the !Kung and central Australians. Dried meat may
have supported regular rendezvous with other bands on some scale.

Blackfoot warfare was a tribal-scale institution. The Blackfeet fought a
chronic guerrilla war against the Shoshone who emerged from the northern Great Basin to hunt bison. Owing to the limited mobility of pedestrian

hunters, most fights were band-scale raids. Nevertheless, informants who lived as young adults in the prehorse days told an early visitor that fights with two hundred warriors on a side sometimes occurred, a fair fraction of the tribe's total force of warriors. Three subtribes of Blackfeet (Piegans, Bloods, and Blackfeet proper), each composed of several bands, were at peace. During the horse era and perhaps earlier, the Blackfeet were allied with two other tribes, the Gros Ventres and the Sarsis, thus maintaining internal peace on a considerable scale.

Commentators on primitive warfare do not always describe the realm over which peace is maintained,[84] yet the scope and quality of internal peace is, perhaps, a more-important index of the strength of tribal institutions than the size and frequency of wars themselves. Logistics limit the size of war parties among foragers, but the realm of peace can, and commonly does, include more people than could ever be assembled in one place. In societies like the Blackfeet, disputes are solved through self-help violence by aggrieved parties. It is testimony to the strength of tribal instincts and their associated cultural institutions that societies lacking formal leadership do not suffer a Hobbesian collapse of social peace.[85]

Even in the horse days, Blackfeet tribal governance was very informal. Anthropologist Christopher Boehm argues that such egalitarian societies have a reverse dominance hierarchy in which followers control the behavior of leaders.[86] Blackfeet band "leaders," so-called peace chiefs, were typically older men with many horses. Generous rich men who lent horses and food to the poor could earn great respect, and only men whose decisions were sound could maintain this regard. Even at that, chiefs could only guide the emergence of a consensus; they could not coerce followers. Errant chiefs were "replaced" whenever popular sentiment came to favor the opinions of another man. Individual families were free to move to other bands if they were dissatisfied with life in their current band. Moreover, groups of families could split off to form a new band. War chiefs, usually younger men than peace chiefs, were entrepreneurs who organized raids on an ad-hoc basis in quest of horses, captives, and glory. War chiefs were not subordinate to peace chiefs or vice versa.

The horse lent the Blackfeet mobility and brought them a wealth of food, but there was little time for the horse era to affect basic institutions. Thus, the earliest horse-era Blackfeet must have been little more than modestly scaled-up, richer versions of pedestrian big-game hunters, with a little more dominance successfully exercised by richer horse owners. It is quite plausible that the range of latest Pleistocene foraging societies encompassed societies of the complexity of the Blackfeet. Of course, how close to the late Pleistocene central tendency they might have been is more difficult to say.

We read the ethnographic evidence as suggesting that many, if not most, Pleistocene societies were multi-level tribal formations in which small residential bands were nested within a larger society. At the simple end of the spectrum were societies something like the Shoshone and !Kung, in which bands were linked into a weakly organized tribal unit. At the other end of the continuum, tribal societies with sufficient resources—rich fishing or hunting grounds—could grow to several thousand people with the aid of sufficiently sophisticated cultural institutions. For example, Nuer tribes ranged from less than ten thousand to more than forty thousand, and they maintained a modicum of unity on this scale with a highly extended kinship ideology and other modest institutions.[87] Most likely, no Pleistocene societies reached this size. More likely, the modal Pleistocene society living in relatively provident temperate environments was something like the Blackfeet, in which relatively limited tribal institutions organized many hundreds or perhaps a few thousand people to cooperate in subsistence and in warfare. If this argument is correct, the dependence of the big-mistake hypothesis on kin and reciprocity seems insufficient to account for the scale of social organization typical of the late Pleistocene.

Modern institutions are based on tribal social instincts

Adaptationist reasoning usually runs "forward in time"—we predict contemporary behavior from a knowledge of past environments. The recent radical changes in human environments and the inadequacy of the archaeological record make this strategy difficult in the case of human social behavior. However, adaptationist reasoning can also be run "backward"—we can predict past environments from present behavior. In this enterprise, the radical changes in the environment work for us. You can think of the evolution of complex societies in the Holocene as a giant field experiment in which the social instincts adapted to smaller-scale societies are subjected to a wide range of new environmental conditions. How does cultural evolution engineer ancient Rome or modern Los Angeles starting with human raw material originally designed for societies, at most, on the scale of the cattle camps of the southern Sudan? The size, degree of division of labor, and degree of hierarchy and subordination of Rome and Los Angeles are orders of magnitude beyond the range of the most complex foraging societies. If either the big-mistake or tribal instincts hypothesis is correct, the structure of our evolved psychology should have left tracks all over the resulting constructions.

The past ten thousand years have seen a race toward ever larger and more-complex societies. In favorable circumstances, foraging can support fairly large, sedentary, hierarchical societies, but in most environments the social complexity of foragers is limited. Foraging was probably the only option during the Pleistocene, because climates during that epoch were hostile to agriculture—dry, low in atmospheric CO_2, and extremely variable on quite short timescales. The warm, moist, stable climates of the last 11,500 years have made agriculture, and therefore larger, more-complex societies, possible over much of the earth. Once they were possible, the race was on. Larger societies can usually marshal larger military units and defeat smaller societies in military competition. Size allows economies of scale, and division of labor generates greater economic productivity. These also contribute to political and military success, and attract imitators and immigrants. Nuer-Dinka style conquest-absorptions are evident from the beginning of the written historical record. The result was a steady increase in social scale and complexity that continues today.[88]

The increase in the size and complexity of human societies has probably not been accompanied by significant changes in our social instincts. While natural selection can sometimes lead to substantial genetic change in a few thousand years, most biologists think that important changes in complex characters take much longer to assemble. Our innate social psychology is probably that bequeathed to us by our Pleistocene ancestors.

If we are correct, the institutions that foster hierarchy, strong leadership, inegalitarian social relations, and an extensive division of labor in modern societies are built on top of a social "grammar" originally adapted to life in tribal societies. To function, humans construct a social world that resembles the one in which our social instincts evolved. At the same time, a large-scale society cannot function unless people are able to behave in ways that are quite different from what they would be in small-scale tribal societies. Labor must be finely divided. Discipline is important, and leaders must have formal power to command obedience. Large societies require routine, peaceful interactions between unrelated strangers. These requirements necessarily conflict with ancient and tribal social instincts, and thus generate emotional conflict, social disruption, and inefficiency.

Consequently, social innovations that make larger-scale society possible, but at the same time effectively simulate life in a tribal-scale society, will tend to spread. If we assume that the social instincts have changed little if any since the beginning of the Holocene, then the evolutionary job of creating complex societies will have to have been done entirely by institutional "work-arounds" that have alternately taken advantage of and finessed our social instincts. People will prefer such arrangements and will adopt them given a choice. Societies with such institutions will suffer less internal conflict and will, all else being equal, be more effective in competition with other groups. To put the idea a little differently, to the extent possible, institutions buttressed by the ancient and tribal social instincts will be used as building blocks in the evolution of complex societies.

However, these building blocks are not especially well suited to the task. For example, the command and control institutions necessary for large-scale cooperation inevitably generate inequality as those in high positions acquire a disproportionate share of society's rewards. Our social instincts do not prepare us to submit to command or tolerate inequality. As a result, our social institutions should resemble a well-broken-in pair of badly fitting boots. We can walk quite a ways in the institutions of complex societies, but at least some segments of society hurt for the effort.

In the section that follows, we describe what seem to us to be the main work-around mechanisms, and the conflicts, compromises, and modes of failure that each entails.

Command backed up by force is necessary but not sufficient

To make a complex society a going concern, the moralistic punishment of tribal societies has to be supplemented with institutionalized coercion. Otherwise, individuals, organized predatory bands, and classes or castes with special access to means of coercion would entirely expropriate the benefits of cooperation, coordination, and division of labor. However, institutionalized coercion *creates* roles, classes, and subcultures with the power to turn coercion to their own narrow advantage. Social institutions of some sort must police the police so that they will act in the larger interest. Such policing is never perfect and, in the worst cases, can be very poor. That elites always advantage themselves shows that narrow interests, rooted in individual selfishness, kinship, and, often, the tribal solidarity of the elite, exert their predictable influence.

While coercive institutions are common enough, there are two reasons to suspect that they are not, by themselves, sufficient to sustain a complex society. First, the elite class itself must be a complex, cooperative venture. Additionally, the tribal instincts and the institutions built on them often give classes quite a high degree of social solidarity. The importance of the military in the politics of so many countries is an example of how highly organized even a highly coercive institution must be to maintain control of a complex society. Weakly organized coercive elites lead to warlordism, and as we now see in Somalia, Afghanistan, Colombia, Zaire/Congo, and some successor republics to the U.S.S.R., this can lead to near anarchy.

The second problem with pure coercion is that defeated and exploited peoples seldom accept subjugation as a permanent state of affairs without costly protest. The instability of dictatorships is evidence that even highly organized coercion is not sufficient in the long term. Deep feelings of injustice generated by manifestly inequitable social arrangements move people to desperate acts, driving the cost of dominance to levels that cripple societies in the short run and cannot be sustained in the long run.[89] Durable conquests, such as those leading to the modern European national states, Han China, or the Roman Empire, leaven raw coercion with more-prosocial institutions. The Confucian system in China and the Roman legal system in the West were far more sophisticated and group-functional institutions than the highly coercive systems that they replaced.

Hierarchies are segmented

Top-down control is generally exerted through a segmentary hierarchy that is adapted to preserve nearly egalitarian relationships at the face-to-face level. As we have argued, late Pleistocene societies probably linked residential bands into larger ethnolinguistic units that served social functions without much formal political organization. The same principle is used in complex societies to deepen and strengthen the hierarchy of command and control. The trick is to construct a formal nested hierarchy of offices, using various mixtures of ascription and achievement principles to staff the offices. Each level of the hierarchy replicates the structure of a hunting and gathering band. A leader at any level interacts mainly with a few near-equals at the next level down in the system and collaborates with peers across the hierarchy. New leaders are usually recruited from the ranks of subleaders, often tapping informal leaders at that level. Bonds of individual reciprocity and small-group esprit leaven tendencies to arbitrary authority deriving from status in the larger hierarchy. Even high-ranking leaders in modern hierarchies typically adopt much of the humble headman's deferential approach to leadership.[90] Charismatic individuals such as Bill Clinton have a gift for reducing their subjective distance from people far beneath them in the official chain of command. As Max Weber so famously argued, bureaucratic institutions attempt by training, symbolic means, and legalistic regulations to routinize charisma in order to legitimize the command-and-control system.[91]

The imperfect fit of institutions and social instincts often makes segmentary hierarchies painfully inefficient. Selfishness and nepotism—corrupt sergeants, incompetent aristocrats, vainglorious generals, power-hungry bureaucrats—degrade the effectiveness of social organizations. Leaders in complex societies must convey orders downward, not just seek consensus among their comrades. Only very careful attention to detail can make subordinates responsive to leaders without destroying the illusion that the same arrangements would have arisen by egalitarian consensus. The chain of command is necessarily long in large complex societies, and remote leaders are not normally able to exercise personal charisma over a mass of subordinates. Devolving substantial leadership responsibility to subleaders far down the chain of command is necessary to create small-scale leaders with face-to-face legitimacy. However, delegation potentially generates friction if lower-level leaders have different objectives than the upper leadership or are seen by followers as helpless pawns of remote superiors. Stratification often creates rigid boundaries so that natural leaders are denied promotion above a certain level, resulting in inefficient use of human resources and a fertile source of resentment to fuel social discontent.

In-group symbols create a sense of solidarity in complex social systems

In complex societies, high population density, division of labor, and improved communication give rise to symbolic systems adapted to simulate the badges and rituals of tribal membership, sometimes on a huge scale, as in modern nationalism.[92] The development of monumental architecture in which to stage mass ritual performances is one of the oldest archaeological markers of complex societies. Usually an established religious organization supports a complex society's institutions. At the same time, complex societies make use of the symbolic in-group instinct to delimit a diverse array of culturally defined subgroups, within which a good deal of cooperation is routinely achieved. Military organizations generally mark a set of middle-level, tribal-scale units with conspicuous badges of membership. A squad or platoon's solidarity can rest on bonds of reciprocity reinforced by prosocial leadership, but ship's companies, regiments, and divisions are made real by symbolic marking. These kind of ethnic group–like sentiments are most strongly reinforced in units that number between one thousand and ten thousand men (British and German regiments, U.S. divisions), groups on the same scale as the tribal societies from which we believe our tribal instincts evolved.[93] In civilian life, symbolically marked units include regions, tribal institutions, ethnic diasporas, castes, large economic enterprises, religions, civic organizations, and, of course, universities.[94]

The evolutionary properties of symbolically marked subgroups give rise to many problems and conflicts in complex societies. Marked subgroups often have enough tribal cohesion to organize at the expense of the larger social system, as when lower-level military units arrange informal truces with the enemy or ideologies of elite superiority support highly exploitative institutions. "Special interests" organize to warp policy in directions favoring their ideology or material well-being. Charismatic innovators regularly launch new belief and prestige systems, which sometimes make radical claims on the allegiance of new members, make large claims at the expense of existing institutions, and grow explosively. The worldwide growth of fundamentalist faiths that challenge the institutions of modern states is a contemporary example.[95] On the other hand, larger loyalties can arise for better or worse, as in the case of modern nationalism and Islam.

Societies often have legitimate institutions that command broad support

At their most functional, institutions create the sense that laws and customs are fair. Rationally administered bureaucracies, lively markets, protection of socially beneficial property rights, widespread participation in public affairs, and the like often combine to provide public and private goods efficiently, and preserve individual liberties and village-scale autonomy to a certain degree. Individuals in modern societies often feel part of culturally labeled tribal-scale groups, such as local political parties, that have influence up through a hierarchy on the remotest leaders. In older complex societies, village councils, local notables, tribal chieftains, or religious leaders often hold courts open to humble petitioners, and these local leaders in turn represented their communities to higher authorities. As long as most individuals feel that existing institutions are reasonably legitimate and that reform can be achieved through ordinary political activities, considerable scope exists for collective social action, including deliberate evolution of new social institutions.

On the other hand, the many unavoidable flaws in the evolving institutions of complex societies make legitimacy a difficult thing to sustain. Individuals who do not accept the legitimacy of the current institutional order are liable to band together in resistance organizations, such as the contemporary fundamentalist and tribal groups that view secular modernism as illegitimate. Stubbornly tribal people such as the Pathans of Afghanistan and Pakistan have effectively resisted incorporation into larger social systems for millennia. Trust varies considerably in complex societies, and variation in trust is the main cause of differences in happiness across societies.[96] Even the most efficient legitimate institutions are prey to manipulation by small-scale organizations and cabals, the so-called special interests of modern democracies.[97]

Conclusion: Coevolution weaves cultural and genetic causes into a single cloth

The main point of this chapter is that the cultural part of the gene-culture coevolutionary processes has played an important role in the evolution of human social institutions. In the short run, cultural evolution, partly driven by ancient and tribal social instincts and partly by selection among cultur-ally variable groups, gave rise to the institutions we observe. In the longer run, cultural evolutionary processes created an environment that led to the evolution of uniquely human social instincts.

This hypothesis provides a theoretically coherent account of the evolu-tion of complex human societies, and is consistent with much empirical evidence. It explains the undeniable elements of functional design in hu-man social institutions *and* the manifest crudity of complex societies in the same theoretical framework. Without the ancient social instincts, we can't explain the many features of our social systems that we share with other primates. Without the tribal social instincts, we can't explain why our so-cieties are so different from those of other primates, the emotional salience of tribal-scale human groups, or their importance in social organization and social conflict. The social instincts of both sorts, acting as biases shap-ing evolution of social institutions, account for the peculiar form of hu-man societies, for the timescales over which institutions evolve, and for the patterns of conflict that routinely plague human societies. The institutions of complex societies are manifestly built on ancient and tribal instincts and have predictable imperfections deriving from cultural evolutionary processes.

While we are quite proud of this hypothesis, we know that it skips lightly over many details. Surely, for example, future discoveries will even-tually yield a better picture of how cultural and genetic processes are in-tegrated in the brain. Social psychologists will be able to tell us how this integration plays out in the everyday social interactions that are the foundation of social institutions. Sociologists, anthropologists, and historians will better map out how our evolved psychology, acting through ongoing cultural evolution, generates the actual social institutions we observe. Nonetheless, we believe a better explanation will retain a number of the essential elements of the hypothesis given here. In particular, it will (1) synthesize genetic and cultural causes, (2) be an evolutionary explanation, and (3) explain the intricate mix of function and dysfunctional conflict in human societies.

Chapter Seven

Nothing About Culture Makes Sense Except in the Light of Evolution

> Nothing in biology makes sense except in the light of evolution.
>
> —Theodosius Dobzhansky, 1973

When Dobzhansky penned our epigraph in the 1970s, relatively few biologists devoted themselves to the study of evolution, and today evolutionary biologists are vastly outnumbered by molecular biologists, physiologists, developmental biologists, ecologists, and all the rest. Nonetheless, evolution plays a central role in biology, because it provides answers to *why* questions. Why do humans have big brains? Why do horses walk on the tips of their toes? Why do female spotted hyenas dominate males? The answers to these questions draw on all parts of biology. To explain why horses walk on their toes, we need to connect the ecology of Miocene grasslands, the developmental biology of the vertebrate limb, the genetics of quantitative characters, the molecular biology and biophysics of keratin, and much more. Because evolution provides the ultimate explanation for why organisms are the way they are, it is the center of a web of biological explanation that links the work of all the other areas of biology into a single, satisfying, explanatory framework. As Dobzhanzky put it, without the light of evolution, biology "becomes a pile of sundry facts, some of them interesting or curious but making no meaningful picture as a whole."[1]

We believe that evolution can play the same role in explaining human culture. The ultimate explanation for cultural phenomena lies in understanding the genetic *and* cultural evolutionary processes that generate them. Genetic evolution is important because culture is deeply intertwined with other parts of human biology. The ways we think, the ways we learn, and the ways we feel shape culture, affecting which cultural variants are learned, remembered, and taught, and thus which variants persist and spread. Parents love their own children more than those of siblings or friends, and this must be part of the explanation for why marriage systems persist. But *why* do people value their own children more than others? Obviously, an important part of the answer is that such feelings were favored by natural selection in our evolutionary past.

Cultural evolution is also important for understanding the nature of culture. Because culture is transmitted, it is subject to natural selection. Some cultural variants persist and spread because they cause their bearers to be more likely to survive and be imitated. The answer to why mothers and fathers send their sons off to war is probably that social groups having norms that encourage such behavior outcompete groups that do not have such norms.

Finally, genetic and cultural evolution interact in complex ways. We saw that social psychologists and experimental economists, working from very different research traditions, have produced compelling evidence that people have prosocial predispositions that cause us to act altruistically. But *why* do we have such predispositions in the first place? Evolutionary theory and the lack of large-scale cooperation in other primates suggest that selection directly on genes is unlikely to produce such predispositions. So, why did they evolve? We think cultural evolutionary processes constructed a social environment that caused individual natural selection to favor empathetic altruism. Our specific explanation may be in error; you seldom get it straight on the first try. The important point is that evolving culture, certainly in theory and probably in fact, has a fundamentally important role in shaping our species.

Is dual inheritance theory the proper theory of cultural evolution?

Of course, to agree that evolutionary theory is a valuable tool in the human sciences is not necessarily to agree that the approach we propose is the right one. Karl Popper, the famous philosopher of science, said that science trades only in conjectures not (yet) refuted. But some issues cease to be debated because the evidence becomes so overwhelming. In our lifetimes, the propositions that genes are DNA and that seafloor spreading causes continental drift have passed from doubtful speculations to textbook conventions. Will the Darwinian theory of cultural evolution be one of those currently controversial ideas that become standard textbook fare in the early twenty-first century? In this chapter, we gather the threads of the case we have laid out in this book to allow readers to answer this question for themselves. We are, of course, partisans of this endeavor, but we hope that the fair-minded skeptic will find the evidence strong and the issues worth pursuing.

Evolutionary biologist E. O. Wilson recently revived the notion of "consilience,"[2] introduced by the nineteenth-century polymath William Whewell. The idea, which was a favorite of Darwin's, holds that seemingly disparate phenomena in the world are in fact connected. For instance, nuclear physics is "remote" scientifically from the social sciences, yet nuclear reactions in the sun are the most important source of energy on earth; nuclear decay in the earth's interior drives seafloor spreading, which in turn shapes terrestrial ecology; and nuclear weapons profoundly altered the shape of international politics. Nematologists will remind you that if the rest of the biosphere suddenly disappeared, nematodes would trace out a ghostly outline of it all. Nothing, then, is in principle irrelevant to the study of the human species. Since this is so, scientific theories are vulnerable to disproof in all the realms of phenomena where they apply.

Evolutionary theories apply to highly consilient phenomena. You will have noticed that our examples have sprawled across a considerable territory. Let us remap the territory in terms of five sorts of investigations where evolutionary theory is vulnerable: logical coherence, investigations of proximate mechanisms, microevolutionary studies, macroevolutionary studies, and patterns of adaptation and maladaptation. This is just a taxonomy of convenience, but most evolutionary investigations fall into one category or another. It is a useful device for depicting the wide-ranging consilience of evolutionary phenomena. Any given evolutionary hypothesis can usually fail in several if not all of these domains.

Logical consistency

We have devoted a lot of effort to making mathematical models of cultural processes. Although we have spared you the details here, the models play a very important role in our story, because they ensure that our arguments are deductively sound.[3] Critics of mathematical models often recoil at their simplicity, yet simple models are an effective prosthesis for a mind that is poor at following intricate, quantitative causal pathways—tools to help us think a little more clearly about complex problems. Without such models we would be forced to rely entirely upon verbal arguments and intuitions whose logical consistency is difficult to check.

Mathematical models stand behind all our explanations of cultural evolution and gene-culture coevolution. In chapter 3, we argued that a style of modeling borrowed from population biology, with suitable modification to reflect the very real differences between culture and genes, can be used to test the logical cogency of cultural evolutionary hypotheses. In chapter 4, we described several models that investigate the basic adaptive properties of cultural transmission, leading to the hypothesis that culture was initially an adaptation to variable environments. In chapter 5, we sketched the results of models that show how adaptive cultural mechanisms systematically lead to the spread of maladaptive cultural variants. Finally, in chapter 6, we outlined models of cultural group selection, a process that might explain our quite unusual and phenomenally successful social systems. These models may be wrong, but they are (probably) deductively sound.

Proximate mechanisms

In chapter 4, we described evidence comparing social learning in humans and other animals. While many animals have rudimentary capacities for social learning, these are uniquely hypertrophied in humans. In late infancy, a suite of behaviors emerges in humans that make us very efficient imitators compared to any other animal. These capacities might underlie language, though the dominant school of linguists insists that language learning is a special-purpose capacity. Regardless of these controversial details, humans are clearly capable of transmitting vast quantities of information by imitation, instruction, and verbal communication. Humans have the capacity to form a large cultural repertoire, and the evidence surveyed in chapter 2 shows that much of our extraordinary behavioral variation stems from differences in cultural traditions. Human populations are characterized by durable traditions that result in different behaviors even in the same environments.

Two other plausible mechanisms explain variation in human behavior among groups: genetic differences and individual adaptation to environmental differences. Genetic differences cannot be very important, as borne out in the most direct data bearing on this issue, the results of cross-cultural adoptions. The evidence indicates that children raised by parents of another culture behave like the members of their adoptive culture, not their natal culture, in all important respects. Until a few thousand years ago, all humans lived in quite simple societies. Since then, most of us have come to live in much more complex ones, albeit some of us much more recently than others. Human behavior, under the influence of evolving cultural traditions, can change enormously without any appreciable genetic evolution. Whatever average innate differences might exist between human populations, they must be small compared to cultural differences.[4]

The importance of individual behavioral versus cultural adaptation to local environments is a more difficult issue. Humans are adaptable and inventive creatures, no doubt. However, if individual behavioral adaptation to local conditions is the primary force generating behavioral differences between groups, then people living in the same environment should all behave in more or less the same way, but we know they often don't. Farmers with Lutheran German, Anabaptist German, and Yankee roots living side by side in the American Midwest behave quite differently, confirming that cultural tradition often has a powerful impact on behavior.

Microevolution

In chapter 3, we built a case for culture being an evolutionary phenomenon susceptible to analysis using Darwinian tools. The heart of Darwinism is the close study of evolutionary processes on the generation-to-generation timescale that allows precise observation and controlled experiment. Such microevolutionary studies contrast with macroevolutionary investigations of the grand results of evolution on timescales of tens of generations and longer. Macroevolutionists normally have to work without the benefit of direct observation and experiment and must rely upon the scrappy fossil record and comparative study of extant forms. Most cultural change is relatively gradual, and is apparently the result of modest innovations spreading by diffusion from their point of origin to other places. Such patterns were well documented by anthropologists in the nineteenth century. In the twentieth century, "diffusionism" fell into disrepute for being atheoretical and merely descriptive.

A Darwinian theory provides the tools needed to analyze the process of invention and diffusion in a rigorous way. Cultural evolution is a population phenomenon. Individuals invent, and they observe the behavior of others. Imitation by discriminating observers selectively retains and spreads innovations which in turn accumulate and eventually yield complex technology and social organization. Darwin described such patterns of change as "descent with modification." The theoretical and empirical tools designed by evolutionary biologists to study genes are well suited to describing cultural evolution *given suitable modification.* The examples we use to illustrate most points about cultural evolutionary processes are micro-evolutionary. For example, in chapter 5 we reviewed evidence that several different processes, operating against the background of the expanding influence of nonparental relative to parental transmission of culture, have successively come to influence attitudes toward family size and family-planning technology.

Macroevolution

Understanding what regulates the rate of evolution in different times and places is one of the main tasks of macroevolutionary studies and one that is none-too-well advanced, even in biology. The large-scale and comparative evidence suggest that cultural evolution has an important role to play in understanding the major events in human evolution. In chapter 4, we reviewed the basic adaptive properties of the cultural system of information transmission. The theoretical models tell us that a system of social learning is likely to have been favored initially as an adaptation to variable environments. Paleoclimatologists tell us that the environments of the last couple of million years have become highly variable on timescales that our models suggest ought to strongly favor a cultural animal. We can even make a stab at guessing why only humans have this adaptation. In chapter 6, we proposed a hypothesis, based on models of cultural group selection, to explain the highly unusual form of human social organization. We explained how the coevolution of genes and culture could create innate psychological dispositions that could never evolve by genes alone.

The macroevolutionary record is a stern test of explanatory hypotheses because the explanation has to get the timescale right. For example, the emergence of complex societies over the last five thousand years can't have been the result of genetic change, because it happened too fast. On the other hand, it happened far too slowly to be explained by purely individual adaptation, be it by rational choice or any other individual-level psychological process. Some factor that has just the right amount of historical inertia is required to explain the moderately rapid growth of social complexity over the last five thousand years. Cultural traditions change on the appropriate timescale, adding credence to the theory. The next question is, can we ferret out what kinds of traditions are the rate-limiting step in such progressive sequences? Many scholars argue that the rate of evolution of social institutions is the rate-limiting step due to the difficulty of observing foreign social institutions and the difficulty of experimenting with any novel institution.[5]

Patterns of adaptation and maladaptation

Humans adapt quickly and efficiently to variable environments using technology, and they evolve variable, often complex, social institutions producing unusual amounts of cooperation, coordination, and division of labor. Much of the diversity of human behavior in time and space results from adaptive microevolutionary processes shaping complexes of technology and social organization that suit us to live in most of the terrestrial and littoral habitats on earth. Other organisms must speciate in order to occupy novel environments, whereas humans rely mostly upon culture. Modern humans apparently have spread out of Africa to the rest of the world in the last one hundred thousand years, relying on their ability to generate complex cultural adaptations suited to virtually every habitat on earth.[6]

Cultural maladaptations are a more pointed test of our approach. The same is true of Darwinian theory generally. Although divine creation accounted for *adaptation,* Darwin's theory has the edge in that it also accounts for vestigial organs and other maladaptations. Maladaptations are plausible byproducts of a messy natural process of descent with modification, but are an embarrassment to the work of an omniscient Creator. Contemporary population geneticists have discovered interesting organic maladaptations that result from the peculiarities of the genetic inheritance system.

Darwinian models of cultural evolution make specific predictions about classes of maladaptations that we should observe with fair frequency. In chapter 5, we presented the case that selfish cultural variants should be reasonably common. The existence of many adaptive cultural traits and the costs of evaluating the utility of different ideas put the sophisticated social learner on the horns of a dilemma. Impressionable observers risk imitating poorly adapted cultural variants, while conservative observers may miss out on valuable new techniques and social arrangements. The human cultural psychology seems adapted to balance these costs and opportunities. We have various forms of "fast and frugal" transmission biases that give us a good chance of sweeping up good ideas and rejecting bad ones. The chance that such biases will find a better variant to favor goes up with the number of models surveyed. But the tendency for selfish cultural variants to be favored by natural selection increases as the influence of nonparental models increases. A certain frequency of maladaptation inevitably results from design tradeoffs confronting an advanced cultural creature.

We presented evidence in chapter 5 that some very common features of human societies, such as the modern transition to low fertility, are caused by selfish cultural variants. Modern societies, by vastly enlarging the scope for nonparental transmission, have also magnified the chance of choosing maladaptive memes. On the one hand, modern technology and social organizations produce a cornucopia of adaptations. On the other, we face a barrage of well-advertised innovations that have the net effect of causing birthrates to plummet in modern societies, thereby reversing the normal correlation of economic and reproductive success. Anabaptist societies show how relentlessly discriminating a culture must be to adopt modern innovations that increase economic efficiency but still retain traditional cultural values that maintain birthrates at fitness-maximizing levels.

In chapter 6, we discussed cultural group selection, another engine for generating maladaptations from the narrow genes'-eye point of view. Human societies are crude superorganisms. One of the human species' main social *adaptations* is the ability to organize cooperation, coordination, and a division of labor on a much larger scale than the typical primate kin group. Yet cultural group selection remains in conflict with selection on genes that may continue to favor only small-scale, family-oriented, and reciprocal cooperation. The dilemma of cooperation exists at the evolutionary level as well as the personal. Selection on genes can't favor large-scale cooperation even if every individual is on average better off if they cooperate. Even taking into account coevolutionary selection pressures for more-docile genes, selection on genes still tends to favor people who look out for themselves, their families, and their partners in significant measure. Human social institutions, particularly those of the really large-scale societies of the past five thousand years, have developed work-arounds to resolve the inevitable conflicts built into our social psychology.

The debate over whether culture is adaptive, maladaptive, or just neutral has gone on for a century. The theory outlined here predicts what the empirical evidence tells us—culture is sometimes adaptive, sometimes maladaptive, and sometimes neutral. It adds the nuance that what is maladaptive from the gene's-eye point of view may result from selection acting on cultural variation. Then, genes adapt secondarily to a world with culturally evolved institutions, so that genes come to support cultural adaptations. In a broader sense, human genes have also on average benefited from cultural adaptations even though natural selection directly on genes never favored large-scale cooperation! The soap opera messiness of human life accords well with the idea that multilevel selection has built conflict into our instincts and our institutions.

The Darwinian theory of cultural evolution and gene-culture coevolution does not fail in any of the five domains, while theories that invoke only genes and individual decision-making have problems with every one. The tracks of culture are all over human behavior.

We need a synthetic theory of human behavior

Consider for a moment how biology is taught to undergraduates. Although students know that biology is composed of many subdisciplines—ecology, molecular biology, genetics, and so forth—it is taught as an integrated subject up through the first course in college. Good instructors take care to present the unifying themes of biology—genetics, basic metabolic principles, and evolutionary processes. They do not do this because they value a general education for its own sake. Rather, they know that all of these levels of organization are linked in a causal web. Biology has many subdisciplines, yet the boundaries between them, and between biology and the other natural sciences, are porous. Some of the most creative scientific work is done by harnessing findings or methods in one field to problems posed in others. The classic example is the importation of chemistry into biology to create a succession of new disciplines—physiology, biochemistry, and molecular biology. Moreover, many of the early molecular biologists were actually trained in physics.[7] Later, Richard Lewontin's pioneering application of biochemists' methods for studying molecular variation almost at the level of the gene[8] led to his discovery that a surprising number of gene loci are polymorphic. This finding attracted a generation of students to the problems of evolution at the level of genes and launched the still-vibrant field of molecular evolution.

When we first began exploring the social sciences, we were struck by how isolated they are from one another, as well as from the natural sciences. The problem begins with educational traditions. Psychology, sociology, economics, linguistics, history, and political science all teach proprietary first-year introductory courses. Students are encouraged to think that the study of humans can be divided into isolated chunks corresponding to these historical disciplines. Why is there no *Homo sapiens* 1 course based on the model of Bio 1, a complete introduction to the whole problem of understanding human behavior? Even in anthropology, where students in traditional programs typically take introductory courses in biological, sociocultural, archaeological and linguistics subfields, efforts to link the subfields are limited (and have become more unfashionable in recent decades).[9]

One reason, perhaps, is that the key integrative fields have not yet developed in the social sciences. If so, a proper evolutionary theory of culture should make a major contribution to the unification of the social sciences. Not only does it allow a smooth integration of the human sciences with the rest of biology; it also provides a framework for linking the human sciences to one another. Much of human psychology is concerned with acquiring and managing culturally acquired information, and the variation in psychology among different groups of people is mainly a cultural phenomenon. The rational-choice disciplines of economics and game theory need theories of constraints and preferences, many of which are cultural in origin. Anthropology, sociology, political science, linguistics, and history have long relied on cultural explanations to account for changes in human behavior and to explain diversity. In this book, we have drawn upon empirical work from all of these disciplines to understand the nature of cultural evolution. We have advanced cultural evolutionary hypotheses to explain interesting phenomena that social scientists have documented, such as the surprising reversal of the correlation between wealth and reproductive success that has gradually spread from society to society over the last two centuries. We don't expect all of these hypotheses to stand the test of time; perhaps none will. Our use of this immense and valuable body of data, we hope, illustrates the relevance of the social sciences to evolutionary questions.

We also hope we have demonstrated to your satisfaction how cultural-evolutionary analyses integrate data from disparate disciplines and schools within the human sciences. Several questions that have excited enormous controversy in the social sciences seem to us to have natural resolutions in the evolutionary framework.

Methodological individualism versus methodological collectivism

The social sciences have long been bedeviled by the "micro-macro problem."[10] If, like economists, you start with a theory based on individual behavior, how can you ever get to a proper account of society-scale phenomena like social institutions? If you start with collective institutions, like many sociologists and anthropologists do, how do you make room for individuals? A distinguished sociologist once astounded us with the claim that it had been *proved* that you had to pick one or the other and that it was a logical certainty that the two approaches could never be unified.

Actually, Darwinian concepts provide a neat account of the relations between individual and collective phenomena. Darwinian tools were *invented* to integrate levels. The basic biological theory includes genes, individuals, and populations. In these models, what happens to *individuals* (for example, natural selection) affects the *population's* properties (for example, the frequencies of genes), even as individuals are the prisoners of the gene pool they draw upon. Many other links between individuals and the populations they live in are possible, and the addition of culture creates still more. We have considered examples such as conformist transmission, where the frequency of a cultural variant, a *population* property, affects its probability of being imitated by *individuals*. Darwinian tools help us build linkages between phenomena at different levels as given problems require. Individuals seem to be hapless prisoners of their institutions because, in the short run, individual decisions don't have much effect on institutions. But, in the long run, accumulated over many decisions, individual decisions have a profound effect on institutions. Evolutionary theory gets right the basic structure of the relationship between individuals and the collective properties of their societies.

History versus science

Historians and historically minded social scientists sometimes argue that the actual evolution of social institutions and the like is produced by a myriad of concrete events peculiar to a particular place and time. Generalizations or hypotheses derived from general models like those used in economics or psychology add nothing to the history of these concrete events, and are often positively misleading because they focus attention on a-priori concerns to the detriment of understanding the actual events of the case at hand.

Historical contingency is as important in the biology of other organisms as it is in our own species. Every species is unique, after all, and derives from the highly contingent events of its evolutionary history. The convergences of plants and animals on similar adaptations in similar but isolated environments are often striking, but equally striking differences remain. The Darwinian theoretical tool box furnishes bits of canned logical analysis applicable to such phenomena. Our empirical methods are similarly tuned in the first instance to the accurate depiction of concrete historical trajectories and the local causal processes that drive them. Students of a particular case should sort through the tool box to select the apt tools for the problem at hand. In the event some models do prove to apply to many cases, and empirical generalizations sometimes have great power. Hamilton's theory of inclusive fitness turns out to apply very broadly; cooperation in animal societies is almost always organized along family lines, although the diversity within that generalization is certainly considerable. Inclusive fitness theory itself accounts for much but by no means all of this diversity.[11] Humans are a partial exception to Hamilton's generalizations, and we showed how a theory of cultural group selection might explain our exceptional level of cooperation. The cultural group selection model is in the same spirit as Hamilton's, with a sharp tweak to fit our unique case. We thus submit that our model building and the kinds of empirical studies we champion are acutely sensitive to the details of the human case. Everything from evolutionary biology has to be rethought in the light of the massive importance of culture in our species, leading to a tool box specifically tailored for the unique features of human evolution.

Models of modestly general applicability and empirical generalizations of modest scope are extremely valuable for two reasons. First, individuals are quite stupid compared to the complexity of the problems we aspire to solve. Well-studied models and well-tested empirical generalizations embody the collective wisdom of one's fellow scientists. An isolated individual thinker has no chance against a problem of any complexity. As teachers we know, for example, that even the simplest population-level process, exponential growth, flummoxes the untutored. Second, most concrete cases are so complex that no one investigator can hope to study in detail every dimension of the problem. In actual historical investigations, many important processes and events will not enter the record at all, and the problem is necessarily simplified, often drastically, by the investigator. Empirical generalizations and theories help to make this inevitable simplification transparent. Sensible evolutionists know they have left out much and know that any conclusions they reach are vulnerable as a consequence. All anyone can hope to do is to make canny simplifications that do minimum damage to understanding.

Ironically, the evolutionary tool box helps explain why historical contingency plays the large role that it does. For example, evolutionary game theory shows how easily multiple evolutionarily stable strategies arise even in rather simple games. For example, in the standard model of reciprocity, the iterated prisoner's dilemma, *any* behavior from never cooperate to always cooperate and everything in between is favored by selection once it becomes common enough. Historians have everything to gain and nothing to lose by using appropriate evolutionary tools for their job.

Functional versus symbolic elements of culture

The relationship between the functional and symbolic elements of culture is a bit intricate, but by no means intractable. Human scientists interested in the symbolic aspects of culture sometimes claim that symbolic considerations rule out functional interpretations of culture.[12] Some evolutionary functionalists claim that a strict separation exists between stylistic elements, like the decorations on a pot, that evolve by random processes, and functional elements, like its size and shape, that evolve by selection.[13] Evolutionary analyses confirm[14] what some social scientists have claimed for a long time:[15] stylistic differences have functions even when the precise form of a style has no function. The pot's decoration may serve to advertise its maker's group membership or status within the group.[16] Evolutionary theory and some good data suggest that symbols are used as badges of group membership, as badges of roles within groups, and as the means to assert personal status. Stylistic displays often convey useful information to potential imitators.[17] On the other hand, the evolution by the runaway process can generate maladaptive exaggeration of style. We considered how status-motivated consumption races may play a role in the demographic transition.

Function and dysfunction

The sources of human happiness and human misery are evolutionary. Take social institutions as an example. Some simple societies lack effective systems of dispute resolution, whereas others have quite effective ones.[18] Levels of trust, happiness, and satisfaction with life differ greatly within western European countries, quite independently of per-capita wealth.[19] People evidently find some sets of social institutions more congenial than others. Since individual decision-making and collective decision-making institutions act as forces in cultural evolution, we may be said to affect our own evolution. However, we are also the prisoners of the culture and genes we inherit.

Aggregating individual decisions to make collective ones is a formidable problem in theory and in practice.[20] In our discussion of the work-arounds that make complex societies possible, we took pains to point out that each functional work-around has its evil twin; emphasizing one element at the expense of the other is a recipe for error. Utopians meet defeat after defeat in attempts to persuade people to slip their chains, and attempts at revolution often fall victim to a combination of impossible dreams and cabals of the selfish, vicious, and power hungry. On the other hand, corrupt regimes must be repressive because they always face resistance from altruistically motivated moralists advocating reform. Societies that are unwilling or unable to change subject their people to much the same vices as failed revolutions. Low-trust societies controlled by authoritarian political institutions look much the same no matter their origins. The modern evolution of technology shows that the rate of evolution can be enormously accelerated, in largely desirable directions, if things such as property-rights institutions are favorable.[21] The evolution of social institutions is the tougher nut to crack, but the capacity of open political systems to build the interpersonal trust that in turn serves as the basis for desirable innovations in social arrangements is fairly impressive. No doubt, if we understood the nature of social evolution better, we could improve the process.

If our general argument is correct, the reason that these classic problems led to intractable debates rather than scientific progress is simply that Darwinian concepts and methods are appropriate to the problems of organic and cultural evolution. Without these tools, you just cannot think straight about problems involving cultural evolution, and problems of cultural evolution are fundamental to understanding human behavior.

The theory is an engine for generating new questions

From the scientist's point of view, the most important function of a scientific theory is productivity. Does it point research in a useful direction? Does it create more new and interesting problems than it solves? A sociologist once remarked to us that Darwinian theories of cultural evolution looked to him like conventional social science done with a different slant. That we have been able to use much of conventional social science to make a case for the theory lends weight to this critique. But cultural evolutionists advocate adding new evolutionary tools, not carrying on as always.

Many cultural scientists of our acquaintance are infected with a certain ennui. They seem to feel that the late "Great Men" of their fields, Marx, Weber, Durkheim, Parsons, and so on, said most of what can be said about the human condition. Contemporary scholars can mine thin ore that the Great Men passed over; they can slice, dice, and recombine old arguments to get interesting but not very novel new variants; or they can abandon science entirely for personalized accounts of human behavior in the humanistic vein. We believe that social scientists should not be discouraged. We really know very little about how cultural evolution works. Some of you may have concluded that this is because cultural evolution is beyond scientific understanding, at least of the sort we advocate. But we believe that thinking about culture using Darwinian tools opens many new avenues for investigation.

Our knowledge of the basic patterns of cultural variation is grossly incomplete, and understanding patterns is often the key to understanding process. While we have argued that many patterns of variation in human behavior are inconsistent with genetic and environmental explanations and quite consistent with cultural ones, high-quality, systematic studies are very few. Most descriptions of cultural variation are qualitative rather than quantitative. While the ethnographic record is a splendid body of knowledge, the study of the processes of cultural evolution needs more precise description. Some studies based on qualitative data are rather sophisticated,[22] but many opportunities to do better work exist. We need to characterize cultural variation in the same quantitative detail as genetic variation. Recent work in cross-cultural psychology[23] and in the use of economic games to investigate norms of fairness cross-culturally[24] will open a new era of quantitative ethnography that will revolutionize our understanding of human behavioral variation.

Cultural variation in time is also poorly quantified. Archaeologists and historians have very clearly documented cultural change in the long run. However, their impulse is usually to attempt to reconstruct the societies that lived in the past. An inherently simpler task is to use the best parts of the sketchy records available to estimate rates of change. Often, inferences about evolutionary processes imply quite different rates of evolution, and the archaeological and historical records are the best places to test these inferences. For example, the rise of literacy should allow increased rates of evolution by creating a form of memory less limited and less prone to error than that in human brains.[25] Impressionistically, over the last five thousand years, evolutionary rates do seem to accelerate with the development and expansion of literacy. Would such a hypothesis withstand quantitative test? Do other processes or variables have a comparable impact?

The evolutionary processes that operate on culture are poorly understood. In this book, we have used a taxonomy of evolutionary forces acting on cultural variation that was developed in our previous book.[26] We are partial to this taxonomy, but it is surely incomplete. The trend in evolutionary biology has been to subdivide general categories of evolutionary processes into many distinctive subtypes, usually because the dynamic behavior of populations under their influence is distinctive. In chapter 4, we introduced the concept of imitating the successful and one of its subtypes, the imitation of those with prestige. But prestige is itself a complex social construction. Some prestige derives from personal charisma, some from institutionalized office. Some kinds of prestige may be recognized by nearly everyone in a society, whereas other forms may be highly local. We have no idea how many distinct varieties of prestige-based selective imitation there might be. We have little doubt that cultural evolution a complex and diverse set of phenomena, though we can only dimly imagine complexity from our present vantage point.

The quantitative roles of the various forces in concrete cases of evolution are scarcely known. In selecting studies to include in this book to illustrate the processes of cultural evolution, we have usually been reduced to examples where a single process, such as natural selection or one of the decision-making forces, is arguably dominant. In general, several forces are liable to simultaneously affect the evolution of any given bit of culture we choose to focus on. For example, innate, learned, and culturally acquired dispositions, often acting in different directions, are liable to simultaneously affect whether certain religious beliefs or innovations increase or decrease in frequency. Much of evolutionary science can be boiled down to estimating the strength of various effects on the trajectory of evolution in a sufficiently large number of cases to obtain some empirical generalizations. The gold-standard study of organic evolution is one in which the investigator estimates the strength of natural selection and other forces in an evolving population.[27] In the case of culture, such studies are still very few.[28]

Conclusion: Nothing about culture makes sense except in the light of evolution

In 1982, the pioneering evolutionary economists Richard Nelson and Sidney Winter remarked that among the interesting intellectual challenges in their discipline, "certainly none is more worthy of attention than that of understanding the great complex of cumulative change in technology and economic organization that has transformed the human situation in the last few centuries."[29] Historians and sociologists would nominate the rise of complex societies beginning five millennia ago and their subsequent development as another paramount question. Anthropologists would nominate the origins of agriculture eleven millennia ago and paleoanthropologists the origins of modern humans that culminated with the first complex cultural systems some one hundred or more millennia ago. At the other end of the spectrum, political scientists would nominate the emergence of new political institutions and public policies, and how these rule systems affect political and economic development on the timescale of a few election cycles. What contemporary humans *are* is a product of such past and ongoing evolutionary events.

Evolutionary processes are thus at the crux of the most interesting questions about our species. How do we find ourselves in the early twenty-first century in the particular state we are in? The cultural evolutionary events of the centuries that came before have everything to do with that. Why do we have the social predispositions that we do? The coevolution of genes and culture over a million or more years has much to do with that. Can we influence the current evolution of human societies in desirable directions? As humans, we are unusually active agents in our own evolution, because we each choose which cultural variants to adopt and which to neglect.[30] Moreover, we organize institutions ranging from a simple tribal council to highly complex modern ones, such as the research university and the political party, that are designed to direct the course of cultural evolution.[31] Yet, cultural evolution is a very big dog on the end of our leash. Even cultural heroes leading great political movements typically have modest effects. Gandhi could not prevent the Muslims from leaving India, nor could he persuade Hindus to reform the caste system. Only by attending properly to the population-level processes can we arrive at a proper picture of cultural evolution. With a reasonable picture of cultural evolution in hand, we could begin to understand how we might humanize processes that often exact savage costs in the currency of human misery.

In this book, we have made the case for using Darwinian methods to understand cultural evolution. Culture is stored in populations, so understanding human brains and how populations change requires population thinking. Darwinian accounts are one part bookkeeping—a quantitative description of cultural variation and its change through time. In addition, they are one part quantitative budget analysis—a systematic attribution of changes to causal processes. If you are going to study cultural evolution in a serious way, you are going to be driven to Darwinian methods of analysis. You have to be able to describe change and you have to be able to account for change. Several research programs in social sciences have independently converged on the Darwinian methods. The sociolinguists' microevolutionary studies of dialect evolution are a particularly sophisticated example; elsewhere we note others.[32]

Our own particular analyses may be maladroit. Borrowing tools from biology and remodeling them for culture has the attraction of capitalizing on the sophistication of evolutionary biology, but it may well introduce distortions. What is more, we have just argued that the Darwinian work to date is at best seriously incomplete. We make no apology for this. Science is an error prone, one-step-at-a-time procedure, and the story shall remain incomplete for a long time if not forever.[33] The only thing about the project that we care to assert with utter conviction is that the Darwinian approach is *worth pursuing.*[34] Those who engage in the pursuit will take proper delight in remedying our generation's errors and omissions!

Much of the objection to applying Darwinian tools to the human case seems to come from a visceral dislike of picturing us as just "another unique species."[35] From the evolutionist's point of view, human exceptionalism is a major problem. As long as humans stand outside the Darwinian synthesis, as long as human culture is said to be superorganic, the whole Darwinian project has a potentially fatal gap. Darwin feared that attacks on the *Descent of Man* would be used as a platform for attacks on the whole edifice of his theory. In this he was not disappointed. As the *Quarterly Review*'s commentator, probably the long hostile and devoutly Catholic St. George Mivart, gloated, the *Descent* "offers a good opportunity for reviewing his whole position" (and rejecting it).[36] The modern secular Science Wars critics evolved from the superorganic version of human exceptionalism that we critiqued in chapter 1, and their objection to science being applied to humans has generally come to be accompanied by a hostility toward science in general. Of course, the religious version persists, too, in fundamentalist circles. Doc Watson sings, "Man came from monkey, so some folks say, but the Good Book don't quite tell it that way. If you believe the monkey business, some people do, then I'd rather be that monkey's brother than you."[37] If humans are outside the bounds of science, then no doubt other things are, too. Science is *bound by its charter* to pursue explanations of human evolution!

Darwinians generally feel more bemused than beleaguered by their critics. Scientists very commonly have humanistic interests. They paint, read novels, write history. So many older scientists try their hand at philosophy that it can practically be regarded as a normal sign of aging. Many are politically active. On the religious side, most scientists will admit to a belief in a god if a sufficiently broad definition is used.[38] Far from feeling a conflict between their science, their religion, and their humanistic impulses, most scientists find their science suffused with the beautiful and the sublime.[39]

Darwin ended *On the Origin of Species* with a lyrical paragraph reading in part as follows:

> It is interesting to contemplate an entangled bank, clothed with many plants of many kinds, with birds singing on the bushes, with various insects flitting about, and with worms crawling through the damp earth, and to reflect that these elaborately constructed forms, so different from each other in so complex a manner, have all been produced by laws acting around us. . . . There is grandeur in this view of life, with its several powers, having originally breathed into a few forms or into one; and that, whilst this planet has gone on cycling on according to the fixed law of gravity, from so simple a beginning endless forms most beautiful and most wonderful have been, and are being, evolved.

Scientific methods are a lot like Zen meditation—arduous and exacting practices that allow the practitioner to win some lovely, if fragile and fallible, truths, eyeball to eyeball with the great mystery. Scratch many a scientist, and a nature mystic bleeds. We feel so about our subject. Peoples and their cultures are wondrous and diverse. The study of human diversity highlights how much humanity we share with the most exotic of our fellows. Darwin believed that anyone whose heart had not been hardened by some specious ideology would feel sympathy for the sufferings of any other human. His description of his feelings about slavery, aroused by his experience of Brazil's treatment of slaves, is the most passionate passage he ever wrote.[40] On the other hand, cultural differences are profound and profoundly interesting. We don't subscribe to an extreme form of cultural relativism (Nazism, after all, was not quaint German folklore). However, the anthropologists' practice of refusing the easy pleasures of ethnocentrism in favor of reserving judgment about other societies—at least until you understand them well—has much to recommend it. Stubbornly anachronistic peoples such as the Anabaptists and the Nuer command respect—even admiration. Though few of us would care to join such societies, we can understand why those brought up in them are proud and successful human beings.

Mathematical models are, as we have said, deliberately shorn of all the rich detail that makes people themselves so interesting. Foolish indeed are the mathematical modelers who confuse their abstractions with reality. But when used properly, mathematics schools our intuition in ways that no other technique can. It is a form of meditation upon nature without peer.

We are constantly struck by the way our naive intuitions are confounded and then rebuilt along new lines by the results of models. Bit by bit, models can be used to dissect the logic of complex systems. The sharp contrast between the difficulty of making good models and their manifest simplicity compared to the phenomena they seek to understand is a humbling, even spiritual, experience. We followed the development of adding social learning to individual learning in simple evolutionary models in chapter 4. We saw that Alan Rogers's very simple model in which social learning evolved without being adaptive led to some real insights into exactly what properties are needed for culture to be adaptive. Good models produce diamond-clear deductive insights into the logic of evolutionary processes. The aesthetic dimension of models is something their critics, unfortunately, never experience. Modelers love a well-designed, well-analyzed representation, as with other artifacts whose beauty lies in their elegant minimalist functionality. We experience when teaching how taking up a nice, old model after a length of time brings on a nice, warm feeling. When it comes to subject areas like evolution, you cannot think straight without them, just like you can't hike for long over rough ground without a good pair of boots. You don't have to *be* a modeler to appreciate models. Much like in any other art form, educated connoisseurs can get a lot out of them.

A good set of data also is a beautiful thing to behold. Foolish, of course, is the empiricist who thinks that even the most beautiful set of data captures any complex phenomenon completely, especially one who thinks that the data from his own case applies without exception to a diverse system such as human culture.[41] However, data are the ultimate arbiter. More than just testing hypotheses, data often start us thinking in the first place. The great pioneer of mathematical population genetics, J. B. S. Haldane, said, "the world is not only queerer than we suppose, but queerer than we can suppose."[42] In chapter 2, we reviewed beautiful studies documenting the existence of cultural variation. Many scholars poke fun at cultural explanations for their supposed lack of sophistication, and argue cogently that innate information, rational calculation, and ecological variation are quite plausible alternatives to cultural explanations. In any given case, perhaps such alternatives are correct, but as general arguments against culture, the empirical data are clear enough. Cultural scientists have developed a considerable body of elegantly compelling, even if largely qualitative, data. The importance of cultural variation in the human species is hardly more dubious than role of gravity in the motions of the planets. As with models, the empirical picture gets built bit by bit, gradually constraining the range of plausible explanations with ever better data.

Some data are so sublime they completely transform our picture of the world in a most surprising way. Data from ice and ocean cores collected over the last decade document the extreme variability of climate during the last ice age, giving us a stunningly surprising picture of the sort of world in which our cultural system arose. We barely dared to imagine that such data would come to light, even though our models suggested that such variability is a plausible engine driving the evolution of our capacities for culture. More surprises in both past and future climates are virtually a certainty.[43] The world is so complex that without sound empirical data the theorists are blind. Those who claim to study unquantifiable complexity are being unreasonable, for quantifying is precisely what we do when things get complicated.

With that thought, we rest our case.

Notes

Chapter One

1. Nisbett and Cohen 1996.

2. Nisbett and Cohen's analysis is restricted to European Southerners.

3. Mayr 1982, 45–47.

4. People have used the word *culture* for a variety of phenomena with many dimensions and varieties. There is widespread agreement that culture in the sense of a body of socially transmitted traditions and practices is important. However, there has been no consensus about how to conceptualize it, or whether this conception of culture is sufficient to explain the phenomena. While definitions of culture such as ours, which emphasizes its individual/psychological aspects, are well known in anthropology, many other kinds of definitions exist (Kroeber and Kluckhohn 1952; Fox and King 2002). We don't think arguing about whether our definition or some other is the "correct" definition of *culture* is worth much effort. Complex natural phenomena such as culture are exceedingly difficult to capture with simple definitions, and quarrelling over which of the many sensible definitions is best does not seem to us a useful exercise. Rather, the question should be, does it generate useful theory?

5. Today an important fraction of culture is stored in written (and electronic, film, etc.) form (Donald 1991), and some has probably always been carried in the form of artifacts of various kinds. This fact has no doubt substantially affected cultural evolution during the last few thousand years.

6. Kroeber 1948, 62.

7. This idea goes back to the pioneers of sociology and anthropology at the turn of the twentieth century. Ingold 1986, 223, discusses three different senses of "superorganic" used

by social scientists over the years, about which he summarizes, "the superorganic has become a banner of convenience under which have paraded anthropological and sociological philosophies of the most diverse kinds."

8. Indeed, anthropologists long interpreted much of culture in adaptive terms (e.g., Steward 1955).

9. Alexander 1974, 1979; Wilson 1975; Symons 1979; Chagnon and Irons 1979; and Barash 1977. See Segerstråle 2000 for a history of the controversy.

10. See Laland and Brown 2002 for a judicious comparative review of the several research traditions, including the one we work from, that together comprise evolutionary social science.

11. Roughly 0.05% of live births in the United States show some form of hand or arm reduction, and some fraction of these cases may be due to exposure to environmental factors (Centers for Disease Control 1993). About 0.2% of live births have more than five digits on either the hands or feet. Many cases of polydactyly seem to be caused by rare mutant alleles.

12. We mean the role of environment in shaping development. Environment plays an ultimate role in natural selection.

13. The distinction between proximate and ultimate causation is Ernst Mayr's 1961.

14. Richard Alexander 1979, 75–81, is quite clear on this point. Evolutionary thinkers disagree about the specificity of these psychological mechanisms. Human behavioral ecologists tend to hold that the psychological mechanisms are what cause humans to act, to a decent first approximation, as general-purpose genetic fitness maximizers. Culture, as defined here, has a strictly secondary role, and for most practical purposes it can be neglected (Smith, Borgerhoff Mulder, and Hill 2001). Many evolutionary psychologists are nativists who believe that the mind has a large collection of rather narrowly specialized, gene-based, content-rich algorithms that can solve a series of narrow problems that had confronted Pleistocene foragers. Contemporary environments have changed so radically that it is vain to hope that behavior will be fitness maximizing today. Evolution is too slow to have readapted the human mind significantly in the last few thousand years (Tooby and Cosmides 1992).

15. The most famous critique is Gould and Lewontin's 1979. The alternatives put forward by such critics have not enjoyed much success (e.g., Carroll 1997) compared, say, to the Darwinism of Campbell 1965, Dawkins 1989, Dennett 1995, Cziko 1995, and Sober and Wilson 1998.

16. Distinguished evolutionists have emphasized both dimensions of culture. Richard Dawkins 1976 coined the term *meme* to emphasize its the genelike properties of culture, while Richard Alexander 1979, 75–78, emphasizes how much culture has in common with individual learning.

17. The climate of the last 11,500 years has been strikingly uniform compared to the previous 100,000. It is unclear whether all warm interglacial episodes have been similar.

18. Edward O. Wilson (Lumsden and Wilson 1981; Wilson 1998) differs from most other evolutionists we are discussing in thinking that culture is very important. He nevertheless thinks that in the final analysis, genetic "leashes" make it possible to reduce culture to ultimate genetic imperatives.

19. Darwin 1874.

20. Galef 1996.

21. Alland 1985; Richerson 1988.

22. Hodgson 2004; Richards 1987; Richerson and Boyd 2001a.

23. Atran 2001; Aunger 1994; Boyer 1998; Cavalli-Sforza and Feldman 1981; Durham 1991; Bowles and Gintis 1998; Gil-White 2001; Henrich and Boyd 1998; Henrich and Gil-White 2001; Henrich 2001; McElreath, Boyd, and Richerson 2003; McElreath, in press; Lumsden and Wilson 1981; Pulliam and Dunford 1980; Sperber 1996.

Chapter Two

1. Some economists have embraced population models as a mechanism for bringing culture into economic theory (Bowles 2004; Schotter and Sopher, 2003).

2. Alexander 1979, 30.

3. Buss 1999, 407.   4. Betzig 1997, 17.

5. Odling-Smee et al. (2003) argue that many organisms in effect construct their environments. Humans are an extreme case. Londoners enjoy an elaborate and efficient urban rail transport network while Angelenos enjoy a highly developed freeway system, both mainly constructed by past generations. Such "niche constructions" clearly influence behavior in important ways.

6. For example J. T. Bonner (1980) wrote a book, *The Evolution of Culture in Animals*, that is mostly about phenotypic flexibility in general, not about social transmission of behaviors.

7. Salamon 1985, 329.

8. Salamon 1984, 334.

9. Salamon 1984, 1980; Salamon and O'Reilly 1979; Salamon, Gegenbacher, and Penas 1986.

10. Kelly 1985.

11. Glickman 1972.

12. Edgerton 1971. This work was part of a larger project initiated and planned by Walter Goldschmidt.

13. Steward 1955.

14. Edgerton 1971, 271. Even at the supratribal level of Bantu (Hehe, Kamba) and Kalenjin (Pokot, Sebei), there are significant effects of tribal history. Several of the differences are related to economically important variables, and should therefore reflect environment according to the environmental hypothesis. Kalenjin are more confident of military success, but less interested in land ownership and in industriousness than Bantu.

15. McElreath, in press.

16. Paciotti 2002.

17. See Knauft 1993 for an excellent example of the diversity that anthropologists routinely find among peoples within culture areas where both environment and general cultural background are similar.

18. Greeley and McCready 1975.

19. Putnam, Leonardi, and Nanetti 1993.

20. Hofstede 1980.

21. LeVine 1966.

22. Finney 1972; Epstein 1968; Pospisil 1978.

23. Harris 1979.   26. Boehm 1983.

24. Rogers 1983.   27. Brooke 1994.

25. Handelman 1995.   28. Salamon 1984.

29. Gil-White 2001.

30. Ruth Benedict 1934 and Margaret Mead 1935 were the most famous spokespersons for this hypothesis in psychological anthropology. Mussen et al. 1969 represent developmental psychology in the period when a version of the Boasian hypothesis was extremely influential.

31. Eaves, Martin, and Eysenck 1989.

32. The genes and IQ debate of the 1960s and '70s sensitized behavioral geneticists to the flaws of early studies of heritable factors in human behavior. In an effort to correct flaws in the more-primitive investigations, the modern studies are based on large samples, and the analysis is sophisticated. The focus shifted from IQ to a broader array of traits, especially personality traits, that developmental psychologists have suggested should depend strongly on the home "environment" (meaning also cultural effects).

33. Feldman and Lewontin 1975; Feldman and Otto 1997.

34. Labov 1973.

35. Scarr 1981.

36. Lydens's unfortunately unpublished Ph.D. dissertation 1988 has a good literature review. See also Andujo 1988 and Altstein and Simon 1991.

37. Our discussion is taken mainly from Hallowell 1963 and Heard 1973.

38. Indian "adoptees" of Anglo Americans in the frontier period were much less successful because of the heavy burden of racism of that time. Probably few or no Indians were fully adopted in this period. Indians raised at boarding schools typically showed highly conflicted ethnic identities.s

39. Gibbs and Grant 1987.

40. The beak of the medium ground finch (*Geospiza fortis*), the species in question, is about 20% smaller than that of its congener, the large ground finch (*Geospiza magnirostris*). The medium ground finch is about 75% lighter. Grant calculates that it would have taken between thirty-six and forty years for the medium ground finch to achieve the beak size of the large ground finch at the rate observed during the 1976 drought.

41. Roe 1955; Oliver 1962.

42. Tooby and Cosmides 1992, 115–16.

43. Gallistel 1990.

44. See Hirschfeld and Gelman 1994.

45. Boyer 1994.

46. Atran et al. 1999; Atran 1990.

47. Bickerton 1990 argues that new languages called creoles arose in slave communities when the children's main linguistic input was an almost syntax-free pidgin. The complex syntax of the creole arises, according to Bickerton, from children's evolved language learning machinery. Other linguists believe that the syntax mainly comes from other languages that the children's parents speak. See for example Thomason and Kaufman 1988.

48. Richerson, Boyd, and Bettinger 2001.

49. Pinker 1997, 209.

50. For a similar view, see Tooby and Cosmides 1992, 119–20.

51. Gould 1977.

52. Boyd and Richerson 1996.

53. Labov 1994, 2001.

54. Nisbett and Ross 1980; Tversky and Kahneman 1974; Tooby and Cosmides 1992; Simon 1979; Gigerenzer and Goldstein 1996. These authors do not agree on the significance and interpretation of human cognitive limitations, but they all agree that the accuracy and comprehensiveness of individual human decision making is strictly limited.

55. Some readers, such as Sperber 1996, have misunderstood the weight that Darwinian cultural evolutionists, beginning with psychologist Donald Campbell's pioneering contributions (1960, 1965, 1975), have put on psychological forces *in a population context.*

56. Basalla 1988.

57. Keller 1931, practically the only early to mid twentieth-century social scientist who is a Darwinian in any deep sense, went so far in his disparagement of the efforts of individual innovators as to discount them entirely.

58. Petroski 1992.   60. Needham 1979.

59. Sobel 1995.   61. Needham 1987.

62. Iannaccone 1994; Finke and Stark 1992; Marty and Appleby 1991.

63. Brooke 1994.

64. Diamond 1978, 1997. See also Henrich 2004.

Chapter Three

1. Attributed to Charles Anderson Dana (1819–97), longtime editor of the New York *Sun* and one of the creators of the modern American newspaper. Dana is the source of another quotation which we admire: "Fight for your opinions, but do not believe that they contain the whole truth, or the only truth."

2. Burrow (1966) provides a classic account. See also Richerson and Boyd 2001a.

3. White 1949; Sahlins, Harding, and Service 1960.

4. Steward 1955; Sahlins, Harding, and Service 1960; Harris 1979.

5. Johnson and Earle 2000; Carneiro 2003.

6. See Sahlins, Harding, and Service 1960 and Steward 1955 for two approaches to dealing simultaneously with the complexity of the evolution of particular traditions and the general trend. For an authoritative modern treatment of this kind of evolutionism, see Johnson and Earle 2000.

7. E.g., Cohen 1977 on the origin of agriculture. Harris 1977, 1979 and Johnson and Earle 2000 make population pressure the engine of cultural evolution.

8. Richerson, Boyd, and Bettinger 2001; Richerson and Boyd 2001c.

9. Blurton-Jones and Konner 1976.

10. Merlin Donald 1991 takes the invention of literacy and similar information technologies to be one of the major revolutions in the origins of the modern mind, greatly in-

creasing the accuracy and volume of information that individuals can access. We don't mean to discount information technology! The pioneering evolutionary economists Richard Nelson and Sidney Winter 1982 used firms as their unit of analysis and the routines of firms as the unit of culture. In the fifth chapter of their book they give the best discussion we know of on the means by which culture can be carried outside individuals' heads.

11. See Griffiths 1997 and Wierzbicka 1992 for the case that the scientific study of emotions has been handicapped by culture-specific concepts. Richard Nisbett 2003 presents considerable evidence that Asians think quite differently from Americans.

12. See Baum 1994 for an evolutionarily sophisticated version of behaviorism and Pinker 1997 for the cognitivist approach.

13. Gallistel 1990 on mental representations, Churchland 1989 on why not.

14. Jackendoff 1990, commentary on Pinker and Bloom 1990.

15. To be more precise, social learning is itself a concept with several subconcepts, only some of which would support imitation-based culture in the human sense (not to say that humans don't sometime use simpler sorts of social learning). See Galef 1988 for an introduction to these complexities.

16. The use of such toy models for didactic purposes is a common practice in some disciplines (e.g., economics, evolutionary biology) but not others (e.g., anthropology, history).

17. Atran 2001; Boyer 1998; Sperber 1996.

18. Salamon 1992, 172.

19. The intent of the distinction between transmission and forces is analytical, not ontological. It is often convenient to assume that there is one step in the life history in which perfect transmission occurs, followed by another step in which the mind applies biases to select among perfectly learned cultural variants. The staged life history is a trick to simplify the structure and analysis of models of evolution that are borrowed from evolutionary biology. The facts may be quite different; the bias may be applied at the point of learning to distort the cultural variant as it is learned. Under most conditions, slight differences in structure don't affect the outcome of models, so we claim that the step-structure approach is usually an innocent simplification. In theory, and no doubt occasionally in practice, there will be cases where a more-realistic psychology of transmission is absolutely necessary. A failure to distinguish between tactical analytical simplifications and truth claims has led a number of unwary critics of this theory to unwarranted conclusions. For example, it may seem "reductionistic" to analyze the no doubt very complex events occurring in farming communities in the Midwest in terms of two cultural variants and two forces. We claim no more than that the highly simplified picture we present is a tolerably good first approximation to that complex phenomenon. Additional variants and forces would be necessary to explain even Salamon's data, much less all the facts of the case, assuming (counterfactually) that they could all be put on the table. In truth, no empirical or theoretical study can manage more than a modest fraction of all the processes ongoing in particular cases of evolution. One is stuck with a choice among alternative simple models (or simple experimental designs) and between doing analysis and practicing mysticism. At least in favorable cases, a few things do dominate the evolutionary process and our analysis leads to great insights. We hasten to add we imply no objection to mysticism. Many excellent "hard" scientists become mystics after two beers; Darwin's last paragraph of the *Origin* is a first-class example.

In unfavorable cases there is little to do but be in awe of the complexity of the tangled bank. In Richerson and Boyd 1987 we outline and defend the simple-models-of-complex-phenomena strategy employed by evolutionary biologists, economists, and engineers, among many others.

20. For a technical discussion see Boyd and Richerson 1985, chap. 5.

21. Ryan and Gross 1943. Rogers 1983 surveys this literature, counting 3,085 studies from 10 different disciplines as of that date.

22. Rogers with Shoemaker 1971 showed that perceived advantage was one of the commonest effects in studies of the diffusion of innovations. This book did a primitive meta-analysis of some fifteen hundred diffusion-of-innovation studies. Henrich 2001 shows how a quantitative analysis of such adoption data can be used to estimate the influence of the various forces of evolution.

23. Wiessner and Tumu 1998; Yen 1974. See Crosby 1972, 1986 for a discussion of the rapid spread of many New World crops in the Old World following the voyages of Columbus, and of Old World plants and animals in the New.

24. Labov 1994 discusses the principles internal to the structure of language that help drive linguistic evolution.

25. Durham 1991.

26. Lindblom 1986, 1996.

27. Alexander 1979; Lumsden and Wilson 1981.

28. As Melvin Thorpe, the fictional governor of Texas in *The Best Little Whorehouse in Texas,* puts it: "Ooo . . . I love to dance the little sidestep. Now they see me now they don't, I've come and gone. And . . . Ooo I love to sweep a-round a wide step, cut a lit-tle swath and lead the peo-ple on" (from "The Side Step," lyrics by Carol Hall). Labov 1994 describes many cases in which language change due to psychological factors decreases communication efficiency.

29. The idea that ideas compete and that the results of this competition drive human history was elaborated by sociologist Gabriel Tarde 1903 at the turn of the twentieth century.

30. Castro and Toro 1998 discuss the potential importance of teaching as opposed to simple imitation in the evolution of some important features of human cultures.

31. Janssen and Hauser 1981.

32. McEvoy and Land 1981.   33. Eaves, Martin, and Eysenck 1989.

34. Some social scientists propose that an explanation of cases like this in terms of beliefs, desires, and intentions is sufficient. We disagree. The explanation that assistant professors work hard because they intend to get tenure and tenured faculty vote against those who are slack because they intend to maintain the quality of the department has great intuitive appeal to our folk psychology. How, then, do we explain why professors tend to prefer writing papers to having children, while rural Africans have quite a different set of intentions? At best, beliefs, desires and intentions are proximal explanations themselves in need of an ultimate evolutionary explanation. See Rosenberg 1988 for a critique of scientific explanations in terms of folk psychology.

35. Hamilton 1967; Dawkins 1982; Jablonka and Lamb 1995; Rice 1996.

36. Cavalli-Sforza and Feldman 1981; Dawkins 1976; Durham 1991.

37. "Cornpone Opinions," Twain 1962, 24.

38. See Blackmore 1999 for a review of the work done using the meme concept. Richard Dawkins's foreword to Blackmore's book gives a particularly clear example of how important the high fidelity of transmission is taken to be by Dawkins at least. See Durham and Weingart 1997 for a discussion of alternative proposals for the unit of cultural inheritance. Dennett 1995 in *Darwin's Dangerous Idea* provides an extended argument in favor of the idea that replicators are necessary for cumulative adaptation.

39. See Aunger 2002 for an elaboration and critique of this view.

40. Sperber 1996.

41. Bynon 1977, characterizing scholars like Chomsky and Halle 1968.

42. Note that this phenomenon may take some of the bite out of Chomsky's argument from the poverty of the stimulus. Perhaps in the case of grammar, all native American English speakers don't all have the same rules in their heads. Perhaps learners adopt the first rule that they stumble across that generates grammatical sentences an acceptably large percentage of the time. There may be more than one rule that does so, so no one is really speaking exactly the same language. Individual speakers certainly do have small differences in their speech called ideolects. According to sociolinguists, ideolectual variation is the raw material out of which language evolution grows, a quite Darwinian notion (Labov 2001; Wardhaugh 1992). It is not so clear whether ideolect includes grammatical rules, but if it does, the sociolinguist's picture of the evolution of phonology may extend to syntax.

43. Bynon 1977.

44. Sperber 1996, chap. 5.

45. Sperber 1996; Boyer 1998, 1994; Atran 2001.

46. Burke and Young 2001. In addition to the 1:1 and 2:1 contracts, they also observed a small number of 3:2 contracts, and, even among the highly market-oriented farmers of Illinois, virtually no other shares. Burke and Young also show that farmers don't adjust shares by varying other inputs such as fertilizer or pesticides.

47. Bloom 2001.   50. Mallory 1989.

48. Spelke 1994.   51. Sperber 1996.

49. Tomasello 1999.   52. Bynon 1977.

53. Cavalli-Sforza and Feldman 1976, 1981; Karlin 1979; Lande 1976.

54. Hallpike 1986, 46.

55. Thomason and Kaufman 1988. See also Thomason 2001.

56. Thomason and Kaufman 1988.

57. Ibid.

58. Ibid.

59. Welsch, Terrell, and Nadolski 1992.

60. Jorgensen 1980; Hodder 1978.

61. Dumézil 1958; Hallpike 1986; Mallory 1989, chap. 5.

62. Brown 1988. Vayda 1995 argues that such explanations are much to be preferred to the general process accounts that we shall focus upon.

63. Boyd and Richerson 1992a.

64. Darwin "wasted" his college years following his dogs across the countryside, shooting birds, collecting beetles, and speculating about geology under the guidance of

Adam Sedgwick. The contextual detail such a naturalist commands certainly rivals that of ethnographers and historians. Some naturalists write and speak lyrically about the pleasure they get from looking the complexity and diversity of nature in the eye. E. O. Wilson's 1984 celebration of the naturalist's craft, *Biophilia,* is an excellent example. So is Darwin's last paragraph of the *Origin* and many passages in his *Journal of Researches* (*Voyage of the Beagle*). W. D. Hamilton was before all else an intrepid and perpetually entranced naturalist, according to those who knew him best. The same can be said of recently departed evolutionary theorist, John Maynard Smith. One of us (Richerson) has spent quite a lot of effort trying to understand the ecology of lakes, one of the simplest sorts of ecological systems, and will trade stories of complexity and diversity with any human scientist who cares to defend the idea that our species is any more complex than the average ecosystem. An accessible description of how evolutionary biologists immerse themselves in the detail of their chosen "system" is Jonathan Weiner's 1994 book *The Beak of the Finch,* describing Peter and Rosemary Grant's wonderful study of the evolution of Darwin's finches on the Galapagos Islands. This is a high-end study, to be sure, but every serious field study of evolution at least aspires to something like its resolution of the concrete events that are eventually summarized as selection of a certain strength on a certain trait.

65. Other authors, using vague or different arguments, have tried to make the case that something different about cultural and genetic evolution has led to cultural evolution being properly studied with methods quite different from genetic evolution. We think that, because of the general similarity of the evolution of the two systems, they all fall prey to this "sauce for the goose, sauce for the gander" analysis. One outfit or the other is doing something wrong! See Sober 1991 and Marks and Staski 1988.

66. See Boyd and Richerson 1985 and Cavalli-Sforza and Feldman 1981 for a full mathematical treatment of the issues. We refer extensively to these and other formal theoretical studies in later chapters.

67. For example, archaeologists often use population pressure to explain phenomena on very long timescales, such as the origins of agriculture. A little elementary modeling of demographic and evolutionary timescales suggests that the changes in subsistence that led up to and then resulted in domestication of plants and animals happen so *slowly* that demographic processes cannot explain either their occurrence in time nor their rate of change (Richerson, Boyd, and Bettinger 2001). As in organic evolution, population pressure does play an important role in our explanations. The Malthusian propensity of populations to grow *rapidly* to environmental limits generates the competition between variants that in turn drives selection. In essence, at the typical evolutionary timescale we assume that demographic processes acting on a shorter timescale generate some average level of population pressure and that therefore such fast-acting processes are not the rate-limiting steps in the evolutionary process. In a generation-by-generation microevolutionary context such assumptions will be violated and models may have to be adjusted accordingly.

68. Endler's 1986 analysis of the patterns of strength of natural selection in the wild is a nice example of the sorts of generalizations we get in the face of diversity and complexity. Selection is often rather strong, stronger more often than evolutionists' intuitions before Endler's review typically supposed. We also know from the analysis of cross-cultural data (pioneers include Murdock 1949, 1983 and Jorgensen 1980) that cultural variation is not without pattern.

69. The foregoing owes much to the work of Wimsatt 1981.

70. Sober 1991.

71. Ibid., 18.

72. Boyd and Richerson 1985; Cavalli-Sforza and Feldman 1981.

73. For the long version of this argument, see Richerson and Boyd 1987.

Chapter Four

1. Boyd and Richerson 1985; Tooby and DeVore 1987; Rosenthal and Zimmerman 1978; Brandon and Hornstein 1986; Pinker and Bloom 1990.

2. A similar game, Guess the Gadget, is played with the audience on the program *Home Matters,* which airs on the Discovery Channel.

3. See Stephens and Krebs 1987 for many examples.

4. Gould and Lewontin 1979.

5. Nilsson 1989.

6. E.g., Tomasello, Kruger, and Ratner 1993.

7. E.g., McGrew 1992.

8. Lefebvre and Palameta 1988; see Moore 1996 for an analysis restricted to the evolution of imitation.

9. Wrangham 1994; Whiten et al. 1999; McGrew 1992.

10. McGrew 1992.

11. The Tasmanian tool kit is unusually reduced, a subject to which we return later in this chapter. It is also important to note that this kit may have included many artifacts that have not been preserved in the archaeological record.

12. Van Schaik and Knott 2001.   14. Moore 1996.

13. Rendell and Whitehead 2001.   15. McComb et al. 2001.

16. Marler and Peters 1977; Baker and Cunningham 1985; Baptista and Trail 1992.

17. Galef 1996.

18. Lefebvre and Palameta 1988.

19. Lachlan, Crooks, and Laland 1998.

20. Most laboratory investigators have strong doubts about attributions of culture based on field studies. See commentaries on Rendell and Whitehead's 2001 paper for an introduction to this controversy. In the absence of controlled experiments, experimentalists argue, it is impossible to know whether observed behaviors are transmitted culturally. Field-workers feel equally strongly that laboratory environments don't provide much opportunity for animals to show off their best tricks, and that animals with seemingly complex behavior, such as chimpanzees and killer whales, are the hardest to deal with in the laboratory. They argue that the circumstantial evidence for sophisticated culture is strong.

21. Chou and Richerson 1992; Terkel 1995; Zohar and Terkel 1992.

22. Galef 1988.

23. Slater, Ince, and Colgan 1980; Slater and Ince 1979.

24. Imitation sometimes connotes rote copying of motor patterns, but we use it to label any form of learning in which individuals learn how to perform a behavior by observ-

ing others. So, for example, learning grammatical rules from hearing others speak is imitation in our usage.

25. Galef 1988; Visalberghi and Fragazy 1991; Whiten and Ham 1992.

26. Tomasello and Ratner 1993.

27. Galef 1988; Whiten and Ham 1992; Tomasello and Ratner 1993; Visalberghi 1993; Visalberghi and Fragazy 1991. But see Heyes 1996 for a quite different take on the evidence.

28. Custance, Whiten, and Fredman 1999 and Tomasello 1996.

29. Whiten 2000 thinks the difference between imitation and emulation is quantitative. Clearly, the exact differences between human and chimpanzee social learning skills are not yet precisely mapped.

30. Heyes and Dawson 1990; Voelkl and Huber 2000.

31. Herman 2001.

32. Pepperberg 1999; Moore 1996; Connor et al. 1998; Heyes 1993; Dawson and Foss 1965; Van Schaik ad Knott 2001; Russon and Galdikas 1995.

33. Darwin and some of his early followers thought that accurate imitation characterized even insects; the low level of sophistication of social learning in so many species would have surprised him (Richerson and Boyd 2001a; Galef 1988).

34. Rogers 1989; Boyd and Richerson 1995 show that Rogers's result generalizes considerably beyond his simple model.

35. Kameda and Nakanishi 2002.

36. Kameda and Nakanishi 2002; Lefebvre and Geraldeau 1994.

37. Basalla 1988; Petroski 1992.

38. Although there is some evidence that molecular-epigenetic systems may exhibit analogous dynamics. See Jablonka and Lamb 1995.

39. As we saw in chapter 3, when there are multiple attractors, bias and guided variation may tug in different directions; and when this is the case, guided variation can cause the population to evolve so that every individual has beliefs near one of the attractors. Weak selection can still be important if there is no strong bias among these attractors.

40. Boyd and Richerson 1988b, 1989b.

41. Todd and Gigerenzer 2000.

42. Henrich and Boyd 1998. See Boyd and Richerson 1985, 1996 and Kameda and Nakanishi 2002 for other treatments.

43. Myers 1993; Sherif and Murphy 1936.

44. Another classic is Asch 1956. The social psychology literature under the heading of conformity is extensive and rich. Conformist cultural transmission in our sense is that part of "informational conformity" that results in relatively durable changes in attitudes, beliefs, skills, and the like. Aronson, Wilson, and Akert 2002 (chap. 8) provide an accessible and up-to-date summary. In chapter 7 we describe some models of punishment without, however, tying these to the sort of coercive socialization of deviants that social psychologists observe in their experiments.

45. Boyd and Richerson 1985, 223–27.

46. Jacobs and Campbell 1961.

47. Myers 1993, chap. 7.

48. Some people like call such effects emergent properties. We are not fond of this term, because the whole-part relations of different kinds of systems are so diverse. The weather is a notoriously difficult-to-understand result of the Newtonian mechanics of flow in compressible fluids. The physics of turbulence has only the most distant analogies to the ecology and biology that drive evolution. To gather the phenomena of hurricanes and adaptations under one term doesn't seem very useful to us.

49. Manufacturer of Hanes brand underwear.

50. Henrich and Gil-White 2001.

51. Boyd and Richerson 1985, 223–27.

52. Ryckman, Rodda, and Sherman 1972.

53. Rogers 1983.   55. Brandon 1990.

54. Labov 2001.   56. Kaplan et al. 2000.

57. Byrne 1999. But some birds do as well or better! See Hunt 1996 and Weir, Chappell, and Kacelnik 2002 on the amazing tool use abilities of the New Caledonian crow.

58. Kaplan et al. 2000.

59. In fact it is not at all clear that humans can learn individually a whole lot faster or better than baboons. The human individual learning mechanism is adapted to manipulate culture. It is likely for this reason to be more general purpose than that of animals that can't depend to anywhere near the same extent on highly specialized cultural adaptations. If Shirley Strum had made a contest of it and transplanted some humans along with her baboons, then perhaps we'd know whether humans can learn any faster than baboons. What humans would do is rapidly transmit the successes of any one individual to the whole group, and in this way we might well best the monkey even if we're not any smarter individual by individual.

60. Lamb 1977; Alley 2000; Partridge et al. 1995; Bradley 1999; National Research Council, Committee on Abrupt Climate Change 2002.

61. Opdyke 1995.

62. Anklin et al. 1993; Lehman 1993; Ditlevsen, Svensmark, and Johnsen 1996.

63. Allen, Watts, and Huntley 2000; Dorale et al. 1998; Frogley, Tzedakis, and Heaton 1999; Hendy and Kennett 2000; Schulz, von Rad, and Erlenkeuser 1998.

64. Lamb 1977; Fagan 2002; Grove 1988.

65. Broecker 1996.

66. Jerison 1973.

67. Opdyke 1995; Klein 1999; deMenocal 1995.

68. Eisenberg 1981, 235–36.

69. Aiello and Wheeler 1995. Also see Martin 1981.

70. Reader and Laland 2002.

71. Kaplan and Robson 2002.

72. The argument works in reverse, too. An environment that favors longer life spans will make a prolonged juvenile period less costly, and therefore will also favor increased behavioral flexibility and larger brains. Kaplan and Robson argue that large brains were favored in Oligocene primates, because arboreal life reduced predation pressure and therefore selected for longer life spans.

73. Boyd and Richerson 1996.

74. Cheney and Seyfarth 1990, 277–30; Tomasello 2000.

75. Diamond 1978.

76. Humphrey 1976; Whiten and Byrne 1988, 1997; Kummer et al. 1997; Dunbar 1992, 1998.

77. Boyd and Richerson 1992a.   80. Susman 1994.

78. Wood and Collard 1999.   81. Povinelli 2000.

79. Toth et al. 1993.   82. Dean et al. 2001.

83. Cavalli-Sforza and Feldman 1981; Shennan and Wilkinson 2001.

84. For a contrary view, see Mithen 1999.

85. Typically, potassium-argon dating methods cannot be used for sites less than about 500,000 years old, and carbon-14 methods cannot be used for sites older than about 40,000 years. In the last couple of decades new methods such as thermoluminescence (TL) and electron spin resonance (ESR) allow sites from this period to be dated. However, many Middle Pleistocene sites were excavated before such methods were available.

86. McBrearty and Brooks 2000.

87. Brooks et al. 1995.

88. Thieme 1997.

89. Henshilwood et al. 2002; Henshilwood et al. 2001.

90. Ingman et al. 2000; Kaessmann and Pääbo 2002; Underhill et al. 2000.

91. Hofreiter et al. 2001.

92. Templeton 2002.

93. Falk 1983; Holloway 1983.

94. Laitman, Gannon, and Reidenberg 1989; Lieberman 1984.

95. Shennan and Steele 1999.

96. Donald 1991.

97. Dunbar 1996; Thompson 1995.

98. Sperber 1996; Atran 2001; Castro and Toro 1998.

99. Henrich 2004.

100. Our picture of hominid evolution will be slowly improved by new paleoanthropological finds and methods. However, the most important advances in the short term are liable to be the paleoclimatologists'. As we write this, the high-frequency variation in climate that is key to our analysis in this chapter exists only for the last glacial and the Holocene. We have only the most speculative ideas on the rest of the relevant record. Is the highly variable climate ancient, implying long lags between climatic events and the evolution of culture? Or does the record of brain size increase and tool sophistication reflect ongoing changes in high-frequency climate variation during the Pleistocene? Are some of the oddities of ancient hominids attributable to as-yet unappreciated oddities in environmental variation? Paleoclimatologists are hot on the trail of data that will bear on these questions, fuelled by the fear that ongoing anthropogenically induced climate change is a serious threat.

101. Tooby and Cosmides 1989.

102. Odling-Smee 1995.

103. Maynard Smith and Szathmáry 1995; Corning 1983.

Chapter Five

1. Sahlins 1976a, 1976b; Hallpike 1986.

2. Among other anthropologists, Marvin Harris 1977, 1972, 1979 was a prominent spokesman for the point of view of taking delight in proposing functional explanations for exotic cultural practices like Aztec cannibalism and Indian sacred cows.

3. Bongaarts and Watkins 1996; United Nations Population Division 2002.

4. Irons 1979; Borgerhoff-Mulder 1988a, 1988b.

5. Kaplan and Lancaster 1999.

6. Gould 2002; Levinton 2001; Carroll 1997; Alcock 1998, 2001.

7. Cronin 1991.

8. Land and Nilsson, 2002.

9. Martindale 1960.

10. Dawkins 1982, 1976. We (Richerson and Boyd 1976, 1978) made a similar argument at about the same time.

11. Hamilton 1967.

12. As occurs today between organisms and their endosymbionts, see Werren 2000.

13. For a simple mathematical treatment, see Boyd and Richerson 1985, chap. 6.

14. We resolved not to do any math in this book. However, we found the ideas in this paragraph especially hard to express in words. So, as an example of the greater precision and clarity that math affords, here is an alternative explanation. The fertility-reducing variant will spread if

$$(1 - A)p + At > 0,$$

where A is a number between zero and 1 that measures the relative influence of teachers. The term $1 - A$ measures the relative influence of parents. If A is near 1, children tend to acquire the beliefs of their teachers—parents aren't important. If A is close to zero, teachers have little influence. The parameter t is the difference between the probability that a person holding the late marriage meme becomes a teacher and the probability that a random person becomes a teacher, divided by the probability that a random person becomes a teacher. p is the difference between the probability that a person holding the late marriage meme becomes a parent and the probability that a random person becomes a parent, divided by the probability that a random person becomes a parent. This quantity is called the selection differential in population genetics. First, notice that if $A = 0$, this expression is negative. This makes sense. If ideas are acquired only from parents, an idea that reduces fertility is a sure loser. On the other hand, if teachers do have some influence, the belief in late marriage can spread. Note that this can occur even if parents have a bigger influence on their children's most basic attitudes than do teachers. That would mean that A would be smaller than $1 - A$. However, the process of attaining social roles like that of teacher can be very selective. Marrying young and starting a family might be a quite big handicap in obtaining the education to become a teacher. If so, t will be bigger in absolute value than p. If this effect is large enough to overcome the differential in the importance of parents in teaching basic family values, the whole expression can easily be positive, causing the late-marriage norm to spread. For the full development of the model, see Boyd and Richerson 1985, chap. 6.

15. Alexander 1979, 1974; Irons 1979; Durham 1976, 1991.

16. Parker and Maynard Smith 1990.

17. Pterosaurs of the genus *Quetzalcoatlus* were probably the heaviest flying creatures that have ever lived on earth. These soaring reptiles had eleven-meter wingspans and are estimated to have weighed a pig-sized one hundred kilograms or so. The greater oxygen content of the atmosphere during the later Mesozoic probably facilitated flight in such large creatures, both because it allowed higher metabolic levels and because denser air made flight easier (Dudley 2000).

18. Boyd and Richerson (1985, 53–55, 180) list a number of studies documenting the importance of vertical, horizontal, and oblique cultural transmission. Also see Harris 1998.

19. Feldman and Otto (1997) maintain that models with explicit terms for cultural transmission suggest that culture plays a larger role than typical behavior genetics studies suggest.

20. Labov 2001, chap. 13.

21. Hewlett and Cavalli-Sforza 1986.

22. Rogers 1983, 217–18.

23. Tooby and Cosmides 1992, 104.

24. Atlatls are also called spear throwers. A piece of light wood, bone, or ivory about the length of the forearm is equipped with a hook at one end that engages the butt of a light spear, the dart. Holding the end opposite the hook, the thrower uses the length of the atlatl to increase the velocity of his throw. The hitting power of such a fast, relatively heavy dart is greater than that of an arrow, a consideration when the prey is a large, blubber-padded sea mammal.

25. Arima 1975, 1987. While this was being written on a sunny May day in Berlin, the US Weather Service was reporting ten-foot seas and thirty-knot winds in the Bering Sea.

26. Tooby and Cosmides 1992, 104–8.

27. Rogers 1983, 231–32.

28. "Own interests" here refers to whatever outcomes are favored by within-group processes. Boyd and Richerson 1982, 1985; Soltis, Boyd, and Richerson 1995; and Sober and Wilson 1998 present the case for cultural group selection.

29. Stark 1997 argues that early Christianity increased rapidly following a scenario much like we outline, and Wilson 2002 makes a similar case for Calvinism during the Reformation.

30. Stark 2003, chap. 3, provides an interesting perspective on the function of the Inquisitions as institutions of social control in the early modern period. He dismisses much of the conventional wisdom about them as anti-Catholic propaganda.

31. Darwin 1874.

32. Fisher 1958 [1930]. Fisher could not explain why exaggerated traits could persist at equilibrium, but recently Iwasa and Pomiankowski 1995 and Pomiankowski, Iwasa, and Nee 1991 describe two different mechanisms that solve this problem.

33. Eberhard 1990.

34. Boyd and Richerson 1985, chap. 8.

35. Boyd and Richerson 1987; Richerson and Boyd 1989a.

36. Boyd and Richerson 1985, chap. 8.

37. Costly signaling provides another process with a maladaptive runaway inherent in it. We think of this as the "rocks in your packsack" hypothesis. If Rob can demonstrate his physical prowess relative to Pete's by routinely carrying heavier burdens, he can advertise his better genes (or his better training regime). If Rob gets the girls or the cultural prestige by carrying one rock, Pete will respond by carrying two, and Rob will have to increase his load to three. The competition will only come to a halt when both of us are carrying a huge load of useless rocks, but one can carry just a bit more than the other. The only way an observer can be sure which competitor is better is if we can carry enough rocks to make a serious contest. In an evolutionary context, wasteful displays will evolve to guarantee the veracity of signals, much as in the Fisherian process. After some rounds of evolutionary bidding in the prestige auction, the benefit from having good genes or good culture will be entirely consumed in the wasteful display to advertise that you have them. In other words, Hummer II owners are likely to have pretty good incomes, but also likely be up to their big pocketbooks in debt. The economist Spence 1974 originally proposed this idea. Biologists later applied the same logic to sexual selection (Zahavi 1975; Zahavi and Zahavi 1997; Grafen 1990a, 1990b), and Smith and Bliege Bird 2000 have applied these ideas to explain a variety of exaggerated displays in small-scale human societies. Ryan 1998 describes a third explanation based on the idea that exaggerated characters are byproducts of selection on female sensory systems. In all three hypotheses the exaggeration is maladaptive. For an excellent nonmathematical treatment of modern sexual selection theory, see Miller 2000.

38. Richerson and Boyd 1989b. See Boyer 1994 for a discussion of the role of abductive reasoning in providing support for religious beliefs.

39. Pascal 1660, §233.

40. There is the problem, if you want to accept Pascal's side of the wager, of deciding exactly what beliefs God intends to reward and punish in the afterlife! The Pope believed that Jansenism was a grave danger to the souls of its practitioners. Perhaps God really rewards all the humble seekers after the truth with heaven, even agnostic scientists, and sends all dogmatists who presume to know his mind to hell, Pascal and Pope alike.

41. As Max Weber 1951 noted, the rationalist approach to ultimate questions, especially to the elaboration of theistic notions that are so prone to doubts, is more pronounced in Christianity than in other religions. Nevertheless, other religions, even quite "primitive" ones, have evolved in their appeal to hard thinkers, not just the credulous. See, e.g., Barth 1990. Stark 2003, chap. 2, notes that many pioneering scientists like Newton had theological agendas comparable to Pascal's.

42. Campbell 1974.

43. Boyer 1994.

44. That is: A implies B. Observe B is true. Therefore A is true.

45. Sloan, Bagiella, and Powell 1999 provide a skeptical review of the psychological health literature. On community-level functions see Stark 2003 and Wilson 2002.

46. Schwartz 1999.

47. Knauft 1985a.

48. Stark 2003, chap. 3.

49. Rabinowitz 2003; see also Linder 2003. As of this date, the reasoning of those who still believe that ritual child abuse did occur in these cases is easy to find on the Web.

50. Lindert 1985 describes the population cycles in England from the late medieval to modern times. The Malthusian pattern obtained until the Industrial Revolution was well under way.

51. Reducing the importance of nonparental transmission reduces the *risk* of maladaptive cultural variants evolving, much as reducing the size of organelle genomes reduces the risk of sex ratio distortion. Some evidence suggests that premodern populations were frequently the victims of maladaptive reductions of fertility. Bruce Knauft 1986 argues that relatively high birthrates and relatively high death rates in preindustrial cities made them demographic black holes. The prestige system of the Roman Empire and early modern England made such places a magnet for country folk and sustained themselves by recruiting from the countryside despite an excess of deaths over births in Rome and London. Coale and Watkins 1986 (14–22, and chap. 3) and their colleagues uncovered some rural as well as urban populations in Europe in which fertility fell below replacement levels well in advance of the main demographic transition. These included a number of rural cases in which one- and two-child families became the norm, while death rates remained at preindustrial levels. Such populations shrank rapidly, selecting against the norms that supported such practices.

52. Coale and Watkins 1986, chap. 1.

53. Skinner 1997.

54. The following discussion is based on Alter 1992; Pollack and Watkins 1993; Kirk 1996; Bongaarts and Watkins 1996; and Borgerhoff-Mulder 1998.

55. Coale and Watkins 1986.

56. Rogers 1990a.

57. Kaplan et al. 1995.

58. Becker 1983.

59. Robinson and Godbey 1997.

60. E.g., the well-known account in Schor 1991.

61. This change in TV viewing is especially interesting, because it is among the lowest rated leisure activities. Indeed, in most surveys many work activities, including child care, rank as high as TV viewing in reported enjoyment. TV's constant availability, low cost, and curiously addictive hold on our attention apparently allow it to crowd out more highly rated activities, such as social activities away from the home.

62. Easterlin, Schaeffer, and Macunovich 1993.

63. Kasarda, Billy, and West 1986, chap. 6.

64. Blake 1989.

65. Belonging to a pronatalist faith reduces the effect of family size. Catholics, particularly high-status, educated ones, show a smaller effect of sibship size on educational attainment than Protestants, but the family-size effect remains significant.

66. Hill and Stafford 1974; Lindert 1978.

67. Witkin and Berry 1975; Witkin and Goodenough 1981; Werner 1979.

68. Inkeles and Smith 1974; Jain 1981.

69. Kohn and Schooler 1983.

70. Bongaarts and Watkins 1996.

71. Rogers 1983.

72. Westoff and Potvin 1967.

73. Roof and McKinney 1987.

74. Bongaarts and Watkins 1996.

75. Our discussion based on Peter 1987; Hostetler 1993; Kraybill and Olshan 1994; Kraybill and Bowman 2001.

76. Nonaka, Miura, and Peter 1994.

77. Even in ordinary modern societies, the effect of families on the propensity to have children is marked. Psychologist Lesley Newson collected some very interesting data relevant to childbearing and patterns of cultural transmission in contemporary Britain. Questionnaire data indicate that men and women with relatively more contact with relatives marry earlier. Women with such contact have earlier first pregnancies and have more children. In a role-playing experiment, Newson asked mature women to write down what sort of advice they thought an older woman would give to a younger woman (either her daughter or a younger friend) in one of four situations. In each of the scenarios, women who imagined giving advice to their daughters were more likely to advise behavior consistent with reproductive success (Newson 2003). Comparable data from an Anabaptist community would be most interesting.

78. See *The Devil's Playground,* an excellent documentary film on *rumspringa* made by Lucy Walker, at http://www.wellspring.com/devilsplayground/.

79. Labov 1973.

80. The first verse:

> Some folk built like this, some folk built like that
> But the way I'm built, you shouldn't call me fat
> Because I'm built for comfort, I ain't built for speed
> But I got everything all the good girls need

81. Tooby and Cosmides 1989, 34–35.

82. Borgerhoff Mulder 1988a and 1988b.

83. Laland, Kumm, and Feldman 1995 provide a model and a test case, societies with much female infanticide and apparently a sex ratios at birth skewed in favor of males. However, Skinner and Jianhua's 1998 data from China, where female-biased infanticide has long been practiced, suggest that the actual sex ratio at birth is not skewed and that statistics to the contrary result from concealed infanticide, not genetic changes. The model is nevertheless indicative of what can happen.

Chapter Six

1. Simoons 1970, 1969. Durham (1991, chap. 5) reviews and reanalyzes the data on adult lactose absorption.

2. Cavalli-Sforza, Menozzi, and Piazza 1994; Holden and Mace 1997.

3. Paul Ehrlich and Peter Raven (1964) coined the term *coevolution* to describe the evolutionary relationship between butterflies and plants. Caterpillars prey on plants, and the plants in turn evolve chemical defenses to mitigate the damage of insect attack, which leads to the evolution of caterpillar detoxification capacities. Since then its meaning has been extended to any case in which two distinct evolutionary systems interact in interesting ways.

4. Another way to think about gene-culture coevolution is in terms of "niche construction" (Odling-Smee et al. 2003). Whenever an organism modifies its environment, natural

selection will result from the effects of the modified environment. For example, beavers construct dams and are much modified for aquatic life in the resulting ponds. In this way of thinking, the products of culture become part of the selective environment of genes, just as the products of genes become part of the selective environment of culture. One just has to be careful to understand what is acting as an inheritance system and what is not. Beaver dams cannot reproduce themselves; the information about how to construct dams is encoded in beaver genes, not in the dams themselves, although in principle beavers could learn dam building by observing the dams of other beavers.

5. Klein 1999, 474–76; Berger and Trinkhaus 1995.

6. For a model of this process see Richerson and Boyd 1989b; Laland 1994; and Laland, Kumm, and Feldman 1995.

7. Lumsden and Wilson 1981, 303. See also E. O. Wilson 1998.

8. Corning 2000, 1983 discusses the evolutionary consequences of synergy in some detail.

9. Maynard Smith and Szathmáry 1995.

10. Margulis 1970.

11. Kaplan et al. 2000.

12. There are two common objections to the term *instinct*. First, some critics say that the term is hollow. A pattern of behavior exists, and merely labeling it an instinct adds nothing to our understanding. To this we answer that we want to distinguish between influences on behavior that are genetic and those that are cultural. Second, some would restrict the term *instinct* to innate patterns of behavior that are little modified by environmental contingencies or culture. Wilson 1975, 26–27, notes that this sense of the term applies only to extreme cases and so endorses the usage we adopt here.

13. We are well aware that anthropologists have used the term *tribe* in such diverse ways that many feel that the term has become hopelessly muddled. Common English usage is also quite polysemous. We use it here in a minimalist sense. Tribes are a unit of social organization that incorporates people of relatively low degrees of biological relatedness into a common social system without depending on formal authority. Extended kinship, sentiment, and informal institutions animate tribes, rather than formal law and leadership with formal powers of coercion. Birdsell's 1953 classic study estimated that the average Australian hunter-gather tribe incorporated about five hundred people. The creation of social units composed of many distantly related families, usually not coresident in hunter-gatherers, is unique to humans. Usually, descent from a common ancestor, often fictitious, honorific, or metaphorical, forms the core of the ideology enjoining feelings of solidarity, which are in turn the main wellspring of common action. Some restrict the term *tribal* to a range of societies of intermediate size and complexity usually characterized by sizes of a few thousand, with fairly elaborate formal political institutions but still no specialized full-time leaders with coercive authority (Service 1962). We believe that even the societies like the Shoshone, Steward's (1955, chap. 6; see esp. p. 109) illustration of an approximation to his "family band" ideal type, are normally part of a multiband community that functions to maintain local peace, resist incursions by other tribes, and provide aid in subsistence crises, even if in extreme cases these functions are rather limited. Murphy and Murphy 1986 and Thomas et al. 1986 argue that Steward's characterization of the Shoshone as family band societies underestimates their social complexity, even taking his caveats into account. In

any case, the Shoshone adaptation to the arid Great Basin is very late, highly derived, and rather sophisticated in its very minimalism (Robert Bettinger, personal communication). No ethnographically known societies lack some form of integration into units considerably larger than the family or coresident band. Simpler societies vary continuously along several dimensions regarding social organization (e.g., Jorgensen 1980), and clean classification is a vain hope. The emergence of social bonds among noncoresident, distantly related people requires a convenient label, and the choice is *tribal* or an awkward neologism.

14. Boehm 1992; Rodseth et al. 1991.

15. Hamilton 1964. The "derivation" we gave in the previous paragraph is in the spirit of that paper.

16. The great population geneticist J. B. S. Haldane gave what is perhaps the pithiest summary of this principle. When asked by a reporter whether the study of evolution had made it more likely that he would give up his life for a brother, Haldane is supposed to have answered, "No, but I would give up my life to save two brothers or eight cousins." We can't resist another Haldane anecdote here, even though it has nothing to do with the subject of this book. Haldane was also asked by a reporter, maybe even the same one, whether the study of evolution had taught him anything about the mind of the Creator, to which Haldane is said to have replied, "He has an inordinate fondness for beetles."

17. Silk 2002; Keller and Chapuisat 1999; Queller and Strassmann 1998; Queller 1989.

18. As opposed to *Nature,* where it figures prominently.

19. Hammerstein manuscript.

20. See Axelrod and Dion 1988 and Nowak and Sigmund 1993, 1998a and 1998b for reciprocity in small groups; Boyd and Richerson 1988a, 1989a and Joshi 1987 for larger groups. Glance and Huberman (1994) present a model in which reciprocity evolves in large groups, but this result depends on constraints on their choice of a set of possible strategies. Simple unconditional defection invades their cooperative Evolutionarily Stable Strategy.

21. E.g., Binmore 1994.

22. Trivers 1971.

23. Boyd et al. 2003 and Boyd and Richerson 1992b.

24. Wynne-Edwards 1962.

25. Maynard Smith 1964; Williams 1966; Lack 1966.

26. Price 1972, 1970.

27. The Price approach has been very fruitful, generating a much clearer understanding of many evolutionary problems—for example, Alan Grafen's 1984 work on kin selection and Steven Frank's 2002 work on the evolution of the immune system, multicellularity, and related issues. This approach can also be used to study cultural evolution. See Henrich, in press, and Henrich and Boyd 2002.

28. Sober and Wilson 1998.

29. Eshel 1972; Aoki 1982; Rogers 1990b.

30. See Boyd and Richerson 1990 for details.

31. There is also a very interesting interaction between conformism and moralistic punishment. If there is a widely held norm of moralistic punishment, it may be that most

people cooperate. This in turn means that it is difficult to know whether punishing is individually advantageous (because nonpunishers are punished) or not (because nonpunishers take a free ride on the police work of others). Recall that when it is difficult to determine the relative merit of alternative variants, then decision-making forces like content bias are relatively weak, which in turn implies that even weak conformist transmission can be important. In this case, conformism can maintain a moral norm that holds that people should engage in moralistic punishment. Such a norm can generate lots of costly punishment that maintains group-beneficial behavior. See Henrich and Boyd 2001 for more details.

32. Darwin 1874, 178–79. Of course, Darwin did not understand organic inheritance, though he did use concepts closely related to the modern notion of culture. The subtleties of the differences between genes and culture were lost on him, but he did understand that selection would normally favor selfish behavior. See Richards 1987 and Richerson and Boyd 2001a.

33. Palmer, Fredrickson, and Tilley 1997.

34. See Cavalli-Sforza and Feldman 1981 for models of cultural drift, and Coyne, Barton, and Turelli 2000 and Lande 1985 for rates at which populations shift from one equilibrium to another due to genetic drift.

35. Keeley 1996; Otterbein 1985; Jorgensen 1980.

36. Wiessner and Tumu 1998.

37. Boyd and Richerson 2002.   38. Stark 1997.

39. Johnson 1976, 75, quoted in Stark 1997, 84.

40. Stark 1997, 83–84.

41. See Rogers 1995 on the reason these properties are necessary for easy diffusion of innovations.

42. Barth 1969, 1981; Cohen 1974.

43. Rappaport 1979.   44. Barth 1981.

45. McElreath, Boyd, and Richerson 2003.

46. See Boyd and Richerson 1987 for mathematical details.

47. Logan and Schmittou 1998 offer the art of the Great Plains Crow as an example of such a process.

48. E.g., van den Berghe 1981; Nettle and Dunbar 1997; Riolo, Cohen, and Axelrod 2001.

49. Harpending and Sobus 1987.

50. See Ostrom's 1990 discussion of punishment in the context of managing public goods. See also Gruter and Masters 1986 on ostracism and Paciotti 2002 for an African tribal system with very sophisticated punishment.

51. Cognitive psychologists like Boyer 1998 would say that we have a "naïve ontology" in which symbolically marked groups are a default category.

52. Kelly 1995; Richerson and Boyd 1998, 2001b; Richerson, Boyd, and Henrich 2003.

53. The idea that much criminal behavior in modern societies is a product of impulsive and otherwise socially maladroit personalities is a classic criminological hypothesis. Scholars differ about the reasons some people attract more punishment than others, but the data suggesting that prison inmates and other delinquents tend to be more impulsive than the average person is rather strong (Caspi et al. 1994; Raine 1993).

54. Pinker 1994, 111–12.

55. Steward 1955, chaps. 6–8; Kelly 1995.

56. Mansbridge's 1990 edited volume gives an excellent sampler.

57. Ghiselin 1974, 247.

58. Batson 1991.

59. Camerer 2003; Henrich et al. 2004.

60. Boehm 1993; Eibl-Eibesfeldt 1989, 279–314; Insko et al. 1983; Salter 1995.

61. Fehr and Gächter 2002.

62. The syndrome of ethnocentrism has received much attention from sociologists, from the work of William Graham Sumner early in the twentieth century onward. Notable summaries include those of Robert LeVine and Donald Campbell 1972 and Nathan Glazer, Daniel Moynihan, and Corinne Schelling's edited volume 1975.

63. Wiessner 1984, 1983.

64. Bettinger 1991.

65. Tajfel 1982, 1981, 1978; Robinson and Tajfel 1996.

66. Turner 1984; Turner, Sachdev, and Hogg 1983.

67. Gil-White 2001 and personal communications.

68. Paldiel 1993.

69. Brewer and Campbell 1976.

70. Alexander 1987, 1979; Cosmides and Tooby 1989; Dunbar 1992.

71. This hypothesis was first, and perhaps most clearly, articulated by Pierre van den Berghe (1981).

72. Some people have interpreted the work of Nowak and Sigmund 1998a, 1998b as showing that indirect reciprocity can lead to helping in larger groups. That conclusion is problematical. First, the Nowak and Sigmund model had significant technical flaws (Leimar and Hammerstein 2001). Second, the corrected model still allows the evolution of indirect reciprocity, but under much-restricted conditions. Also see Panchanathan and Boyd 2003. Third and most important, this model is still limited to pairwise interactions. It does not explain the evolution of public-good provision.

73. Jorgensen 1980.

74. Price and Brown 1985.

75. R. L. Bettinger, University of California, Davis, personal communication.

76. Wiessner 1983, 1984.

77. Yengoyan 1968.

78. Knauft 1987.

79. Knauft 1985a; Otterbein 1968.

80. Keeley 1996, 28.

81. Arnold 1996; Price and Brown 1985.

82. It is also true that the institutions of small-scale societies vary for reasons that have no discernable correlation with ecological circumstances. Among the work cited here, Knauft 1985b, 1993 and Jorgensen 1980 describe the considerable degree of variation that exists in simple societies, apparently independent of environment.

83. For example, a trader first visited Blackfeet of the northwestern Plains in 1787, during the second generation of the horse era, and at that time a few elderly people experienced with pedestrian hunting were still alive to give him an impression of that life (Ewers 1958).

84. Otterbein 1968; Boehm 1984.

85. Service 1966, 54–61.   86. Boehm 1993.

87. Evans-Pritchard 1940; Kelly 1985, chap. 4.

88. Richerson, Boyd, and Bettinger 2001; Richerson and Boyd 2001c.

89. Kennedy 1987. Insko et al.'s 1983 elegant experiments in social evolution showed dramatic resistance that coercive dominance generates compared to leadership that is perceived as more legitimate. They also show how domination and resistance to domination weaken the productivity of the group as a whole.

90. Eibl-Eibesfeldt 1989, 314.

91. Salter 1995 provides a detailed analysis of how the institutions of dominance in complex societies function to manipulate our evolved psychology.

92. Benedict Anderson 1991 argues that nations came to be the dominant actors on the political stage when mass literacy and newspapers allowed cultural-political writers to appeal to the whole of the community speaking a vernacular. We imagine that the ritual systems centered on dramatic public buildings we so admire as ruins were the analogs in ancient city-states. The Mayans and the Greeks that participated in the construction of such complexes and in the ceremonies that took place in them could easily imagine themselves to be part of a common community. Today, the Muslim hajj (pilgrimage to Mecca) is the largest extant ritual and probably plays a real role in giving Muslims a sense of a common community despite the huge size of that community (Peters 1994).

93. Kellett 1982, 112–17.

94. Garthwaite 1993; Curtin 1984; Gadgil and Malhotra 1983; Srinivas 1962; Fukuyama 1995; Putnam, Leonardi, and Nanetti 1993; Light and Gold 2000; Light 1972.

95. Marty and Appleby 1991; Roof and McKinney 1987; Juergensmeyer 2000.

96. Inglehart and Rabier 1986.

97. We have elsewhere reviewed two sets of comparative cases, World War II armies and village-scale commons management institutions, in the light of this taxonomy of workarounds and their problems (Richerson, Boyd, and Paciotti 2002; Richerson and Boyd 1999).

Chapter Seven

1. Dobzhansky 1973, 129.

2. Wilson 1998.

3. For an elementary treatment of cultural evolutionary models, see Richerson and Boyd 1992. For advanced treatments see Cavalli-Sforza and Feldman 1981 and Boyd and Richerson 1985.

4. Human geneticists also tell us that total human genetic variation is modest; that

most of it is expressed within, not between, populations; and that Africans are altogether more variable than the rest of our species (Harpending and Rogers 2000).

5. North and Thomas 1973; Bettinger and Baumhoff 1982.

6. Klein 1999, chap. 7. We of course do not deny that biological adaptations like skin color, body form, and disease resistance alleles are important in human adaptation to new environments.

7. Weiner 1999.

8. Lewontin and Hubby 1966.

9. Donald Campbell's 1969, 1979, 1986a cheerleading for interdisciplinary studies more than a generation ago shows that recognition of the problem goes back a long ways.

10. Alexander 1987.

11. For instance, Keller 1995 and Keller and Ross 1993 describe some fascinating social systems in ants. Both our personal kitchen ants here in California are Argentine ants, a recent invader that lacks genetic diversity for colony odor and violates inclusive fitness expectations even more massively than humans. The species forms giant supercolonies that have about a two-fold advantage over its competitors: because colonies cannot recognize strangers, they do not fight. (Genetic relatedness within subcolonies is practically zero.) It has driven most competing ant species out of the habitats that are suitable to it (Holway, Suarez, and Case 1998).

12. Sahlins 1976a.

13. Dunnell 1978.

14. Bettinger, Boyd, and Richerson 1996.

15. Cohen 1974.

16. Bettinger, Boyd, and Richerson 1996.

17. Henrich and Gil-White 2001.

18. Edgerton 1992; Knauft 1985a.

19. Inglehart and Rabier 1986.

20. Arrow 1963.

21. North and Thomas 1973.

22. E.g., Jorgensen 1980.

23. Nisbett 2003; McElreath, in press.

24. Henrich et al. 2004.

25. Donald 1991.

26. Boyd and Richerson 1985.

27. Endler (1986) utilizes many such studies of natural selection in the wild in his meta-analysis of the strength of natural selection.

28. For an experimental example see Insko et al. 1983; for an observational approach see Cavalli-Sforza et al. 1982 and McElreath, in press.

29. Nelson and Winter 1982, 3.

30. Other organisms are also active in their own evolution through "niche construction"; culture is just a particularly efficient mechanism for doing so. For a more-general theory, see Odling-Smee et al. 2003.

31. Richerson and Boyd 2000.

32. Labov 2001; Weingart et al. 1997, 292–97.

33. Recall Vannevar Bush's 1945 characterization of science as the endless frontier. If the frontier truly is endless, the story will never be complete.

34. As the philosopher of science John Beatty 1987 notes, that is about the best you can say for any research program.

35. Thanks to Robert Foley for this phrase.

36. Anonymous [St. George Mivart] 1871. Ironically, Mivart later ran afoul of Catholic orthodoxy and was excommunicated (http://www.newadvent.org/cathen/10407b.htm).

37. The song is "That's All" from the CD *Elementary Doc Watson,* Collectables, 1997.

38. Easterbrook 1997.

39. Kiester's 1996/97 essay on the aesthetics of biodiversity is interesting in this regard.

40. Darwin 1902. His paean against slavery begins (561–63):

On the 19th of August, we finally left the shores of Brazil. I thank God I shall never again visit a slave country. To this day, if I hear a distant scream, it recalls with vivid painfulness my feelings when, passing a house near Pernambuco I heard the most pitiable moans, and could not but suspect that some poor slave was being tortured, yet knew that I was as powerless as a child even to remonstrate.

And ends:

It makes one's blood boil, yet heart tremble, to think that Englishmen and our American descendants with their boastful cry of liberty, have been and are so guilty: but it is a consolation to reflect that we have made a greater sacrifice than ever made by any nation to expiate our sin. [Britain freed the slaves in all her colonies in 1838.]

41. PJR is a sometime limnologist, a student of lakes. Limnologists have a saying that is inevitably truer than it should be: "Everyone sees the world from the shores of their own lake."

42. Haldane 1927, 286.

43. A recent National Academy of Sciences committee entitles their report *Abrupt Climate Change: Inevitable Surprises* (National Research Council 2002).

References and Author Index

Aiello, L. C., and P. Wheeler. 1995. The expensive-tissue hypothesis: The brain and the digestive system in human and primate evolution. *Current Anthropology* 36:199–221. [270n69]

Alcock, John. 1998. Unpunctuated equilibrium in the Natural History essays of Stephen Jay Gould. *Evolution and Human Behavior* 19:321–36. [272n6]

———. 2001. *The triumph of sociobiology*. Oxford: Oxford Univ. Press. [272n6]

Alexander, J. C., B. Giesen, R. Münch, and N. J. Smelser. 1987. *The micro-macro link*. Berkeley and Los Angeles: Univ. of California Press.

Alexander, Richard D. 1974. The evolution of social behavior. *Annual Review of Ecology and Systematics* 5:325–83. [18, 72, 260n9, 272n15]

———. 1979. *Darwinism and human affairs; The Jessie and John Danz lectures*. Seattle: Univ. of Washington Press. [260n9, 260n14, 260n16, 261n2, 265n27, 272n15, 280n70]

———. 1987. *The biology of moral systems*. Hawthorne, NY: A. de Gruyter. [280n70, 282n10]

Alland, Alexander. 1985. *Human nature, Darwin's view*. New York: Columbia Univ. Press. [261n21]

Allen, J. R. M., W. A. Watts, and B. Huntley. 2000. Weichselian palynostratigraphy, palaeovegetation and palaeoenvironment: The record from Lago Grande de Monticchio, southern Italy. *Quaternary International* 73/74:91–110. [270n63]

Alley, Richard B. 2000. *The two-mile time machine: Ice cores, abrupt climate change, and our future*. Princeton, NJ: Princeton Univ. Press. [270n60]

Alter, G. 1992. Theories of fertility decline: A nonspecialist's guide to the current debate.

In *The European experience of declining fertility, 1850–1970: The quiet revolution,* ed. J. R. Gillis, L. A. Tilly, and D. Levine, 13–27. Cambridge, MA: Blackwell. [275n54]

Altstein, H., and R. J. Simon. 1991. *Intercountry adoption: A multinational perspective.* New York: Praeger. [262n36]

Anderson, Benedict R. O'G. 1991. *Imagined communities: Reflections on the origin and spread of nationalism.* Rev. and extended ed. London: Verso. [281n92]

Andujo, E. 1988. Ethnic identity of transethnically adopted hispanic adolescents. *Social Work* 33:531–35. [262n36]

Anklin, M., J. M. Barnola, J. Beer, T. Blunier, J. Chappellaz, H. B. Clausen, D. Dahljensen, et al. 1993. Climate instability during the last interglacial period recorded in the GRIP ice core. *Nature* 364: 203–7. [270n62]

Anon. [St. George Mivart, The Wellesly Index]. 1871. Review of the *Descent of Man and Selection in Relation to Sex* by Charles Darwin. *The Quarterly Review* 131:47–90.

Aoki, K. 1982. A condition for group selection to prevail over counteracting individual selection. *Evolution* 36:832–42. [278n29]

Arima, Eugene Y. 1975. *A contextual study of the Caribou Eskimo kayak.* Ottawa: National Museums of Canada. [273n25]

———. 1987. *Inuit kayaks in Canada: A review of historical records and construction, based mainly on the Canadian Museum of Civilization's collection.* Ottawa: Canadian Museum of Civilization. [273n25]

Arnold, J. E. 1996. The archaeology of complex hunter-gatherers. *Journal of Archaeological Method and Theory* 3:77–126. [280n81]

Aronson, Elliot, Timothy D. Wilson, and Robin M. Akert. 2002. *Social psychology.* 4th ed. Upper Saddle River, NJ: Prentice-Hall. [269n44]

Arrow, Kenneth J. 1963. *Social choice and individual values.* 2nd ed. New Haven, CT: Yale Univ. Press. [282n20]

Asch, Solomon E. 1956. Studies of independence and conformity: I. A minority of one against a unanimous majority. *Psychological Monographs* 70:1–70. [122, 269n44]

Atran, Scott. 1990. *Cognitive foundations of natural history: Towards an anthropology of science.* Cambridge: Cambridge Univ. Press. [262n46]

———. 2001. The trouble with memes—Inference versus imitation in cultural creation. *Human Nature—An Interdisciplinary Biosocial Perspective* 12:351–81. [45, 261n23, 264n17, 266n45, 271n98]

Atran, Scott, D. Medin, N. Ross, E. Lynch, J. Coley, E. U. Ek, and V. Vapnarsky. 1999. Folkecology and commons management in the Maya lowlands. *Proceedings of the National Academy of Sciences of the United States of America* 96:7598–603. [83, 262n46]

Atran, Scott, D. Medin, N. Ross, E. Lynch, V. Vapnarsky, E. U. Ek, J. Coley, C. Timura, and M. Baran. 2002. Folkecology, cultural epidemiology, and the spirit of the commons— A common garden experiment in the Maya lowlands, 1991–2001. *Current Anthropology* 43:421–50.

Aunger, Robert. 1994. Are food avoidances maladaptive in the Ituri Forest of Zaire? *Journal of Anthropological Research* 50:277–310. [261n23]

———. 2002. *The electric meme: A new theory of how we think.* New York: Free Press. [266n39]

Axelrod, R., and D. Dion. 1988. The further evolution of cooperation. *Science* 242:1385–90. [278n20]

Baker, M. C., and M. A. Cunningham. 1985. The biology of bird-song dialects. *Behavior and Brain Science* 8:85–133. [268n16]

Bandura, Albert. 1977. *Social learning theory.* Englewood Cliffs, NJ: Prentice-Hall.

———. 1986. *Social foundations of thought and action: A social cognitive theory.* Prentice-Hall Series in Social Learning Theory. Englewood Cliffs, NJ: Prentice-Hall.

Baptista, L. F., and P. W. Trail. 1992. The role of song in the evolution of passerine birds. *Systematic Biology* 41:242–47. [268n16]

Barash, D. P. 1977. *Sociobiology and behavior: The biology of altruism.* New York: Elsevier. [260n9]

Barth, Fredrik. 1956. Ecologic relationships of ethnic groups in Swat, North Pakistan. *American Anthropologist* 58:1079–89.

———, ed. 1969. *Ethnic groups and boundaries. the social organization of culture difference.* Boston: Little, Brown and Co. [279n42]

———. 1981. *Features of person and society in Swat: Collected essays on Pathans.* London: Routledge & Kegan Paul. [279n42, 279n44]

———. 1990. Guru and the conjurer: Transactions in knowledge and the shaping of culture in Southeast Asia and Melanesia. *Man* 25:640–53.

Basalla, G. 1988. *The evolution of technology.* Cambridge: Cambridge Univ. Press. [263n56, 269n37]

Batson, C. Daniel. 1991. *The altruism question: Toward a social psychological answer.* Hillsdale, NJ: Lawrence Erlbaum Associates. [217, 218, 280n58]

Baum, William B. 1994. *Understanding behaviorism: Science, behavior, and culture.* New York: HarperCollins. [264n12]

Beatty, John. 1987. Natural selection and the null hypothesis. In *The latest on the best: Essays on evolution and optimality,* ed. J. Dupre. Cambridge, MA: MIT Press. [282n34]

Becker, Gary. 1983. Family economics and macro behavior. *American Economic Review* 78:1–13. [175, 275n58]

Benedict, Ruth. 1934. *Patterns of culture.* Boston: Houghton Mifflin Co. [262n30]

Berger, T. D., and E. Trinkhaus. 1995. Patterns of trauma among Neanderthals. *Journal of Archaeological Science* 22:841–52. [277n5]

Bettinger, R. L. 1991. *Hunter-gatherers: Archaeological and evolutionary theory.* New York: Plenum Press. [280n64]

Bettinger, R. L., and M. A. Baumhoff. 1982. The numic spread: Great Basin cultures in competition. *American Antiquity* 47:485–503. [282n5]

Bettinger, R. L., R. Boyd, and P. J. Richerson. 1996. Style, function, and cultural evolutionary processes. In *Darwinian Archaeologies,* ed. H. D. G. Maschner, 133–64. New York: Plenum Press. [282n14, 282n16]

Betzig, Laura L. 1997. *Human nature: A critical reader.* New York: Oxford Univ. Press. [19, 261n4]

Bickerton, Derek. 1990. *Language and species.* Chicago: Univ. of Chicago Press. [262n47]

Binmore, Kenneth G. 1994. *Game theory and the social contract.* Cambridge, MA: MIT Press. [278n21]

Birdsell, J. B. 1953. Some environmental and cultural factors influencing the structuring of Australian aboriginal populations. *The American Naturalist* 87 : 171–207. [277n13]

Blackmore, Susan. 1999. *The meme machine.* Oxford: Oxford Univ. Press. [81, 266n38]

Blake, Judith. 1989. *Family size and achievement.* Berkeley and Los Angeles: Univ. of California Press. [178, 275n64]

Bloom, Paul. 2001. *How children learn the meanings of words.* Cambridge, MA: MIT Press. [87, 266n47]

Blurton-Jones, Nicholas, and M. Konner. 1976. !Kung knowledge of animal behavior. In *Kalahari hunter-gatherers: Studies of the !Kung San and their neighbors,* ed. R. B. Lee and I. DeVore, 325–48. Cambridge: Cambridge Univ. Press. [263n9]

Boehm, Christopher. 1983. *Montenegrin social organization and values: Political ethnography of a refuge area tribal adaptation.* New York: AMS Press. [262n26]

———. 1984. *Blood revenge: The anthropology of feuding in Montenegro and other tribal societies.* Lawrence: Univ. Press of Kansas. [281n84]

———. 1992. Segmentary "warfare" and the management of conflict: Comparison of East African chimpanzees and patrilineal-patrilocal humans. In *Coalitions and alliances in humans and other animals,* ed. A. H. Harcourt and F. B. M. DeWaal, 137–73. New York: Oxford Univ. Press. [278n14]

———. 1993. Egalitarian behavior and reverse dominance hierarchy. *Current Anthropology* 34 (3): 227–54. [228, 280n60, 281n86]

Bongaarts, John, and Susan C. Watkins. 1996. Social interactions and contemporary fertility transitions. *Population and Development Review* 22 : 639–82. [179, 272n3, 275n54, 275n70, 276n74]

Bonner, John Tyler. 1980. *The evolution of culture in animals.* Princeton, NJ: Princeton Univ. Press. [261n6]

Borgerhoff Mulder, Monique. 1988a. Behavioural ecology in traditional societies. *Trends in Ecology & Evolution* 3 : 260–64. [272n4, 276n82]

———. 1988b. Kipsigis bridewealth payments. In *Human reproductive behaviour: A Darwinian perspective,* ed. L. L. Betzig, M. Borgerhoff Mulder, and P. W. Turke, 65–82. Cambridge: Cambridge Univ. Press. [272n4, 276n82]

———. 1998. The demographic transition: Are we any closer to an evolutionary explanation? *Trends in Ecology & Evolution* 44 : 266–72. [275n54]

Bowles, Samuel. 2004. *Microeconomics: Behavior, institutions, and evolution.* New York: Russell Sage Foundation; Princeton, NJ: Princeton Univ. Press. [261n1]

Bowles, S., and H. Gintis. 1998. The moral economy of communities: Structured populations and the evolution of pro-social norms. *Evolution and Human Behavior* 19 : 3–25. [261n23]

Boyd, Robert, Herbert Gintis, Samuel Bowles, and Peter J. Richerson. 2003. The evolution of altruistic punishment. *Proceedings of the National Academy of Sciences USA* 100 : 3531–35. [278n23]

Boyd, Robert, and Peter J. Richerson. 1982. Cultural transmission and the evolution of cooperative behavior. *Human Ecology* 10 : 325–51. [273n28]

———. 1985. *Culture and the evolutionary process.* Chicago: Univ. of Chicago Press.

[265n20, 267n66, 268n1, 268n72, 269n45, 270n51, 272n13, 272n14, 273n18, 273n28, 273n34, 273n36, 281n3, 282n26]

———. 1987. The evolution of ethnic markers. *Cultural Anthropology* 2:65–79. [268n73, 269n42, 273n35, 279n46]

———. 1988a. The evolution of reciprocity in sizable groups. *Journal of Theoretical Biology* 132:337–56. [278n20]

———. 1988b. An evolutionary model of social learning: The effects of spatial and temporal variation. In *Social learning: Psychological and biological perspectives,* ed. T. Zentall and B. G. Galef, 29–48. Hillsdale, NJ.: Lawrence Erlbaum Associates. [269n40]

———. 1989a. The evolution of indirect reciprocity. *Social Networks* 11:213–36. [273n35, 278n20]

———. 1989b. Social learning as an adaptation. *Lectures on Mathematics in the Life Sciences* 20:1–26. [269n40]

———. 1990. Culture and cooperation. In *Beyond self-interest,* ed. J. J. Mansbridge, 111–32. Chicago: Univ. of Chicago Press. [278n30]

———. 1992a. How microevolutionary processes give rise to history. In *History and evolution,* ed. M. H. Nitecki and D. V. Nitecki, 178–209. Albany: State Univ. of New York Press. [266n63, 271n77]

———. 1992b. Punishment allows the evolution of cooperation (or anything else) in sizable groups. *Ethology and Sociobiology* 13:171–95. [278n23]

———. 1995. Why does culture increase human adaptability? *Ethology and Sociobiology* 16:125–43. [269n34]

———. 1996. Why culture is common but cultural evolution is rare. *Proceedings of the British Academy* 88:73–93. [263n52, 270n73]

———. 2002. Group beneficial norms can spread rapidly in a structured population. *Journal of Theoretical Biology* 215:287–96. [279n37]

Boyer, Pascal. 1994. *The naturalness of religious ideas: A cognitive theory of religion.* Berkeley and Los Angeles: Univ. of California Press. [45, 167, 262n45, 274n38, 274n43]

———. 1998. Cognitive tracks of cultural inheritance: How evolved intuitive ontology governs cultural transmission. *American Anthropologist* 100:876–89. [83, 261n23, 264n17, 266n45, 279n51]

Bradley, R. S. 1999. *Paleoclimatology: Reconstructing climates of the Quaternary.* 2nd ed. San Diego: Academic Press. [270n60]

Brandon, Robert N. 1990. *Adaptation and environment.* Princeton, NJ: Princeton Univ. Press. [127, 270n55]

Brandon, Robert N., and N. Hornstein. 1986. From icons to symbols: Some speculations on the origins of language. *Biology and Philosophy* 1:169–89. [268n1]

Brewer, Marilyn B., and Donald T. Campbell. 1976. *Ethnocentrism and intergroup attitudes: East African evidence.* Beverly Hills: Sage Publications. [280n69]

Broecker, W. 1996. Glacial climate in the tropics. *Science* 272:1902–3. [270n65]

Brooke, John L. 1994. *The refiner's fire: The making of Mormon cosmology, 1644–1844.* Cambridge: Cambridge Univ. Press. [53, 262n27, 263n63]

Brooks, A. S., J. Yellen, E. Corneliesen, M. Mehlman, and K. Stewart. 1995. A Middle Stone

Age worked bone industry from Katanda Upper Semliki Valley, Zaire. *Science* 268: 553–56. [271n87]

Brown, Donald E. 1988. *Hierarchy, history, and human nature: The social origins of historical consciousness.* Tucson: Univ. of Arizona Press. [266n62]

Brown, M. J. 1995. "We savages didn't bind feet." The implications of cultural contact and change in southwestern Taiwan for an evolutionary anthropology. Ph.D. diss., Anthropology, Univ. of Washington, Seattle.

Burke, Mary A., and Peyton Young. 2001. Competition and custom in economic contracts: A case study of Illinois agriculture. *American Economic Review* 91:559–73. [266n46]

Burrow, J. W. 1966. *Evolution and society: A study in Victorian social theory.* Cambridge: Cambridge Univ. Press. [263n2]

Bush, Vannevar. 1945. *Science, the endless frontier. A report to the President.* Washington, DC: U.S. Government Printing Office. [282n33]

Buss, David M. 1999. *Evolutionary psychology: The new science of the mind.* Boston: Allyn and Bacon. [18, 261n3]

Bynon, Theodora. 1977. *Historical linguistics.* Cambridge: Cambridge Univ. Press. [266n41, 266n43, 266n52]

Byrne, Richard W. 1999. Cognition in great ape foraging ecology: Skill learning ability opens up foraging opportunities. In *Mammalian social learning: Comparative and ecological perspectives,* ed. H. O. Box and K. R. Gibson, 333–50. Cambridge: Cambridge Univ. Press. [270n57]

Camerer, Colin. 2003. *Behavioral game theory: Experiments on social interaction.* Princeton, NJ: Princeton Univ. Press. [280n59]

Campbell, Donald T. 1960. Blind variation and selective retention in creative thought as in other knowledge processes. *Psychological Review* 67:380–400. [17, 263n55]

———. 1965. Variation and selective retention in socio-cultural evolution. In *Social change in developing areas: A reinterpretation of evolutionary theory,* ed. H. R. Barringer, G. I. Blanksten, and R. W. Mack, 19–49. Cambridge, MA: Schenkman Publishing Company. [17, 260n15, 263n55]

———. 1969. Ethnocentrism of disciplines and the fish-scale model of omniscience. In *Interdisciplinary relationships in the social sciences,* ed. M. Sherif and C. W. Sherif, 328–48. Chicago: Aldine Publishing Company. [17, 282n9]

———. 1974. Evolutionary epistemology. In *The philosophy of Karl Popper,* ed. P. A. Schilpp, 413–63. LaSalle, IL: Open Court Publishing Co. [274n42]

———. 1975. On the conflicts between biological and social evolution and between psychology and moral tradition. *American Psychologist* 30:1103–26. [263n55]

———. 1979. A tribal model of the social system vehicle carrying scientific knowledge. *Knowledge: Creation, Diffusion, Utilization* 1:181–201. [282n9]

———. 1986a. Science policy from a naturalistic sociological epistemology. In *PSA* 2:14–29. [282n9]

———. 1986b. Science's social system of validity-enhancing collective belief change and the problems of the social sciences. In *Metatheory in the social sciences: Pluralisms and subjectivities,* ed. D. W. Fiske and R. A. Shweder, 108–35. Chicago: Univ. of Chicago Press.

Carneiro, Robert. 2003. *Evolutionism in cultural anthropology.* Boulder, CO: Westview Press. [263n5]

Carpenter, Stephen R. 1989. Replication and treatment strength in whole-lake experiments. *Ecology* 70:1142–52.

Carroll, Robert L. 1997. *Patterns and processes of vertebrate evolution.* New York: Cambridge Univ. Press. [260n15, 272n6]

Caspi, Avshalom, Terrie E. Moffitt, Phil A. Silva, Magda Stouthamer-Loeber, Robert F. Krueger, and Pamela S. Schmutte. 1994. Are some people crime-prone? Replications of the personality-crime relationship across countries, genders, races, and methods. *Criminology* 32:163–195. [279n53]

Castro, Laureano, and Miguel A. Toro. 1998. The long and winding road to the ethical capacity. *History and Philosophy of the Life Sciences* 20:77–92. [265n30, 271n98]

Cavalli-Sforza, Luigi L., and Marcus.W. Feldman, 1976. Evolution of continuous variation: Direct approach through joint distribution of genotypes and phenotypes. *Proc. Natl. Acad. Sci. U.S.A.* 73:1689–92. [266n53]

———. 1981. *Cultural transmission and evolution: A quantitative approach.* Monographs in Population Biology, vol. 16. Princeton, NJ: Princeton Univ. Press. [79, 261n23, 265n36, 266n53, 267n66, 268n72, 271n83, 279n34, 281n3]

Cavalli-Sforza, L. L., M. W. Feldman, K. H. Chen, and S. M. Dornbusch. 1982. Theory and observation in cultural transmission. *Science* 218:19–27. [282n28]

Cavalli-Sforza, L. L., Paolo Menozzi, and Alberto Piazza. 1994. *The history and geography of human genes.* Princeton, NJ: Princeton Univ. Press. [276n2]

Centers for Disease Control. 1993. Surveillance for and comparison of birth defect prevalences in two geographic areas—United States, 1983–88. *CDC Weekly Mortality and Morbidity Report,* vol. 42 (March). [260n11]

Chagnon, Napoleon A., and William Irons. 1979. *Evolutionary biology and human social behavior: An anthropological perspective.* North Scituate, MA: Duxbury Press. [260n9]

Cheney, Dorothy L., and Robert M. Seyfarth. 1990. *How monkeys see the world: Inside the mind of another species.* Chicago: Univ. of Chicago Press. [271n74]

Chomsky, Noam, and Morris Halle. 1968. *The sound pattern of English.* New York: Harper & Row. [266n41]

Chou, L. S., and P. J. Richerson. 1992. Multiple models in social transmission of food selection by Norway rats, *Rattus norvegicus. Animal Behaviour* 44:337–43. [268n21]

Chrislock, C. H. 1971. *The Progressive Era in Minnesota 1988–1918.* St. Paul: Minnesota Historical Society.

Churchland, Patricia Smith. 1989. *Neurophilosophy: Toward a unified science of the mind-brain, computational models of cognition and perception.* Cambridge, MA: MIT Press. [264n13]

Coale, Ansley J., and Susan Cotts Watkins. 1986. *The decline of fertility in Europe.* Princeton, NJ: Princeton Univ. Press. [170, 172, 275n51, 275n52, 275n55]

Cohen, Abner. 1974. *Two-dimensional man: An essay on the anthropology of power and symbolism in complex society.* Berkeley and Los Angeles: Univ. of California Press. [279n42, 282n15]

Cohen, Mark N. 1977. *The food crisis in prehistory: Overpopulation and the origins of agriculture.* New Haven, CT: Yale Univ. Press. [263n7]

Connor, R. C., J. Mann, P. L. Tyack, and H. Whitehead. 1998. Social evolution in toothed whales. *Trends in Ecology and Evolution* 13:228–32. [269n32]

Corning, Peter A. 1983. *The Synergism hypothesis: A theory of progressive evolution*. New York: McGraw-Hill. [271n103, 277n8]

———. 2000. The synergism hypothesis: On the concept of synergy and its role in the evolution of complex systems. *Journal of Social and Evolutionary Systems* 21:133–72. [277n8]

Cosmides, Leda, and John Tooby. 1989. Evolutionary psychology and the generation of culture. 2. Case study: A computational theory of social exchange. *Ethology and Sociobiology* 10:51–97. [146, 189, 280n70]

Coyne, Jerry A., Nicholas H. Barton, and Michael Turelli. 2000. Is Wright's shifting balance process important in evolution? *Evolution* 54:306–17. [279n34]

Cronin, Helena. 1991. *The ant and the peacock: Altruism and sexual selection from Darwin to today*. Cambridge: Cambridge Univ. Press. [272n7]

Crosby, Alfred W. 1972. *The Columbian exchange: Biological and cultural consequences of 1492*. Westport, CT: Greenwood. [265n23]

———. 1986. *Ecological imperialism: The biological expansion of Europe, 900–1900*. Studies in Environment and History. Cambridge: Cambridge Univ. Press. [265n23]

Curtin, Philip D. 1984. *Cross-cultural trade in world history*. Studies in Comparative World History. Cambridge: Cambridge Univ. Press. [281n94]

Custance, D., A. Whiten, and T. Fredman. 1999. Social learning of an artificial fruit task in Capuchin monkeys (*Cebus apella*). *Journal of Comparative Psychology* 113:13–23. [109, 269n28]

Cziko, Gary. 1995. *Without miracles: Universal selection theory and the second Darwinian revolution*. Cambridge, MA: MIT Press. [260n15]

Darwin, Charles. 1874. *The descent of man and selection in relation to sex*. 2nd ed. 2 vols. New York: American Home Library. [254, 260n19, 273n31, 279n32]

———. 1902. *Journal of researches by Charles Darwin*. [2nd] ed. New York: P. F. Collier. [266–67n64, 283n40]

Dawkins, Richard. 1976. *The selfish gene*. Oxford: Oxford Univ. Press. [79, 152, 260n16, 265n36, 272n10]

———. 1982. *The extended phenotype: The gene as the unit of selection*. San Francisco: Freeman. [82, 152, 265n35, 272n10]

———. 1989. *The selfish gene*. New ed. Oxford: Oxford Univ. Press. [260n15]

Dawson, B. V., and B. M. Foss. 1965. Observational learning in budgerigars. *Animal Behaviour* 13:470–74. [269n32]

Dean, C., M. G. Leakey, D. Reid, F. Schrenk, G. T. Schwartz, C. Stringer, and A. Walker. 2001. Growth processes in teeth distinguish modern humans from *Homo erectus* and earlier hominins. *Nature* 414:628–31. [141, 271n82]

deMenocal, P. B. 1995. Plio-Pleistocene African climate. *Science* 270:53–59. [270n67]

Dennett, Daniel C. 1995. *Darwin's dangerous idea: Evolution and the meanings of life*. New York: Simon & Schuster. [260n15, 266n38]

Diamond, Jared. 1978. The Tasmanians: The longest isolation, the simplest technology. *Nature* 273:185–86. [54, 263n64, 271n75]

————. 1992. Diabetes running wild. *Nature* 357:362.

————. 1996. Empire of uniformity. *Discover,* March, 78–85.

————. 1997. *Guns, germs, and steel: The fates of human societies:* New York: W. W. Norton; London: Jonathan Cape/Random House. [54, 263n64]

Ditlevsen, P. D., H. Svensmark, and S. Johnsen. 1996. Contrasting atmospheric and climate dynamics of the last-glacial and Holocene periods. *Nature* 379:810–12. [270n62]

Dobzhansky, Theodosius. 1973. Nothing in biology makes sense except in the light of evolution. *American Biology Teacher* 35:125–29. [237, 238, 281n1]

Donald, Merlin. 1991. *Origins of the modern mind: Three stages in the evolution of culture and cognition.* Cambridge, MA: Harvard Univ. Press. [144, 263–64n10, 271n96, 282n25]

Dorale, J. A., R. L. Edwards, E. Ito, and L. A. Gonzales. 1998. Climate and vegetation history of the midcontinent from 75 to 25 ka: A speleothem record from Crevice Cave, Missouri, USA. *Science* 282:1871–74. [270n63]

Dudley, R. 2000. The evolutionary physiology of animal flight: Paleobiological and present perspectives. *Annual Review of Physiology* 62:135–55. [273n17]

Dumézil, G. G. 1958. *L'Ideologie Tripartie des Indo-Europeens.* Brussels: Colléction Latomus, vol. XXXI, *Latomus—Revue d'études latines.* [93, 266n61]

Dunbar, Robin I. M. 1992. Neocortex size as a constraint on group size in primates. *Journal of Human Evolution* 22:469–93. [271n76, 280n70]

————. 1996. *Grooming, gossip and the evolution of language.* London: Faber. [271n97]

————. 1998. The social brain hypothesis. *Evolutionary Anthropology* 6:178–90. [271n76]

Dunnell, R. C. 1978. Style and function: A fundamental dichotomy. *American Antiquity* 43:192–202. [282n13]

Durham, William H. 1976. The adaptive significance of cultural behavior. *Human Ecology* 4:89–121. [272n15]

————. 1991. *Coevolution: Genes, culture, and human diversity.* Stanford, CA: Stanford Univ. Press. [71, 79, 261n23, 265n25, 265n36, 272n15, 276n1]

Durham, William H., and Peter Weingart. 1997. Units of culture. In *Human by nature: Between biology and the social sciences,* ed. P. Weingart, S. D. Mitchell, P. J. Richerson, and S. Maasen, 300–13. Mahwah, NJ: Lawrence Erlbaum Associates. [266n38]

Easterbrook, G. 1997. Science and God: A warming trend? *Science* 277:890–93. [283n38]

Easterlin, R. A., C. M. Schaeffer, and D. J. Macunovich. 1993. Will the baby boomers be less well off than their parents? Income, wealth, and family circumstances over the life cycle in the United States. *Population and Development Review* 19:497–522. [275n62]

Eaves, L. J., N. G. Martin, and H. J. Eysenck. 1989. *Genes, culture, and personality: An empirical approach.* San Diego: Academic Press. [262n31, 265n33]

Eberhard, William G. 1990. Animal genitalia and female choice. *American Scientist* 78:134–41. [273n33]

Edgerton, Robert B. 1971. *The individual in cultural adaptation: A study of four East African peoples.* Berkeley and Los Angeles: Univ. of California Press. [26, 55, 261n12, 261n14]

————. 1992. *Sick societies: Challenging the myth of primitive harmony.* New York: Free Press. [282n18]

Ehrlich, Paul R., and Peter H. Raven. 1964. Butterflies and plants: A study in coevolution. *Evolution* 18:586–608. [276n3]

Eibl-Eibesfeldt, Irenäus. 1989. *Human ethology.* New York: Aldine De Gruyter. [280n60, 281n90]

Eisenberg, John Frederick. 1981. *The mammalian radiations: An analysis of trends in evolution, adaptation, and behavior.* Chicago: Univ. of Chicago Press. [270n68]

Endler, John A. 1986. *Natural selection in the wild.* Monographs in Population Biology 21. Princeton, NJ: Princeton Univ. Press. [267n68, 282n27]

Epstein, T. S. 1968. *Capitalism, primitive and modern: Some aspects of Tolai economic growth.* Manchester: Manchester Univ. Press. [262n22]

Eshel, I. 1972. On the neighborhood effect and the evolution of altruistic traits. *Theoretical Population Biology* 3:258–77. [278n29]

Evans-Pritchard, E. E. 1940. *The Nuer: A description of the modes of livelihood and political institutions of a nilotic people.* Oxford: Clarendon Press. [281n87]

Ewers, John C. 1958. *The Blackfeet: Raiders on the northwestern Plains.* The Civilization of the American Indian. Norman: Univ. of Okalahoma Press.

Fagan, Brian M. 2002. *The little Ice Age: How climate made history, 1300–1850.* 1st pbk. ed. New York: Basic Books. [270n64]

Falk, D. 1983. Cerebral cortices of East-African early hominids. *Science* 221:1072–74. [271n93]

Fehr, E., and S. Gächter. 2002. Altruistic punishment in humans. *Nature* 415:137–40. [219, 220, 280n61]

Feldman, Marcus W., and Richard C. Lewontin. 1975. The heritability hangup. *Science* 190:1163–68. [262n33]

Feldman, M. W., and S. P. Otto. 1997. Twin studies, heritability, and intelligence. *Science* 278:1383–84. [262n33, 273n19]

Finke, R., and R. Stark. 1992. *The churching of America, 1776–1990: Winners and losers in our religious economy.* New Brunswick, NJ: Rutgers Univ. Press. [263n62]

Finney, Ben R. 1972. Big men, half-men, and trader chiefs: Entrepreneurial styles in New Guinea and Polynesia. In *Opportunity and response: Case studies in economic development,* ed. T. S. Epstein and D. H. Penny, 114–261. London: Hurst. [262n22]

Fisher, Ronald A. 1958. *The genetical theory of natural selection.* Rev. ed. New York: Dover. [88, 164, 273n32]

Foster, George M. 1960. *Culture and conquest: America's spanish heritage.* Viking Fund Publications in Anthropology, no. 27. New York: Wenner-Gren Foundation for Anthropological Research.

Fox, Richard Gabriel, and Barbara J. King. 2002. *Anthropology beyond culture.* Oxford: Berg.

Frank, Steven A. 2002. *Immunology and evolution of infectious disease.* Princeton, NJ: Princeton Univ. Press. [278n27]

Frogley, M. R., P. C. Tzedakis, and T. H. E. Heaton. 1999. Climate variability in northwest Greece during the last interglacial. *Science* 285:1886–89. [270n63]

Fukuyama, Francis. 1995. *Trust: Social virtues and the creation of prosperity.* New York: Free Press. [281n94]

Gadgil, Madhav, and K. C. Malhotra. 1983. Adaptive significance of the Indian caste system: An ecological perspective. *Annals of Human Biology* 10:465–78. [281n94]

Galef, B. G. Jr. 1988. Imitation in animals: History, definition, and interpretation of data from the psychological laboratory. In *Social learning: Psychological and biological perspectives,* ed. T. R. Zentall and B. G. Galef Jr., 3–28. Hillsdale, NJ: Lawrence Erlbaum Associates. [264n15, 268n22, 269n25, 269n27, 269n33]

———. 1996. Social enhancement of food preferences in Norway rats: A brief review. In *Social learning in animals: The roots of culture,* ed. C. M. Heyes and B. G. Galef Jr., 49–64. San Diego: Academic Press. [106, 261n20, 268n17]

Gallardo, Helio. 1993. *500 Año: Fenomenología del Mestizo: Violencia y Resistencia.* 1st ed. San José, Costa Rica: Editorial Departamento Ecuménico de Investigaciones.

Gallistel, C. R. 1990. *The organization of learning: Learning, development, and conceptual change.* Cambridge, MA: MIT Press. [262n43, 264n13]

Garthwaite, Gene R. 1993. Reimagined internal frontiers: Tribes and nationalism—Bakhtiyari and Kurds. In *Russia's Muslim frontiers: New directions in cross-cultural analysis,* ed. D. F. Eickelman, 130–48. Bloomington: Indiana Univ. Press. [281n94]

Ghiselin, Michael T. 1974. *The economy of nature and the evolution of sex.* Berkeley and Los Angeles: Univ. of California Press. [216, 280n57]

Gibbs, H. R., and P. L. Grant. 1987. Oscillating selection on Darwin's finches. *Nature* 327: 511–14. [42, 262n39]

Gigerenzer, G., and D. G. Goldstein. 1996. Reasoning the fast and frugal way: Models of bounded rationality. *Psychological Review* 103:650–69. [263n54]

Gil-White, Francisco J. 2001. Are ethnic groups biological "species" to the human brain? Essentialism in our cognition of some social categories. *Current Anthropology* 42:515–54. [222, 223, 261n23, 262n29, 280n67]

Glance, Natalie S., and Bernardo A. Huberman. 1994. Dynamics of social dilemmas. *Scientific American* 270:58–63. [278n20]

Glazer, Nathan, Daniel P. Moynihan, and Corinne Schelling, eds. 1975. *Ethnicity: Theory and experience.* Cambridge, MA: Harvard Univ. Press. [280n62]

Glickman, Maurice. 1972. The Nuer and the Dinka, a further note. *Man,* n.s. 7:587–94. [25, 261n11]

Gould, Stephen Jay. 1977. The return of the hopeful monster. *Natural History* 86:22–30. [49, 263n51]

———. 2002. *The structure of evolutionary theory.* Cambridge, MA: Harvard Univ. Press. [150, 272n6]

Gould, S. J., and R. C. Lewontin. 1979. The spandrels of San Marco and the panglossian paradigm: A critique of the adaptationist programme. *Proceedings of the Royal Society of London,* ser. B 205:581–98. [102, 103, 260n15, 268n4]

Grafen, Alan. 1984. A geometric view of relatedness. *Oxford Surveys of Evolutionary Biology* 2:28–89. [278n27]

———. 1990a. Biological signals as handicaps. *Journal of Theoretical Biology* 144:517–46. [274n37]

———. 1990b. Sexual selection unhandicapped by the Fisher process. *Journal of Theoretical Biology* 144:473–516. [274n37]

Graham, J. B., R. Dudley, N. M. Aguilar, and C. Gans. 1995. Implications of the Late Palaeozoic oxygen pulse for physiology and evolution. *Nature* 375:117–20.

Greeley, A. M., and W. C. McCready. 1975. The transmission of cultural heritages: The case of the Irish and Italians. In *Ethnicity: Theory and experience,* ed. N. A. Glazer and D. P. Moynihan, 209–35. Cambridge, MA: Harvard Univ. Press. [27, 261n18]

Griffiths, Paul E. 1997. *What emotions really are: The problem of psychological categories, science, and its conceptual foundations.* Chicago: Univ. of Chicago Press. [264n11]

Grousset, René. 1970. *The empire of the steppes: A history of central Asia.* New Brunswick, NJ: Rutgers Univ. Press.

Grove, Jean M. 1988. *The Little Ice Age.* London: Methuen. [270n64]

Gruter, Margaret, and Roger D. Masters. 1986. Ostracism as a social and biological phenomenon: An introduction. *Ethology and Sociobiology* 7 : 149–58. [279n50]

Haldane, J. B. S. 1927. Possible worlds. In *Possible worlds and other essays.* London: Chatto & Windus. [256, 283n42]

Hallowell, A. I. 1963. American Indians, white and black: The phenomenon of transculturalization. *Current Anthropology* 4 : 519–31. [262n37]

Hallpike, C. R. 1986. *The principles of social evolution.* New York: Oxford Univ. Press. [90, 266n54, 266n61, 272n1]

Hamilton, William D. 1964. Genetic evolution of social behavior I, II. *Journal of Theoretical Biology* 7 : 1–52. [198, 278n15]

———. 1967. Extraordinary sex ratios. *Science* 156 : 477–88. [265n35, 272n11]

Handelman, Stephen. 1995. *Comrade criminal: Russia's new Mafiya.* New Haven, CT: Yale Univ. Press. [262n25]

Harpending, H. C., and A. Rogers. 2000. Genetic perspectives on human origins and differentiation. *Annual Review of Genomics and Human Genetics* 1 : 361–85. [281–82n4]

Harpending, H. C., and J. Sobus. 1987. Sociopathy as an adaptation. *Ethology and Sociobiology* 8 (suppl.): 63–72. [279n49]

Harris, Judith R. 1998. *The nurture assumption: Why children turn out the way they do.* New York: Free Press. [273n18]

Harris, Marvin. 1972. *Cows, pigs, wars, and witches: The riddles of culture.* New York: Random House. [272n2]

———. 1977. *Cannibals and kings: The origins of cultures.* New York: Random House. [263n7, 272n2]

———. 1979. *Cultural materialism: The struggle for a science of culture.* New York: Random House. [29, 262n23, 263n4, 263n7, 272n2]

Heard, J. Norman. 1973. *White into red: A study of the assimilation of white persons captured by Indians.* Lanham, MD: The Scarecrow Press, Inc. [41, 42, 262n37]

Hendy, I. L., and J. P. Kennett. 2000. Dansgaard-Oeschger cycles and the California Current system: Planktonic foraminiferal response to rapid climate change in Santa Barbara Basin, Ocean Drilling Program Hole 893A. *Paleooceanography* 15 : 30–42. [270n63]

Henrich, Joseph. 2001. Cultural transmission and the diffusion of innovations: Adoption dynamics indicate that biased cultural transmission is the predominate force in behavioral change. *American Anthropologist* 103 : 992–1013. [124, 261n23, 265n22]

———. 2004. Demography and cultural evolution, why adaptive cultural processes produced maladaptive losses in Tasmania. *American Antiquity* 69 : 197–214. [145, 278n27]

———. 2004. Cultural group selection, coevolutionary processes and large-scale cooperation. *Journal of Economic Behavior and Organization* 53 : 3–35. [263n64, 271n99]

Henrich, Joseph, and Robert Boyd. 1998. The evolution of conformist transmission and the emergence of between-group differences. *Evolution and Human Behavior* 19:215–41. [261n23, 269n42]

———. 2001. Why people punish defectors—Weak conformist transmission can stabilize costly enforcement of norms in cooperative dilemmas. *Journal of Theoretical Biology* 208:79–89. [278–79n31]

———. 2002. On modeling cognition and culture: Why replicators are not necessary for cultural evolution. *Culture and Cognition* 2:67–112. [278n27]

Henrich, J., R. Boyd, S. Bowles, C. Camerer, E. Fehr, H. Gintis, and R. McElreath. 2001. In search of Homo economicus: Behavioral experiments in 15 small-scale societies. *American Economic Review* 91:73–78.

Henrich, J., R. Boyd, S. Bowles, C. Camerer, E. Fehr, H. Gintis. 2004. *Foundations of human sociality: Economic experiments and ethnographic evidence from fifteen small-scale societies.* New York: Oxford Univ. Press. [280n59, 282n24]

Henrich, Joseph, and Francisco J. Gil-White. 2001. The evolution of prestige—Freely conferred deference as a mechanism for enhancing the benefits of cultural transmission. *Evolution and Human Behavior* 22:165–96. [270n50, 282n17]

Henshilwood, Christopher S., Francesco d'Errico, Curtis W. Marean, Richard G. Milo, and Royden Yates. 2001. An early bone tool industry from the Middle Stone Age at Blombos Cave, South Africa: Implications for the origins of modern human behaviour, symbolism and language. *Journal of Human Evolution* 41:631–78. [271n89]

Henshilwood, Christopher S., F. d'Errico, R. Yates, Z. Jacobs, C. Tribolo, G. A. T. Duller, N. Mercier, J. C. Sealey, H. Valladas, I. Watts, and A. G. Wintle. 2002. Emergence of modern human behavior: Middle Stone Age engravings from South Africa. *Science* 295:1278–80. [271n89]

Herman, Louis M. 2001. Vocal, social, and self-imitation by bottlenosed dolphins. In *Imitation in animals and artifacts,* ed. K. D. and C. L. Nehaniv, 63–108. Cambridge, MA: MIT. [269n31]

Hewlett, Barry S., and Luigi L. Cavalli-Sforza. 1986. Cultural transmission among Aka Pygmies. *American Anthropologist* 88:922–34. [157, 273n21]

Heyes, Cecilia M. 1993. Imitation, culture, and cognition. *Animal Behavior* 46:999–1010. [269n32]

———. 1996. Genuine imitation? In *Social learning in animals: The roots of culture,* ed. C. M. Heyes and B. G. Galef Jr., 371–89. San Diego: Academic Press. [269n27]

Heyes, Cecilia M., and G. R. Dawson. 1990. A demonstration of observational learning using a bidirectional control. *Quarterly Journal of Experimental Psychology* 42B:59–71. [269n30]

Hildebrandt, William R., and Kelly R. McGuire. 2002. The ascendance of hunting during the California Middle Archaic: An evolutionary perspective. *American Antiquity* 67:231–56.

Hill, R. C., and F. P. Stafford. 1974. The allocation of time to preschool children and educational opportunity. *Journal of Human Resources* 9:323–41. [275n66]

Hirschfeld, Lawrence A., and Susan A. Gelman. 1994. *Mapping the mind: Domain specificity in cognition and culture.* Cambridge: Cambridge Univ. Press. [262n44]

Hodder, Ian. 1978. *The spatial organisation of culture, new approaches in archaeology.* Pittsburgh: Univ. of Pittsburgh Press. [260n60]

Hodgson, Geoffrey M. 2004. *Reconstructing institutional economics: Evolution, agency and structure in American institutionalism.* London: Routledge. [261n22]

Hofreiter, M., D. Serre, H. N. Poinar, M. Kuch, and S. Pääbo. 2001. Ancient DNA. *Nature Reviews Genetics* 2:353–60. [271n91]

Hofstede, Geert H. 1980. *Culture's consequences: International differences in work-related values.* Beverly Hills, CA: Sage Publications. [28, 261n20]

Holden, C., and R. Mace. 1997. Phylogenetic analysis of the evolution of lactose digestion in adults. *Human Biology* 69:605–28. [276n2]

Holloway, Ralph. 1983. Human paleontological evidence relevant to language behavior. *Human Neurobiology* 2:105–14. [271n93]

Holway, D. A., A. V. Suarez, and T. J. Case. 1998. Loss of intraspecific aggression in the success of a widespread invasive social insect. *Science* 282:949–52. [282n11]

Hostetler, John Andrew. 1993. *Amish society.* 4th ed. Baltimore: Johns Hopkins Univ. Press. [276n75]

Humphrey, Nicolas. 1976. The social function of the intellect. In *Growing points in ethology,* ed. P. P. G. Bateson and R. A. Hinde, 303–17. Cambridge: Cambridge Univ. Press. [271n76]

Hunt, G. R. 1996. Manufacture and use of hook-tools by New Caledonian crows. *Nature* 379:1249–51. [270n57]

Iannaccone, L. R. 1994. Why strict churches are strong. *American Journal of Sociology* 99: 1180–1211. [263n62]

Inglehart, R., and J.-R. Rabier. 1986. Aspirations adapt to situations—but why are the Belgians so much happier the French? A cross-cultural analysis of the subjective quality of life. In *Research on the quality of life,* ed. F. M. Andrews, 1–56. Survey Research Center, Institute for Social Research, Univ. of Michigan. [281n96, 282n19]

Ingman, M., H. Kaessmann, S. Pääbo, and U. Gyllensten. 2000. Mitochondrial genome variation and the origin of modern humans. *Nature* 408:708–13. [271n90]

Ingold, Tim. 1986. *Evolution and social life.* Cambridge: Cambridge Univ. Press. [259–60n7]

Inkeles, A., and D. H. Smith. 1974. *Becoming modern: Individual change in six developing countries.* Cambridge, MA: Harvard Univ. Press. [275n68]

Insko, C. A., R. Gilmore, S. Drenan, A. Lipsitz, D. Moehle, and J. Thibaut. 1983. Trade versus expropriation in open groups: A comparison of two type of social power. *Journal of Personality and Social Psychology* 44:977–99. [280n60, 281n89, 282n28]

Irons, William. 1979. Cultural and biological success. In *Evolutionary biology and human social behavior,* ed. N. A. Chagnon and W. Irons, 257–72. North Scituate, MA: Duxbury Press. [272n4, 272n15]

Iwasa, Y., and A. Pomiankowski. 1995. Continual change in mate preferences. *Nature* 377: 420–22. [273n32]

Jablonka, Eva, and Marion J. Lamb. 1995. *Epigenetic inheritance and evolution: The Lamarckian dimension.* Oxford: Oxford Univ. Press. [265n35, 269n38]

Jackendoff, Ray. 1990. What would a theory of language evolution have to look like? *Behavioral and Brain Sciences* 13:737–38. [62–63, 264n14]

Jacobs, R. C., and D. T. Campbell. 1961. The perpetuation of an arbitrary tradition through several generations of laboratory microculture. *Journal of Abnormal and Social Psychology* 62:649–68. [123, 269n46]

Jain, A. K. 1981. The effect of female education on fertility: A simple explanation. *Demography* 18:577–95. [275n68]

Janssen, S. G., and R. M. Hauser. 1981. Religion, socialization, and fertility. *Demography* 18:511–28. [76, 77, 265n31]

Jerison, H. J. 1973. *Evolution of the brain and intelligence.* New York: Academic Press. [270n66]

Johnson, Allen W., and Timothy K. Earle. 2000. *The evolution of human societies: From foraging group to agrarian state.* 2nd ed. Stanford, CA: Stanford Univ. Press. [263n5, 263n6, 263n7]

Johnson, Paul. 1976. *A history of Christianity.* London: Weidenfeld & Nicolson. [279n39]

Jones, Archer. 1987. *The art of war in the Western world.* Urbana: Univ. of Illinois Press.

Jorgensen, Joseph G. 1980. *Western Indians: Comparative environments, languages, and cultures of 172 western American Indian tribes.* San Francisco: W. H. Freeman. [266n60, 267n68, 277–78n13, 279n35, 280n73, 280–81n82, 282n22]

Joshi, N. V. 1987. Evolution of cooperation by reciprocation within structured demes. *Journal of Genetics* 66:69–84. [278n20]

Juergensmeyer, Mark. 2000. *Terror in the mind of God: The global rise of religious violence.* Updated ed. Berkeley and Los Angeles: Univ. of California Press. [281n95]

Kaessmann, H., and S. Pääbo. 2002. The genetical history of humans and the great apes. *Journal of Internal Medicine* 251:1–18. [271n90]

Kameda, Tatsuya, and Diasuke Nakanishi. 2002. Cost-benefit analysis of social/cultural learning in a nonstationary uncertain environment: An evolutionary simulation and an experiment with human subjects. *Evolution and Human Behavior* 23:373–93. [269n35, 269n36, 269n42]

Kaplan, Hillard S., K. Hill, J. Lancaster, and A. M. Hurtado. 2000. A theory of human life history evolution: Diet, intelligence, and longevity. *Evolutionary Anthropology* 9:156–85. [128, 129, 270n56, 270n58, 277n11]

Kaplan, Hillard S., and Jane B. Lancaster. 1999. The evolutionary economics and psychology of the demographic transition to low fertility. In *Adaptation and human behavior: An anthropological perspective,* ed. L. Cronk, N. Chagnon, and W. Irons, 283–322. New York: Aldine de Gruyter. [149, 272n5]

Kaplan, Hillard, Jane B. Lancaster, J. Bock, and S. Johnson. 1995. Does observed fertility maximize fitness among New Mexico men? A test of an optimality model and a new theory of parental investment in the embodied capital of offspring. *Human Nature* 6:325–60. [173, 275n57]

Kaplan, Hillard S., and A. J. Robson. 2002. The emergence of humans: The coevolution of intelligence and longevity with intergenerational transfers. *Proceedings of the National Academy of Sciences USA* 99:10221–26. [135, 270n71, 270n72]

Karlin, Samuel. 1979. Models of multifactorial inheritance. 1. Multivariate formulations and basic convergence results. *Theoretical Population Biology* 15:308–55. [266n53]

Kasarda, J. D., J. O. G. Billy, and K. West. 1986. *Status enhancement and fertility: Reproductive responses to social mobility and educational opportunity.* New York: Academic Press, Inc. [275n63]

Keeley, Lawrence H. 1996. *War before civilization.* New York: Oxford Univ. Press. [279n35, 280n80]

Keller, A. G. 1931. *Societal evolution: A study of the evolutionary basis of the science of society.* New York: The Macmillan Company. [263n57]

Keller, Laurent. 1995. Social life: The paradox of multiple-queen colonies. *Trends in Ecology & Evolution* 10:355–60. [282n11]

Keller, Laurent, and Michel Chapuisat. 1999. Cooperation among selfish individuals in insect societies. *Bioscience* 49:899–909. [278n17]

Keller, L., and K. G. Ross. 1993. Phenotypic plasticity and "cultural transmission" of alternative social organizations in the fire ant *Solenopsis invicta. Behavioral Ecology and Sociobiology* 33:121–29. [282n11]

Kellett, Anthony. 1982. *Combat motivation: The behavior of soldiers in battle.* Boston: Kluwer. [281n93]

Kelly, Raymond C. 1985. *The Nuer conquest: The structure and development of an expansionist system.* Ann Arbor: Univ. of Michigan Press. [23, 24, 261n10, 281n87]

Kelly, Robert L. 1995. *The Foraging spectrum: Diversity in hunter-gatherer lifeways.* Washington, DC: Smithsonian Institution Press. [279n52, 280n55]

Kennedy, Paul M. 1987. *The rise and fall of the great powers: Economic change and military conflict from 1500 to 2000.* 1st ed. New York: Random House. [281n89]

Khazanov, Anatoly M. 1994. *Nomads and the outside world.* 2nd ed. Madison: Univ. of Wisconsin Press.

Kiester, A. Ross. 1996/1997. Aesthetics of biodiversity. *Human Ecology Review* 3:151–57. [283n39]

Kirk, D. 1996. Demographic transition theory. *Population Studies* 50:361–87. [275n54]

Klein, R. G. 1999. *The Human career: Human biological and cultural origins.* 2nd ed. Chicago: Univ. of Chicago Press. [270n67, 277n5, 282n6]

Knauft, Bruce M. 1985a. *Good company and violence: Sorcery and social action in a lowland New Guinea society.* Studies in Melanesian Anthropology. Berkeley: Univ. of California Press. [168, 274n47, 280n79, 282n18]

———. 1985b. Ritual form and permutation in New Guinea: Implications of symbolic process for socio-political evolution. *American Ethnologist* 12:321–40. [280–81n82]

———. 1986. Divergence between cultural success and reproductive fitness in preindustrial cities. *Cultural Anthropology* 2:94–114. [275n51]

———. 1987. Reconsidering violence in simple human societies. *Current Anthropology* 28:457–500. [280n78]

———. 1993. *South coast New Guinea cultures: History, comparison, dialectic.* Cambridge Studies in Social and Cultural Anthropology 89. Cambridge: Cambridge Univ. Press. [261n17, 280–81n82]

Kohn, Melvin L., and Carmi Schooler. 1983. *Work and personality: An inquiry into the impact of social stratification.* Norwood, NJ: Ablex Pub. Corp. [178, 275n69]

Kraybill, Donald B., and Carl F. Bowman. 2001. *On the backroad to heaven. Old Order Hutterites, Mennonites, Amish, and Brethren.* Edited by G. F. Thompson, Center for American Places, Books in Anabaptist Studies. Baltimore: Johns Hopkins Univ. Press. [276n75]

Kraybill, D. B., and M. A. Olshan. 1994. *The Amish struggle with modernity.* Hanover, NH: Univ. Press of New England. [276n75]

Kroeber, Alfred L. 1948. *Anthropology: Race, language, culture, psychology, pre-history.* New ed. New York: Harcourt, Brace & World. [7, 208, 259n6]

Kroeber, Alfred L., and Clyde Kluckhohn. 1952. *Culture; A critical review of concepts and definitions.* Cambridge, MA: Peabody Museum of American Archæology and Ethnology, Harvard University.

Kummer, Hans, Lorraine Daston, Gerd Gigerenzer, and Joan Silk. 1997. The social intelligence hypothesis. In *Human by nature,* ed. P. Weingart, S. D. Mitchell, P. J. Richerson, and S. Maasen, 157–79. Mahwah, NJ: Lawrence Erlbaum Associates. [271n76]

Labov, William. 1973. *Sociolinguistic patterns.* Philadelphia: Univ. of Pennsylvania Press. [262n34, 276n79]

———. 1994. *Principles of linguistic change: Internal factors.* Oxford: Blackwell. [263n53, 265n24, 265n28]

———. 2001. *Principles of linguistic change: Social factors.* Oxford: Blackwell. [266n42, 270n54, 273n20, 282n32]

Lachlan, R. F., L. Crooks, and K. N. Laland. 1998. Who follows whom? Shoaling preferences and social learning of foraging information in guppies. *Animal Behaviour* 56: 181–90. [268n19]

Lack, David L. 1966. *Population studies of birds.* Oxford: Clarendon. [173, 202, 278n25]

Laitman, J. T., P. J. Gannon, and J. S. Reidenberg. 1989. Charting changes in the hominid vocal-tract-the fossil evidence. *American Journal of Physical Anthropology* 78: 257–58. [271n94]

Laland, Kevin N. 1994. Sexual selection with a culturally transmitted mating preference. *Theoretical Population Biology* 45: 1–15. [277n6]

———. 1999. Exploring the dynamics of social transmission with rats. In *Mammalian social learning: Comparative and ecological perspectives,* ed. H. O. Box and K. R. Gibson, 174–87. Cambridge: Cambridge Univ. Press.

Laland, Kevin N., J. Kumm, and Marcus W. Feldman. 1995. Gene-culture coevolutionary theory: A test case. *Current Anthropology* 36: 131–56. [276n83, 277n6]

Laland, K. N., F. J. Odling-Smee, and M. W. Feldman. 1996. The evolutionary consequences of niche construction: A theoretical investigation using two-locus theory. *Journal of Evolutionary Biology* 9: 293–316.

Lamb, H. H. 1977. *Climatic history and the future.* Princeton, NJ: Princeton Univ. Press. [270n60, 270n64]

Lanchester, F. W. 1916. *Aircraft in warfare; The dawn of the fourth arm.* London: Constable and Company Limited.

Land, Michael F., and Dan-Eric Nilsson. 2002. *Animal eyes.* Oxford: Oxford Univ. Press. [272n8]

Lande, Russell. 1976. The maintenance of genetic variability by mutation in a polygenic character with linked loci. *Genetic Research* 26: 221–35. [266n53]

———. 1985. Expected time for random genetic drift of a population between stable phenotypic states. *Proceedings of the National Academy of Sciences USA* 82: 7641–45. [279n34]

Lefebvre, L., and L.-A. Giraldeau. 1994. Cultural transmission in pigeons is affected by the number of tutors and bystanders present. *Animal Behaviour* 47:331–37. [269n36]

Lefebvre, L., and B. Palameta. 1988. Mechanisms, ecology, and population diffusion of socially-learned, food-finding behavior in feral pigeons. In *Social learning, psychological and biological perspectives,* ed. T. Zentall and J. B. G. Galef, 141–65. Hillsdale, NJ: Lawrence Erlbaum Associates. [104, 268n8, 268n18]

Lehman, S. 1993. Ice sheets, wayward winds and sea change. *Nature* 365:108–9. [270n62]

Leigh, Egbert G. J. 1977. How does selection reconcile individual advantage with the good of the group? *Proceedings of the National Academy of Sciences USA* 74:4542–46.

Leimar, Olof, and Peter Hammerstein. 2001. Evolution of cooperation through indirect reciprocity. *Proceedings of the Royal Society of London,* ser. B 268:745–53. [280n72]

LeVine, Robert Alan. 1966. *Dreams and deeds: Achievement motivation in Nigeria.* Chicago: Univ. of Chicago Press. [262n21]

LeVine, Robert, and Donald T. Campbell. 1972. *Ethnocentrism: Theories of conflict, ethnic attitudes, and group behavior.* New York: Wiley. [280n62]

Levinton, Jeffrey S. 2001. *Genetics, paleontology, and macroevolution.* 2nd ed. Cambridge: Cambridge Univ. Press. [272n6]

Lewontin, Richard C., and J. L. Hubby. 1966. A molecular approach to the study of genetic heterozygosity in natural populations. II. Amount of variation and degree of heterozygosity in natural populations of *Drosophila pseudoobscura. Genetics* 54:595–609. [245, 282n8]

Lieberman, Philip. 1984. *The biology and evolution of language.* Cambridge, MA: Harvard Univ. Press. [271n94]

Light, Ivan H. 1972. *Ethnic enterprise in America; Business and welfare among Chinese, Japanese, and Blacks.* Berkeley and Los Angeles: Univ. of California Press. [281n94]

Light, Ivan H., and Steven J. Gold. 2000. *Ethnic economies.* San Diego: Academic Press. [281n94]

Lindblom, B. 1986. Phonetic universals in vowel systems. In *Experimental phonology,* ed. J. J. Ohala and J. J. Jaeger, 13–44. Orlando, FL: Academic Press. [265n26]

———. 1996. Systemic constraints and adaptive change in the formation of sound structure. In *Evolution of human language,* ed. J. Hurford, 242–64. Edinburgh: Edinburgh Univ. Press. [265n26]

Linder, Douglas. 2003. *Famous trials: The McMartin preschool abuse trials.* Available from http://www.law.umkc.edu/faculty/projects/ftrials/mcmartin/mcmartin.html. [274n49]

Lindert, Peter H. 1978. *Fertility and scarcity in America.* Princeton, NJ: Princeton Univ. Press. [275n66]

———. 1985. English population, prices, and wages, 1541–1913. *Journal of Interdisciplinary History* 15:609–34. [275n50]

Logan, M. H., and D. A. Schmittou. 1998. The uniqueness of Crow art: A glimpse into the history of an embattled people. *Montana: The Magazine of Western History* (Summer): 58–71. [279n47]

Lumsden, Charles J., and Edward O. Wilson. 1981. *Genes, mind, and culture: The coevolu-*

tionary process. Cambridge, MA: Harvard Univ. Press. [72, 194, 260n18, 261n23, 265n27, 277n7]

Lydens, Lois A. 1988. A longitudinal study of crosscultural adoption: Identity development among Asian adoptees at adolescence and early adulthood. Ph.D. diss., Northwestern Univ., Chicago. [39, 40, 42, 262n36]

Mallory, J. P. 1989. *In search of the Indo-Europeans: Language, archaeology, and myth.* New York: Thames and Hudson. [266n50, 266n61]

Mänchen-Helfen, Otto. 1973. *The world of the Huns: Studies in their history and culture.* Berkeley and Los Angeles: Univ. of California Press.

Mansbridge, Jane J., ed. 1990. *Beyond self-interest.* Chicago: Univ. of Chicago Press. [280n56]

Margulis, Lynn. 1970. *Origin of eukaryotic cells: Evidence and research implications for a theory of the origin and evolution of microbial, plant, and animal cells on the Precambrian earth.* New Haven, CT: Yale Univ. Press. [277n10]

Marks, J., and E. Staski. 1988. Individuals and the evolution of biological and cultural systems. *Human Evolution* 3:147–61. [267n65]

Marler, Peter, and S. Peters. 1977. Selective vocal learning in a sparrow. *Science* 189:514–21. [268n16]

Martin, R. D. 1981. Relative brain size and basal metabolic rate in terrestrial vertebrates. *Nature* 293:57–60. [270n69]

Martindale, Don. 1960. *The nature and types of sociological theory.* Boston: Houghton Mifflin. [272n9]

Marty, Martin E., and R. Scott Appleby. 1991. *Fundamentalisms observed. The fundamentalism project,* vol. 1. Chicago: Univ. of Chicago Press. [263n62, 281n95]

Maynard Smith, John. 1964. Group selection and kin selection. *Nature* 201:1145–46. [202, 278n25]

Maynard Smith, John, and Eörs Szathmáry. 1995. *The major transitions in evolution.* Oxford: W. H. Freeman Spektrum. [194, 271n103, 277n9]

Mayr, Ernst. 1961. Cause and effect in biology. *Science* 134:1501–6. [5, 260n13]

———. 1982. *The growth of biological thought: Diversity, evolution, and inheritance.* Cambridge, MA: Harvard Univ. Press.

McBrearty, S., and A. S. Brooks. 2000. The revolution that wasn't: A new interpretation of the origin of modern human behavior. *Journal of Human Evolution* 39:453–563. [271n86]

McComb, K., C. Moss, S. M. Durant, L. Baker, and S. Sayialel. 2001. Matriarchs as repositories of social knowledge in African elephants. *Science* 292:491–94. [268n15]

McElreath, Richard. In press. Social learning and the maintenance of cultural variation: An evolutionary model and data from East Africa. *American Anthropologist.* [27, 261n15, 261n23, 282n23, 282n28]

McElreath, Richard, Robert Boyd, and Peter J. Richerson. 2003. Shared norms and the evolution of ethnic markers. *Current Anthropology* 44:122–29. [261n23, 279n45]

McEvoy, L., and G. Land. 1981. Life-Style and death patterns of Missouri RLDS church members. *American Journal of Public Health* 71:1350–57. [76, 265n32]

McGrew, W. C. 1992. *Chimpanzee material culture: Implications for human evolution.* Cambridge: Cambridge Univ. Press. [105, 268n7, 268n9, 268n10]

McNeill, William Hardy. 1963. *The rise of the West: A history of the human community.* New York: New American Library.

———. 1986. *Mythistory and other essays.* Chicago: Univ. of Chicago Press.

Mead, Margaret. 1935. *Sex and temperament in three primitive societies.* New York: W. Morrow & Company. [262n30]

Miller, Geoffrey F. 2000. *The mating mind: How sexual choice shaped the evolution of human nature.* 1st ed. New York: Doubleday. [274n37]

Mithen, Steven. 1999. Imitation and cultural change: A view from the Stone Age, with specific reference to the manufacture of handaxes. In *Mammalian social learning: Comparative and ecological perspectives,* ed. H. O. Box and K. R. Gibson, 389–99. Cambridge: Cambridge Univ. Press. [271n84]

Moore, Bruce R. 1996. The evolution of imitative learning. In *Social learning in animals: The roots of culture,* ed. C. M. Heyes and B. G. Galef Jr., 245–65. San Diego: Academic Press. [268n8, 268n14, 269n32]

Murdock, George Peter. 1949. *Social structure.* New York: Macmillan Co. [267n68]

———. 1983. *Outline of world cultures.* 6th rev. ed. HRAF manuals. New Haven, CT: Human Relations Area Files. [267n68]

Murphy, Robert F., and Yolanda Murphy. 1986. Northern Shoshone and Bannock. In *Handbook of North American Indians: Great Basin,* ed. W. L. d'Azevedo, 284–307. Washington, DC: Smithsonian Institution Press. [277n13]

Mussen, Paul Henry, John Janeway Conger, and Jerome Kagan. 1969. *Child development and personality.* 3rd ed. New York: Harper & Row.

Myers, D. G. 1993. *Social psychology.* 4th ed. New York: McGraw-Hill, Inc. [269n43, 269n47]

National Research Council, Committee on Abrupt Climate Change. 2002. *Abrupt climate change: Inevitable surprises.* Washington, DC: National Academy Press. [270n60, 283n43]

Needham, Joseph. 1979. *Science in traditional China: A comparative perspective.* Hong Kong: The Chinese Univ. Press. [263n60, 263n61]

———. 1987. *Science and civilization in China.* Vol. 5, pt. 7, *The gunpowder epic.* Cambridge: Cambridge Univ. Press.

Nelson, Richard R., and Sidney G. Winter. 1982. *An evolutionary theory of economic change.* Cambridge, MA: Harvard Univ. Press, Belknap Press. [252, 263–64n10, 282n29]

Nettle, D., and R. I. M. Dunbar. 1997. Social markers and the evolution of reciprocal exchange. *Current Anthropology* 38:93–99. [279n48]

Newson, Lesley. 2003. Kin, culture, and reproductive decisions. Ph.D. diss., Psychology, Univ. of Exeter. [276n77]

Nilsson, D. E. 1989. Vision optics and evolution—Nature's engineering has produced astonishing diversity in eye design. *Bioscience* 39:289–307. [268n5]

Nisbett, Richard E. 2003. *The geography of thought: How Asians and Westerners think differently—And why.* New York: Free Press. [264n11, 282n23]

Nisbett, Richard E., and Dov Cohen. 1996. *Culture of honor: The psychology of violence in the South.* New Directions in Social Psychology. Boulder, CO: Westview Press. [1, 2, 3, 259n1, 259n2]

Nisbett, R. E., K. P. Peng, I. Choi, and A. Norenzayan. 2001. Culture and systems of thought: Holistic versus analytic cognition. *Psychological Review* 108:291–310.

Nisbett, Richard E., and Lee Ross. 1980. *Human inference: Strategies and shortcomings of social judgment.* Englewood Cliffs, NJ: Prentice-Hall. [263n54]

Nonaka, K., T. Miura, and K. Peter. 1994. Recent fertility decline in Dariusleut Hutterites: An extension of Eaton and Mayer's Hutterite fertility study. *Human Biology* 66:411–20. [276n76]

North, Douglass C., and Robert P. Thomas. 1973. *The rise of the Western world: A new economic history.* Cambridge: Cambridge Univ. Press. [282n5, 282n21]

Nowak, Martin A., and Karl Sigmund. 1993. A strategy of win stay, lose shift that outperforms tit-for-tat in the prisoners dilemma game. *Nature* 364:56–58. [278n20]

———. 1998a. The dynamics of indirect reciprocity. *Journal of Theoretical Biology* 194:561–74. [278n20, 280n72]

———. 1998b. Evolution of indirect reciprocity by image scoring. *Nature* 393:573–77. [278n20, 280n72]

Odling-Smee, F. John. 1995. Niche construction, genetic evolution and cultural change. *Behavioural Processes* 35:195–202. [271n102]

Odling-Smee, F. John, Kevin N. Laland, and Marcus W. Feldman. 2003. *Niche construction: The neglected process in evolution.* Ed. S. A. Levin and H. S. Horn. Monographs in Population Biology, vol. 37. Princeton, NJ: Princeton Univ. Press. [261n5, 276–77n4, 282n30]

Oliver, Chad. 1962. *Ecology and cultural continuity as contributing factors in the social organization of the Plains Indians.* Univ. of California Publications in American Archaeology and Ethnology, vol. 48, no. 1. Berkeley and Los Angeles: Univ. of California Press. [262n41]

Opdyke, Neil D. 1995. Mammalian migration and climate over the last seven million years. In *Paleoclimate and evolution, with emphasis on human origins,* ed. E. S. Vrba, G. H. Denton, T. C. Partridge, and L. H. Burckle, 109–14. New Haven, CT: Yale Univ. Press. [270n61, 270n67]

Ostergren, R. C. 1988. *A community transplanted: The trans-Atlantic experience of a Swedish immigrant settlement in the upper Middle West, 1835–1915.* Madison: Univ. of Wisconsin Press.

Ostrom, Elinor. 1990. *Governing the commons: The evolution of institutions for collective action, the political economy of institutions and decisions.* Cambridge: Cambridge Univ. Press. [279n50]

Otterbein, Keith F. 1968. Internal war: A cross-cultural study. *American Anthropologist* 80:277–89. [280n79, 281n84]

———. 1985. *The evolution of war: A cross-cultural study.* New Haven, CT: Human Relations Area Files Press. [279n35]

Paciotti, Brian. 2002. Cultural evolutionary theory and informal social control institutions: The Sungusungu of Tanzania and honor in the American South. Ph.D. diss., Ecology Graduate Group, Univ. of California–Davis. [261n16, 279n50]

Paldiel, Mordecai. 1993. The path of the righteous: Gentile rescuers of Jews during the Holocaust. Hoboken, NJ: Ktav. [280n68]

Palmer, C. T., B. E. Fredrickson, and C. F. Tilley. 1997. Categories and gatherings: Group selection and the mythology of cultural anthropology. *Evolution and Human Behavior* 18:291–308. [279n33]

Panchanathan, Karthik, and Robert Boyd. 2003. A tale of two defectors: The importance of standing for evolution of indirect reciprocity. *Journal of Theoretical Biology* 224:115–26. [280n72]

Parker, George A., and John Maynard Smith. 1990. Optimality theory in evolutionary biology. *Nature* 348:27–33. [273n16]

Partridge, T. C., G. C. Bond, C. J. H. Hartnady, P. B. deMenocal, and W. F. Ruddiman. 1995. Climatic effects of Late Neogene tectonism and vulcanism. In *Paleoclimate and evolution with emphasis on human origins,* ed. E. S. Vrba, G. H. Denton, T. C. Partridge, and L. H. Burckle, 8–23. New Haven, CT: Yale Univ. Press. [270n60]

Pascal, Blaise. 1660. *Pensees.* Trans. W. F. Trotter. 1910 ed. available from CyberLibrary (http://www.leaderu.com/cyber/books/pensees/pensees.htm). [165, 274n39]

Pepperberg, I. M. 1999. *The Alex studies: Cognitive and communicative abilities of grey parrots.* Cambridge, MA: Harvard Univ. Press. [269n32]

Peter, K. A. 1987. *The dynamics of Hutterite society: An analytical approach.* Edmonton, Canada: Univ. of Alberta Press. [276n75]

Peters, F. E. 1994. *The Hajj: The Muslim pilgrimage to Mecca and the holy places.* Princeton, NJ: Princeton Univ. Press. [281n92]

Petroski, Henry. 1992. *The evolution of useful things.* New York: Vintage Books. [263n58, 269n37]

Pinker, Steven. 1994. *The language instinct.* 1st ed. New York: W. Morrow and Co. [280n54]

———. 1997. *How the mind works.* New York: Norton. [48, 263n49]

Pinker, S., and P. Bloom. 1990. Natural language and natural selection. *Behavioral and Brain Sciences* 13:707–84. [264n14, 268n1]

Pollack, R. A., and S. C. Watkins. 1993. Cultural and economic approaches to fertility—Proper marriage or misalliance? *Population and Development Review* 19:467–96. [275n54]

Pomiankowski, A., Y. Iwasa, and S. Nee. 1991. The evolution of costly mate preferences. 1. Fisher and biased mutation. *Evolution* 45:1422–30. [273n32]

Pospisil, Leopold J. 1978. *The Kapauku Papuans of West New Guinea.* 2nd ed. *Case studies in cultural anthropology.* New York: Holt Rinehart and Winston. [262n22]

Povinelli, D. J. 2000. *Folk physics for apes: The chimpanzee's theory of how the world works.* Oxford: Oxford Univ. Press. [271n81]

Price, George R. 1970. Selection and covariance. *Nature* 277:520–21. [202, 278n26]

———. 1972. Extensions of covariance selection mathematics. *Annals of Human Genetics* 35:485–90. [202, 278n26]

Price, T. Douglas, and James A. Brown. 1985. *Prehistoric hunter-gatherers: The emergence of cultural complexity.* Orlando, FL: Academic Press. [280n74, 280n81]

Pulliam, H. Ronald, and Christopher Dunford. 1980. *Programmed to learn: An essay on the evolution of culture.* New York: Columbia Univ. Press. [261n23]

Putnam, Robert D., Robert Leonardi, and Raffaella Nanetti. 1993. *Making democracy work:*

Civic traditions in modern Italy. Princeton, NJ: Princeton Univ. Press. [27, 261n19, 281n94]

Queller, David C. 1989. Inclusive fitness in a nutshell. *Oxford Surveys in Evolutionary Biology* 6:73–109. [278n17]

Queller, David C., and Joan E. Strassmann. 1998. Kin selection and social insects: Social insects provide the most surprising predictions and satisfying tests of kin selection. *Bioscience* 48:165–75. [278n17]

Rabinowitz, Dorothy. 2003. *No crueler tyrannies: Accusation, false witness, and other terrors of our times.* New York: Simon and Schuster. [169, 274n49]

Raine, Adrian. 1993. *The psychopathology of crime.* San Diego: Academic Press. [279n53]

Rappaport, Roy A. 1979. *Ecology, meaning, and religion.* Richmond, CA: North Atlantic Books. [279n43]

Reader, S. M. , and K. N. Laland. 2002. Social intelligence, innovation, and enhanced brain size in primates. *Proceedings of the National Academy of Sciences USA* 99:4436–41. [135, 270n70]

Rendell, Luke, and Hal Whitehead. 2001. Culture in whales and dolphins. *Behavioral & Brain Sciences* 24:309–82. [105, 268n13, 268n20]

Renfrew, C. Camerer. 1988. *Archaeology and language: The puzzle of Indo-European origins.* London: Jonathan Cape.

Rice, W. R. 1996. Sexually antagonistic male adaptation triggered by experimental arrest of female evolution. *Nature* 381:232–34. [265n35]

Richards, Robert J. 1987. *Darwin and the emergence of evolutionary theories of mind and behavior.* Chicago: Univ. of Chicago Press. [261n22, 279n32]

Richerson, Peter J. 1988. Review of "Human Nature: Darwin's View" by Alexander Alland Jr. *BioScience* 38:115–16. [261n21]

Richerson, Peter J., and Robert Boyd. 1976. A simple dual inheritance model of the conflict between social and biological evolution. *Zygon* 11:254–62. [272n10]

———. 1978. A dual inheritance model of the human evolutionary process I: Basic postulates and a simple model. *Journal of Social Biological Structures* 1:127–54. [272n10]

———. 1987. Simple models of complex phenomena: The case of cultural evolution. In *The latest on the best: Essays on evolution and optimality,* ed. J. Dupré, 27–52. Cambridge: MIT Press. [264–65n19]

———. 1989a. A Darwinian theory for the evolution of symbolic cultural traits. In *The relevance of culture,* ed. M. Freilich, 120–42. Boston: Bergin and Garvey.

———. 1989b. The role of evolved predispositions in cultural evolution: Or sociobiology meets Pascal's Wager. *Ethology and Sociobiology* 10:195–219. [274n38, 277n6]

———. 1992. Cultural inheritance and evolutionary ecology. In *Evolutionary ecology and human behavior,* ed. E. A. Smith and B. Winterhalder, 61–92. New York: Aldine De Gruyter. [281n3]

———. 1998. The evolution of human ultrasociality. In *Indoctrinability, ideology, and warfare; Evolutionary perspectives,* ed. I. Eibl-Eibesfeldt and F. K. Salter, 71–95. New York: Berghahn Books. [279n52]

———. 1999. Complex societies—The evolutionary origins of a crude superorganism. *Human Nature—An Interdisciplinary Biosocial Perspective* 10:253–89. [281n97]

————. 2000. Evolution: The Darwinian theory of social change. In *Paradigms of social change: Modernization, development, transformation, evolution,* ed. W. Schelkle, W.-H. Krauth, M. Kohli, and G. Elwert, 257–82. Frankfurt: Campus Verlag. [282n31]

————. 2001a. Built for speed, not for comfort: Darwinian theory and human culture. *History and Philosophy of the Life Sciences* 23:423–63. [261n22, 263n2, 263n8, 269n33, 279n32]

————. 2001b. The evolution of subjective commitment to groups: A tribal instincts hypothesis. In *Evolution and the capacity for commitment,* ed. R. M. Nesse, 186–220. New York: Russell Sage Foundation. [279n52]

————. 2001c. Institutional evolution in the Holocene: The rise of complex societies. In *The origin of human social institutions,* ed. W. G. Runciman, 197–234. Oxford: Oxford Univ. Press. [281n88]

Richerson, Peter J., Robert Boyd, and Robert L. Bettinger. 2001. Was agriculture impossible during the Pleistocene but mandatory during the Holocene? A climate change hypothesis. *American Antiquity* 66:387–411. [263n8, 263n48, 267n67, 281n88]

Richerson, Peter J., Robert Boyd, and Joseph Henrich. 2003. The cultural evolution and cooperation. In *Genetic and cultural evolution of cooperation,* ed. P. Hammerstein, 357–88. Berlin: MIT Press. [279n52]

Richerson, Peter J., Robert Boyd, and Brian Paciotti. 2002. An evolutionary theory of commons management. In *The drama of the commons,* ed. E. Ostrom, T. Dietz, N. Dolsak, P. C. Stern, S. Stonich, and E. U. Weber, 403–42. Washington, DC: National Academy Press. [281n97]

Ridley, Mark. 1993. *Evolution.* Cambridge, MA: Blackwell Scientific Publications.

Riolo, R. L., M. D. Cohen, and R. Axelrod. 2001. Evolution of cooperation without reciprocity. *Nature* 414:441–43. [279n48]

Robinson, J. P. , and G. Godbey. 1997. *Time for life: The surprising ways Americans use their time.* University Park, PA: Pennsylvania State Univ. Press. [275n59]

Robinson, W. P., and Henri Tajfel. 1996. *Social groups and identities: Developing the legacy of Henri Tajfel.* International Series in Social Psychology. Oxford: Butterworth-Heinemann. [280n65]

Rodseth, Lars, Richard W. Wrangham, A. M. Harrigan, and Barbara B. Smuts. 1991. The human community as a primate society. *Current Anthropology* 32:221–54. [278n14]

Roe, Frank Gilbert. 1955. *The Indian and the horse.* 1st ed. Norman: Univ. of Oklahoma Press. [262n41]

Rogers, Alan R. 1989. Does biology constrain culture? *American Anthropologist* 90:819–31. [111, 112, 113, 256, 269n34]

————. 1990a. Evolutionary economics of human reproduction. *Ethology and Sociobiology* 11:479–95. [275n56]

————. 1990b. Group selection by selective emigration: The effects of migration and kin structure. *American Naturalist* 135:398–413. [278n29]

Rogers, Everett M. 1983. *Diffusion of innovations.* 3rd ed.. New York: Free Press. [262n24, 265n21, 270n53, 273n22, 273n27, 275n71]

————. 1995. *Diffusion of innovations.* 4th ed. New York: Free Press. [279n41]

Rogers, Everett M., and F. Floyd Shoemaker. 1971. *Communication of innovations: A cross-cultural approach*. 2nd ed. New York: Free Press. [265n22]

Roof, Wade Clark, and William McKinney. 1987. *American mainline religion: Its changing shape and future*. New Brunswick, NJ: Rutgers Univ. Press. [180, 275n73, 281n95]

Rosenberg, Alexander. 1988. *Philosophy of social science*. Boulder, CO: Westview Press. [265n34]

Rosenthal, Ted L., and Barry J. Zimmerman. 1978. *Social learning and cognition*. New York: Academic Press. [268n1]

Ruhlen, Merritt. 1994. *The origin of language: Tracing the evolution of the mother tongue*. New York: Wiley.

Russon, A. E., and B. M. F. Galdikas. 1995. Imitation and tool use in rehabilitant orangutans. In *The neglected ape*, ed. R. Nadler. New York: Plenum Press. [269n32]

Ryan, Bryce, and Neal C. Gross. 1943. The diffusion of hybrid seed corn in two Iowa communities. *Rural Sociology* 8:15–24. [69, 265n21]

Ryan, M. J. 1998. Sexual selection, receiver biases, and the evolution of sex differences. *Science* 281:1999–2003. [274n37]

Ryckman, R. M., W. C. Rodda, and W. F. Sherman. 1972. The competence of the model and the learning of imitation and nonimitation. *Journal of Experimental Psychology* 88:107–14. [270n52]

Sahlins, Marshall. 1976a. *Culture and practical reason*. Chicago: Univ. of Chicago Press. [148, 272n1, 282n12]

———. 1976b. *The use and abuse of biology*. Ann Arbor: Univ. of Michigan Press. [272n1]

Sahlins, Marshall David, Thomas G. Harding, and Elman Rogers Service. 1960. *Evolution and culture*. Ann Arbor: Univ. of Michigan Press. [263n3, 263n4, 263n6]

Salamon, Sonya. 1980. Ethnic-Differences in farm family land transfers. *Rural Sociology* 45:290–308. [23, 261n9]

———. 1984. Ethnic origin as explanation for local land ownership patterns. In *Focus on agriculture: Research in rural sociology and development*, ed. H. K. Schwarzweller. Greenwich, CT: JAI Press. [22, 23, 34, 261n8, 261n9, 262n28]

———. 1985. Ethnic-Communities and the structure of agriculture. *Rural Sociology* 50:323–40. [21, 261n7]

———. 1992. *Prairie patrimony: Family, farming, and community in the Midwest*. Studies in Rural Culture. Chapel Hill: Univ. of North Carolina Press. [66, 67, 68, 264n18]

Salamon, S., K. M. Gegenbacher, and D. J. Penas. 1986. Family factors affecting the intergenerational succession to farming. *Human Organization* 45:24–33. [23, 261n9]

Salamon, S., and S. M. O'Reilly. 1979. Family land and development cycles among Illinois farmers. *Rural Sociology* 44:525–42. [23, 261n9]

Salter, Frank K. 1995. *Emotions in command: A naturalistic study of institutional dominance*. Oxford: Oxford Univ. Press. [280n60, 281n91]

Scarr, S. 1981. *Race, social class, and individual differences in IQ*. Hillsdale, NJ: Lawrence Erlbaum Associates. [262n35]

Schor, J. B. 1991. *The overworked American: The unexpected decline of leisure*. New York: Basic Books. [275n60]

Schotter, Andrew, and Barry Sopher. 2003. Social learning and coordination conventions in inter-generational games: An experimental study. *Journal of Political Economy* 111: 498–529. [261n1]

Schulz, H., U. von Rad, and H. Erlenkeuser. 1998. Correlation between Arabian Sea and Greenland climate oscillations of the past 110,000 years. *Nature* 393:54–57. [270n63]

Schwartz, Scott W. 1999. *Faith, serpents, and fire: Images of Kentucky Holiness believers.* Jackson: Univ. Press of Mississippi. [274n46]

Segerstråle, Ullica. 2000. *Defenders of the truth: The sociobiology debate.* Oxford: Oxford Univ. Press. [260n9]

Service, Elman R. 1962. *Primitive social organization: An evolutionary perspective.* New York: Random House. [277n13]

———. 1966. *The hunters.* Englewood Cliffs, NJ: Prentice-Hall. [281n85]

Shennan, Stephen J., and James Steele. 1999. Cultural learning in hominids: A behavioural ecological approach. In *Mammalian social learning: Comparative and ecological perspectives,* ed. H. O. Box and K. R. Gibson, 367–88. Cambridge: Cambridge Univ. Press. [144, 271n95]

Shennan, S. J., and J. R. Wilkinson. 2001. Ceramic style change and neutral evolution: A case study from Neolithic Europe. *American Antiquity* 66:577–93. [271n83]

Sherif, Muzafer, and Gardner Murphy. 1936. *The psychology of social norms.* New York: Harper & Brothers. [122, 269n43]

Silk, Joan B. 2002. Kin selection in primate groups. *International Journal of Primatology* 23: 849–75. [278n17]

Simon, Herbert A. 1979. *Models of thought.* New Haven, CT: Yale Univ. Press. [263n54]

Simoons, Fredrick J. 1969. Primary adult lactose intolerance and the milking habit: A problem in biologic and cultural interrelations: I. Review of the medical research. *The American Journal of Digestive Diseases* 14:819–36. [191, 192, 276n1]

———. 1970. Primary adult lactose intolerance and the milking habit: A problem in biologic and cultural interrelations: II. A culture historical hypothesis. *The American Journal of Digestive Diseases* 15:695–710. [191, 192, 276n1]

Skinner, G. William. 1997. Family systems and demographic processes. In *Anthropological demography: Toward a new synthesis,* ed. D. I. Kertzer and T. Fricke, 53–114. Chicago: Univ. of Chicago Press. [171, 275n53]

Skinner, G. W., and Y. Jianhua. Unpublished manuscript. *Reproduction in a patrilineal joint family system: Chinese in the lower Yangzi macroregion.* [276n83]

Slater, P. J. B., and S. A. Ince. 1979. Cultural evolution of chaffinch song. *Behaviour* 71: 146–66. [268n23]

Slater, P. J. B., S. A. Ince, and P. W. Colgan. 1980. Chaffinch song types: Their frequencies in the population and distribution between the repertoires of different individuals. *Behaviour* 75:207–18. [268n23]

Sloan, R. P., E. Bagiella, and T. Powell. 1999. Religion, spirituality and health. *Lancet* 353: 664–67. [274n45]

Smith, Eric A., and Rebecca L. Bliege Bird. 2000. Turtle hunting and tombstone opening: Public generosity as costly signaling. *Evolution and Human Behavior* 21:245–61. [274n37]

Smith, E. A., M. Borgerhoff Mulder, and K. Hill. 2001. Controversies in the evolutionary social sciences: A guide for the perplexed. *Trends in Ecology & Evolution* 16:128–35. [260n14]

Sobel, Dava. 1995. *Longitude: The true story of a lone genius who solved the greatest scientific problem of his time.* New York: Walker. [263n59]

Sober, Elliot. 1991. Models of cultural evolution. In *Trees of life: Essays in philosophy of biology,* ed. P. Griffiths, 17–38. Dordrecht: Kluwer. [96, 267n65, 268n70, 268n71]

Sober, Elliot, and David Sloan Wilson. 1998. *Unto others: The evolution and psychology of unselfish behavior.* Cambridge, MA: Harvard Univ. Press. [260n15, 273n28, 278n28]

Soltis, Joseph, Robert Boyd, and Peter J. Richerson. 1995. Can group-functional behaviors evolve by cultural group election? An empirical test. *Current Anthropology* 36:473–94. [208, 209t, 273n28]

Spelke, Elizabeth. 1994. Initial knowledge: Six suggestions. *Cognition* 50:431–45. [266n48]

Spence, A. Michael. 1974. *Market signaling: Informational transfer in hiring and related processes.* Cambridge, MA: Harvard Univ. Press. [274n37]

Sperber, Dan. 1996. *Explaining culture: A naturalistic approach.* Oxford: Blackwell. [82, 83, 84, 261n23, 263n55, 264n17, 266n40, 266n44, 266n45, 266n51, 271n98]

Srinivas, Mysore N. 1962. *Caste in modern India, and other essays.* Bombay: Asia Publishing House. [281n94]

Stark, Rodney. 1997. *The rise of Christianity: How the obscure, marginal Jesus movement became the dominant religious force in the Western world in a few centuries.* San Francisco: HarperCollins. [210, 273n29, 279n38, 279n39, 279n40]

———. 2003. *For the glory of God: How monotheism led to reformations, science, witch-hunts, and the end of slavery.* Princeton, NJ: Princeton Univ. Press. [168, 273n30, 274n41, 274n45, 274n48]

Stephens, D. W., and J. R. Krebs. 1987. *Foraging theory.* Princeton, NJ: Princeton Univ. Press. [268n3]

Steward, Julian H. 1955. *Theory of culture change: The methodology of multilinear evolution.* Urbana: Univ. of Illinois Press. [260n8, 261n13, 263n4, 263n6, 277n13, 280n55]

Sulloway, Frank J. 1996. *Born to rebel: Birth order, family dynamics, and creative lives.* 1st ed. New York: Pantheon Books.

Susman, R. L. 1994. Fossil evidence for early hominid tool use. *Science* 265:1570–73. [271n80]

Symons, Donald. 1979. *The evolution of human sexuality.* Oxford: Oxford Univ. Press. [260n9]

Tajfel, Henri. 1978. *Differentiation between social groups: Studies in the social psychology of intergroup relations.* European Monographs in Social Psychology 14. London: Academic Press. [221, 222, 280n65]

———. 1981. *Human groups and social categories: Studies in social psychology.* Cambridge: Cambridge Univ. Press. [221, 222, 280n65]

———. 1982. *Social identity and intergroup relations.* Cambridge: Cambridge Univ. Press. [221, 222, 280n65]

Tarde, Gabriel. 1903. *The laws of imitation.* New York: Holt. [265n29]

Templeton, A. R. 2002. Out of Africa again and again. *Nature* 416:45–51. [271n92]

Terkel, Joseph. 1995. Cultural transmission in the black rat—pine-cone feeding. *Advances in the Study of Behavior* 24:195–210. [107, 268n21]

Thieme, H. 1997. Lower Palaeolithic hunting spears from Germany. *Nature* 385:807–10. [271n88]

Thomas, David H., Lorann S. A. Pendleton, and Stephen C. Cappannari. 1986. Western Shoshone. In *Handbook of North American Indians: Great Basin,* ed. W. L. d'Azevedo, 262–83. Washington, DC: Smithsonian Institution Press. [277n13]

Thomason, Sarah Grey. 2001. *Language contact: An introduction.* Washington, DC: Georgetown Univ. Press. [266n55]

Thomason, Sarah Grey, and Terrence Kaufman. 1988. *Language contact, creolization, and genetic linguistics.* Berkeley and Los Angeles: Univ. of California Press. [91, 262n47, 266n55, 266n56, 266n57, 266n58]

Thompson, Nicolas S. 1995. Does language arise from a calculus of dominance? *Behavior and Brain Sciences* 18:387. [271n97]

Todd, Peter M., and Gerd Gigerenzer. 2000. Simple heuristics that make us smart. *Behavioral and Brain Sciences* 23:727–80. [119, 120, 269n41]

Tomasello, Michael. 1996. Do apes ape? In *Social learning in animals: The roots of culture,* ed. C. M. Heyes and B. G. Galef Jr., 319–46. New York: Academic Press. [110, 269n28]

———. 1999. *The cultural origins of human cognition.* Cambridge, MA: Harvard Univ. Press. [266n49]

———. 2000. Two hypotheses about primate cognition. In *The evolution of cognition,* ed. C. Heyes and L. Huber, 165–83. Cambridge, MA: MIT Press. [271n74]

Tomasello, M., A. C. Kruger, and H. H. Ratner. 1993. Cultural learning. *Behavioral and Brain Sciences* 16:495–552. [268n6, 269n26, 269n27]

Tooby, John, and Leda Cosmides. 1989. Evolutionary psychology and the generation of culture. 1. Theoretical considerations. *Ethology and Sociobiology* 10:29–49. [271n101, 276n81]

———. 1992. The psychological foundations of culture. In *The adapted mind: Evolutionary psychology and the generation of culture,* ed. J. Barkow, L. Cosmides, and J. Tooby, 19–136. New York: Oxford Univ. Press. [44, 45, 158, 160, 260n14, 262n42, 263n50, 263n54, 273n23, 273n26]

Tooby, J., and I. DeVore. 1987. The reconstruction of hominid behavioral evolution through strategic modeling. In *Primate models of hominid behavior,* ed. W. Kinzey, 183–237. New York: SUNY Press. [268n1]

Toth, N., K. D. Schick, E. S. Savage-Rumbaugh, R. A. Sevcik, and D. M. Rumbaugh. 1993. Pan the tool-maker—Investigations into the stone tool-making and tool-using capabilities of a bonobo (*Pan paniscus*). *Journal of Archaeological Science* 20:81–91. [271n79]

Trivers, Robert L. 1971. The evolution of reciprocal altruism. *Quarterly Review of Biology* 46:35–57. [200, 278n22]

Turner, J. 1984. Social identification and psychological group formation. In *The Social dimension: European developments in social psychology,* ed. H. Tajfel, C. Fraser, and J. M. F. Jaspars, 518–36. Cambridge: Cambridge Univ. Press. [222, 280n66]

Turner, J. C., I. Sachdev, and M. A. Hogg. 1983. Social categorization, interpersonal attraction and group formation. *British Journal of Social Psychology* 22:227–39. [222, 280n66]

Tversky, Amos, and Daniel Kahneman. 1974. Judgment under uncertainty: Heuristics and biases. *Science* 185:1124–31. [263n54]

Twain, Mark. 1962. *Mark Twain on the damned human race.* Ed. and with an introduction by Janet Smith. 1st ed. New York: Hill and Wang. [80, 266n37]

Underhill, P. A., P. D. Shen, et al. 2000. Y chromosome sequence variation and the history of human populations. *Nature Genetics* 26:358–61. [271n90]

United Nations Population Division. 2002a [cited October 29, 2002]. *Analytical Report,* vol. 3. Available from http://www.un.org/esa/population/publications/wpp2000/wpp2000_volume3.htm. [272n3]

————. 2002b. *World population prospects: The 2000 revision.* New York: United Nations.

van den Berghe, Pierre L. 1981. *The ethnic phenomenon.* New York: Elsevier. [279n48, 280n71]

Van Schaik, Carel P., and Cheryl D. Knott. 2001. Geographic variation in tool use on *Neesia* fruits in orangutans. *American Journal of Physical Anthropology* 114:331–42. [268n12, 269n32]

Vayda, A. P. 1995. Failures of explanation in Darwinian ecological anthropology: Part I. *Philosophy of the Social Sciences* 25:219–49. [266n62]

Visalberghi, Elisabetta. 1993. Ape ethnography. *Science* 261:1754. [269n27]

Visalberghi, E., and D. Fragaszy. 1991. Do monkeys ape? In *"Language" and intelligence in monkeys and apes,* ed. S. T. Parker and K. R. Gibson, 247–73. Cambridge: Cambridge Univ. Press. [269n25, 269n27]

Voelkl, Bernard, and Ludwig Huber. 2000. True imitation in marmosets. *Animal Behaviour* 60:195–202. [269n30]

Wardhaugh, Ronald. 1992. *An introduction to sociolinguistics.* 2nd ed. Oxford: Blackwell. [266n42]

Weber, Max. 1951. *The religion of China: Confucianism and Taoism.* Glencoe, IL: Free Press. [274n41]

Weiner, Jonathan. 1994. *The beak of the finch: A story of evolution in our time.* 1st ed. New York: Knopf. Distributed by Random House. [266–67n64]

————. 1999. *Time, love, memory: A great biologist and his quest for the origins of behavior.* 1st ed. New York: Knopf. [282n7]

Weingart, Peter, Sandra D. Mitchell, Peter J. Richerson, and Sabine Maasen. 1997. *Human by nature: Between biology and the social sciences.* Mahwah, NJ: Lawrence Erlbaum Associates. [282n32]

Weir, A. A. S., J. Chappell, and A. Kacelnik. 2002. Shaping of hooks in New Caledonian crows. *Science* 297:981. [270n57]

Welsch, R. L., J. Terrell, and J. A. Nadolski. 1992. Language and culture on the North Coast of New Guinea. *American Anthropologist* 94:568–600. [266n59]

Werner, Emmy E. 1979. *Cross cultural child development: A view from the planet Earth.* Monterey, CA: Brooks/Cole. [275n67]

Werren, J. H. 2000. Evolution and consequences of Wolbachia symbioses in invertebrates. *American Zoologist* 40:1255. [272n12]

Westoff, C. F., and R. H. Potvin. 1967. *College women and fertility values.* Princeton, NJ: Princeton Univ. Press. [275n72]

White, Leslie A. 1949. *The science of culture, a study of man and civilization.* New York: Farrar Straus. [263n3]

Whiten, Andrew. 2000. Primate culture and social learning. *Cognitive Science* 24:477–508. [109, 110, 269n29]

Whiten, Andrew, and Richard W. Byrne. 1988. *Machiavellian intelligence: Social expertise and the evolution of intellect in monkeys, apes, and humans.* Oxford: Oxford Univ. Press. [271n76]

———. 1997. *Machiavellian intelligence II: Extensions and evaluations.* Cambridge: Cambridge Univ. Press. [271n76]

Whiten, A., J. Goodall, W. C. McGrew, T. Nishida, V. Reynolds, Y. Sugiyama, C. E. G. Tutin, R. W. Wrangham, and C. Boesch. 1999. Cultures in chimpanzees. *Nature* 399:682–85. [268n9]

Whiten, A., and R. Ham. 1992. On the nature and evolution of imitation in the animal kingdom: Reappraisal of a century of research. *Advances in the Study of Behavior* 21:239–83. [269n25, 269n27]

Wierzbicka, A. 1992. *Semantics, culture, and cognition: Human concepts in culture-specific configurations.* New York: Oxford Univ. Press. [264n11]

Wiessner, Polly W. 1983. Style and social information in Kalahari San projectile points. *American Antiquity* 48:253–76. [221, 280n63, 280n76]

———. 1984. Reconsidering the behavioral basis for style: A case study among the Kalahari San. *Journal of Anthropological Archaeology* 3:190–234. [221, 280n63, 280n76]

Wiessner, Polly, and Akii Tumu. 1998. *Historical vines: Enga networks of exchange, ritual, and warfare in Papua New Guinea.* Smithsonian Series in Ethnographic Inquiry. Washington, DC: Smithsonian Institution Press. [265n23, 279n36]

Williams, George C. 1966. *Adaptation and natural selection: A critique of some current evolutionary thought.* Princeton, NJ: Princeton Univ. Press. [202, 278n25]

Wilson, David Sloan. 2002. *Darwin's cathedral: Evolution, religion, and the nature of society.* Chicago: Univ. of Chicago Press. [273n29, 274n45]

Wilson, Edward Osborne. 1975. *Sociobiology: The new synthesis.* Cambridge, MA: Harvard Univ. Press, Belknap Press. [260n9, 277n12]

———. 1984. *Biophilia.* Cambridge, MA: Harvard Univ. Press. [266–67n64]

———. 1998. *Consilience: The unity of knowledge.* New York: Knopf. [194, 239, 260n18, 277n7, 281n2]

Wimsatt, William C. 1981. Robustness, reliability, and overdetermination. In *Scientific inquiry and the social sciences,* ed. D. T. Campbell, M. B. Brewer, and B. E. Collins, 124–63. San Francisco: Jossey-Bass. [267n69]

Witkin, Herman A., and John W. Berry. 1975. Psychological differentiation in cross-cultural perspective. *Journal of Cross-Cultural Psychology* 6:111–78. [275n67]

Witkin, Herman A., and Donald R. Goodenough. 1981. *Cognitive styles, essence and origins: Field dependence and field independence.* Psychological Issues Monograph 51. New York: International Universities Press. [275n67]

Wood, B., and M. Collard. 1999. The human genus. *Science* 284:65–71. [271n78]

Wrangham, Richard W. 1994. *Chimpanzee cultures.* Cambridge, MA: Harvard Univ. Press. [268n9]

Wynne-Edwards, Vero C. 1962. *Animal dispersion in relation to social behaviour.* Edinburgh: Oliver and Boyd. [201, 278n24]

Yen, D. E. 1974. *The sweet potato and Oceania: An essay in ethnobotany.* Honolulu: Bishop Museum Press. [265n23]

Yengoyan, Aram A. 1968. Demographic and ecological influences on aboriginal Australian marriage systems. In *Man the hunter,* ed. R. B. Lee and I. DeVore, 185–99. Chicago: Aldine. [226, 280n77]

Zahavi, Amotz. 1975. Mate selection—A selection for a handicap. *Journal of Theoretical Biology* 53:205–14. [274n37]

Zahavi, Amotz, and Avishag Zahavi. 1997. *The handicap principle: A missing piece of Darwin's puzzle.* New York: Oxford Univ. Press. [274n37]

Zohar, O., and J. Terkel. 1992. Acquisition of pine cone stripping behaviour in black rats (*Rattus rattus*). *International Journal of Comparative Psychology* 5:1–6 [107, 268n21]

General Index

The letter *t* following a page number denotes a table. The letter *f* following a page number denotes a figure.